GENETIC TECHNOLOGY

A New Frontier

About the Book

Genetic Technology: A New Frontier
Office of Technology Assessment

Genetic technologies may contribute to meeting some of the most fundamental human needs—from health care to food supplies to energy. They open up new possibilities for developing vaccines against such intractable diseases as hepatitis and malaria; they can transform inedible biomass into food for humans and animals; and they can aid in food processing. Genetically engineered micro-organisms may also be developed for use in oil recovery, pollution control, and mineral leaching.

Yet, there are problems. Technical constraints and questions about potential effects on human health and the environment are major considerations, particularly in the case of genetically engineered micro-organisms. No evidence exists that any unexpectedly harmful genetically engineered organism has been created; still, few experts believe that molecular genetic techniques are totally without risk, and there is uncertainty about the regulation of production methods using engineered micro-organisms. Perhaps a more perplexing issue is the potential impact of the new technologies on human values.

This book is one of the first comprehensive documents on emerging genetic technologies and their implications for society. The authors discuss the opportunities and problems involved, describe current techniques, and attempt to project some of the economic, environmental, and institutional impacts of those techniques. The issues they raise go beyond those of technology, utility, and economic feasibility. As we gain the ability to manipulate life, we must face basic questions of just what life means and how far we can reasonably—and safely—allow ourselves to go in the pursuit of a better future.

The **Office of Technology Assessment** was created in 1972 as an advisory arm of the U.S. Congress. OTA's basic function is to help legislative policymakers anticipate and plan for the consequences of technological changes and to examine the many ways, expected and unexpected, in which technology affects people's lives. The assessment of technology calls for exploration of the physical, biological, economic, social, and political impacts that can result from applications of scientific knowledge. OTA provides Congress with independent and timely information about the potential effects—both beneficial and harmful—of technological applications.

GENETIC TECHNOLOGY

A New Frontier

Office of Technology Assessment

WESTVIEW PRESS • BOULDER, COLORADO
CROOM HELM • LONDON, ENGLAND

Impacts of Applied Genetics Advisory Panel

J. E. Legates, *Chairman*
Dean, School of Agriculture and Life Sciences, North Carolina State University

Ronald E. Cape
Cetus Corp.

Nina V. Fedoroff
Department of Embryology
Carnegie Institution of Washington

June Goodfield
The Rockefeller University

Harold P. Green
Fried, Frank, Harris, Shriver and Kampelman

Halsted R. Holman
Stanford University Medical School

M. Sylvia Krekel
Health and Safety Office
Oil, Chemical, and Atomic Workers
International Union

Elizabeth Kutter
The Evergreen State College

Oliver E. Nelson, Jr.
Laboratory of Genetics
University of Wisconsin

David Pimentel
Department of Entomology
Cornell University

Robert Weaver
Department of Agricultural Economics
Pennsylvania State University

James A. Wright
Pioneer Hi-Bred International
Plant Breeding Division

Norton D. Zinder
The Rockefeller University

Applied Genetics Assessment Staff

Joyce C. Lashof, *Assistant Director, OTA*
Health and Life Sciences Division

Gretchen Kolsrud, *Program Manager*

Zsolt Harsanyi, *Project Director*

Project Staff

Fred H. Bergmann, *Senior Analyst* **
Marya Breznay, *Administrative Assistant*
Lawrence Burton, *Analyst*
Susan Clymer, *Research Assistant*
Renee G. Ford,* *Technical Editor*
Michael Gough, *Senior Analyst*
Robert Grossmann,* *Analyst*
Richard Hutton,* *Editor*
Geoffrey M. Karny, *Legal Analyst*

Major Contractors

Benjamin G. Brackett, University of Pennsylvania
The Genex Corp.
William P. O'Neill, Poly-Planning Services
Plant Resources Institute
Anthony J. Sinskey, Massachusetts Institute of Technology
Aladar A. Szalay, Boyce-Thompson Institute
Virginia Walbot, Washington University

OTA Publishing Staff

John C. Holmes, *Publishing Officer*

John Bergling*   Kathie S. Boss   Debra M. Datcher
Patricia A. Dyson*   Mary Harvey*   Joe Henson

---

*OTA contract personnel.
**On detail from National Institute of General Medical Sciences, NIH.

Published in 1982 in the United States of America by
  Westview Press, Inc.
  5500 Central Avenue
  Boulder, Colorado 80301

Published in 1982 in Great Britain by
  Croom Helm Ltd
  2-10 St Johns Road
  London SW11

Library of Congress Catalog Card Number: 81-600046
ISBN (U.S.): 0-86531-327-X
ISBN (U.S.): 0-86531-328-8 (pbk.)
ISBN (U.K.): 0-7099-1913-1

Composition for this book was provided by the Office of Technology Assessment.
Printed and bound in the United States of America.

Foreword

This report examines the application of classical and molecular genetic technologies to micro-organisms, plants, and animals. Congressional support for an assessment in the field of genetics dates back to 1976 when 30 Representatives requested a study of recombinant DNA technology. Letters of support for this broader study came from the then Senate Committee on Human Resources and the House Committee on Interstate and Foreign Commerce, Subcommittee on Health and the Environment.

Current developments are especially rapid in the application of genetic technologies to micro-organisms; these were studied in three industries: pharmaceutical, chemical, and food. Classical genetics continue to play the major role in plant and animal breeding but new genetic techniques are of ever-increasing importance.

This report identifies and discusses a number of issues and options for the Congress, such as:

- Federal Government support of R&D,
- methods of improving the germplasm of farm animal species,
- risks of genetic engineering,
- patenting living organisms, and
- public involvement in decisionmaking.

The Office of Technology Assessment was assisted by an advisory panel of scientists, industrialists, labor representatives, and scholars in the fields of law, economics, and those concerned with the relationships between science and society. Others contributed in two workshops held during the course of the assessment. The first was to investigate public perception of the issues in genetics; the second examined genetic applications to animals. Sixty reviewers drawn from universities, Government, industry, and the law provided helpful comments on draft reports. The Office expresses sincere appreciation to all those individuals.

JOHN H. GIBBONS
Director

Contents

Glossary

Aerobic.—Growing only in the presence of oxygen.

Anaerobic.—Growing only in the absence of oxygen.

Alkaloids.—A group of nitrogen-containing organic substances found in plants; many are pharmacologically active—e.g., nicotine, caffeine, and cocaine.

Allele.—Alternate forms of the same gene. For example, the genes responsible for eye color (blue, brown, green, etc.) are alleles.

Amino acids.—The building blocks of proteins. There are 20 common amino acids; they are joined together in a strictly ordered "string" which determines the character of each protein.

Antibody.—A protein component of the immune system in mammals found in the blood.

Antigen.—A large molecule, usually a protein or carbohydrate, which when introduced in the body stimulates the production of an antibody that will react specifically with the antigen.

Aromatic chemical.—An organic compound containing one or more six-membered rings.

Aromatic polymer.—Large molecules consisting of repeated structural units of aromatic chemicals.

Artificial insemination.—The manual placement of sperm into the uterus or oviduct.

Bacteriophage (or phage).—A virus that multiplies in bacteria. Bacteriophage lambda is commonly used as a vector in recombinant DNA experiments.

Bioassay.—Determination of the relative strength of a substance (such as a drug) by comparing its effect on a test organism with that of a standard preparation.

Biomass.—Plant and animal material.

Biome.—A community of living organisms in a major ecological region.

Biosynthesis.—The production of a chemical compound by a living organism.

Biotechnology.—The collection of industrial processes that involve the use of biological systems. For some of these industries, these processes involve the use of genetically engineered microorganisms.

Blastocyst.—An early developmental stage of the embryo; the fertilized egg undergoes several cell divisions and forms a hollow ball of cells called the blastocyst.

Callus.—The cluster of plant cells that results from tissue culturing a single plant cell.

Carbohydrates.—The family of organic molecules consisting of simple sugars such as glucose and sucrose, and sugar chains (polysaccharides) such as starch and cellulose.

Catalyst.—A substance that enables a chemical reaction to take place under milder than normal conditions (e.g., lower temperatures). Biological catalysts are enzymes; nonbiological catalysts include metallic complexes.

Cell fusion.—The fusing together of two or more cells to become a single cell.

Cell lysis.—Disruption of the cell membrane allowing the breakdown of the cell and exposure of its contents to the environment.

Cellulase.—An enzyme that degrades cellulose to glucose.

Cellulose.—A polysaccharide composed entirely of several glucose units linked end to end; it constitutes the major part of cell walls in plants.

Chimera.—An individual composed of a mixture of genetically different cells.

Chloroplast.—The structure in plant cells where photosynthesis occurs.

Chromosomes.—The thread-like components of a cell that are composed of DNA and protein. They contain most of the cell's DNA.

Clone.—A group of genetically identical cells or organisms asexually descended from a common ancestor. All cells in the clone have the same genetic material and are exact copies of the original.

Conjugation.—The one-way transfer of DNA between bacteria in cellular contact.

Crossing-over.—A genetic event that can occur during cellular replication, which involves the breakage and reunion of DNA molecules.

Cultivar.—An organism developed and persistent under cultivation.

Cytogenetics.—A branch of biology that deals with the study of heredity and variation by the methods of both cytology (the study of cells) and genetics.

Cytoplasm.—The protoplasm of a cell, external to the cell's nuclear membrane.

Diploid.—A cell with double the basic chromosome number.

DNA (deoxyribonucleic acid).—The genetic material found in all living organisms. Every inherited characteristic has its origin somewhere in the code of each individual's complement of DNA.

DNA vector.—A vehicle for transferring DNA from one cell to another.

Dominant gene.—A characteristic whose expression prevails over alternative characteristics for a given trait.

Escherichia coli.—A bacterium that commonly inhabits the human intestine. It is a favorite organism for many microbiological experiments.

Endotoxins.—Complex molecules (lipopolysaccharides) that compose an integral part of the cell wall, and are released only when the integrity of the cell is disturbed.

Embryo transfer.—Implantation of an embryo into the oviduct or uterus.

Enzyme.—A functional protein that catalyzes a chemical reaction. Enzymes control the rate of metabolic processes in an organism; they are the active agents in the fermentation process.

Estrogens.—Female sex hormones.

Estrus ("heat").—The period in which the female will allow the male to mate her.

Eukaryote.—A higher, compartmentalized cell characterized by its extensive internal structure and the presence of a nucleus containing the DNA. All multicellular organisms are eukaryotic. The simpler cells, the prokaryotes, have much less compartmentalization and internal structure; bacteria are prokaryotes.

Exotoxins.—Proteins produced by bacteria that are able to diffuse out of the cells; generally more potent and specific in their action than endotoxins.

Fermentation.—The biochemical process of converting a raw material such as glucose into a product such as ethanol.

Fibroblast.—A cell that gives rise to connective tissues.

Gamete.—A mature reproductive cell.

Gene.—The hereditary unit; a segment of DNA coding for a specific protein.

Gene expression.—The manifestation of the genetic material of an organism as specific traits.

Genetic drift.—Changes of gene frequency in small populations due to chance preservation or extinction of particular genes.

Genetic code.—The biochemical basis of heredity consisting of codons (base triplets along the DNA sequence) that determine the specific amino acid sequence in proteins and that are the same for all forms of life studied so far.

Genetic engineering.—A technology used at the laboratory level to alter the hereditary apparatus of a living cell so that the cell can produce more or different chemicals, or perform completely new functions. These altered cells are then used in industrial production.

Gene mapping.—Determining the relative locations of different genes on a given chromosome.

Genome.—The basic chromosome set of an organism—the sum total of its genes.

Genotype.—The genetic constitution of an individual or group.

Germplasm.—The total genetic variability available to an organism, represented by the pool of germ cells or seed.

Germ cell.—The sex cell of an organism (sperm or egg, pollen or ovum). It differs from other cells in that it contains only half the usual number of chromosomes. Germ cells fuse during fertilization.

Glycopeptides.—Chains of amino acids with attached carbohydrates.

Glycoprotein.—A conjugated protein in which the nonprotein group is a carbohydrate.

Haploid.—A cell with only one set (half of the usual number) of chromosomes.

Heterozygous.—When the two genes controlling a particular trait are different, the organism is heterozygous for that trait.

Homozygous.—When the two genes controlling a particular trait are identical for a pair of chromosomes, the organism is said to be homozygous for that trait.

Hormones.—The "messenger" molecules of the body that help coordinate the actions of various tissues; they produce a specific effect on the activity of cells remote from their point of origin.

Hybrid.—A new variety of plant or animal that results from cross-breeding two different existing varieties.

Hydrocarbon.—All organic compounds that are composed only of carbon and hydrogen.

Immunoproteins.—All the proteins that are part of the immune system (including antibodies, interferon, and cytokines).

In vitro.—Outside the living organism and in an artificial environment.

In vivo.—Within the living organism.

Leukocytes.—The white cells of blood.

Lipids.—Water insoluble biomolecules, such as cellular fats and oils.

Lipopolysaccharides.—Complex substances composed of lipids and polysaccharides.

Lymphoblastoid.—Referring to malignant white blood cells.

Lymphokines.—The biologically active soluble factor produced by white blood cells.

Maleic anhydride.—An important organic chemical used in the manufacture of synthetic resins, in fungicides, in the dyeing of cotton textiles, and to prevent the oxidation of fats and oils during storage and rancidity.

Messenger RNA.—Ribonucleic acid molecules that serve as a guide for protein synthesis.

Metabolism.—The sum of the physical and chemical processes involved in the maintenance of life and by which energy is made available.

Mitochondria.—Structures in higher cells that serve as the "powerhouse" for the cell, producing chemical energy.

Monoclonal antibodies.—Antibodies derived from a single source or clone of cells which recognize only one kind of antigen.

Mutants.—Organisms whose visible properties with respect to some trait differ from the norm of the population due to mutations in its DNA.

Mutation.—Any change that alters the sequence of bases along the DNA, changing the genetic material.

Myeloma.—A malignant disease in which tumor cells of the antibody producing system synthesize excessive amounts of specific proteins.

n-alkanes.—Straight chain hydrocarbons—the main constituents of petroleum.

Nif genes.—The genes for nitrogen fixation present in certain bacteria.

Nucleic acid.—A polymer composed of DNA or RNA subunits.

Nucleotides.—The fundamental units of nucleic acids. They consist of one of the four bases—adenine, guanine, cytosine, and thymine (uracil in the case of RNA)—and its attached sugar-phosphate group.

Organic compounds.—Chemical compounds based on carbon chains or rings, which contain hydrogen, and also may contain oxygen, nitrogen, and various other elements.

Parthenogenesis.—Reproduction in animals without male fertilization of the egg.

Pathogen.—A specific causative agent of disease.

Peptide.—Short chain of amino acids.

pH.—A measure of the acidity or basicity of a solution; on a scale of 0 (acidic) to 14 (basic): for example, lemon juice has a pH of 2.2 (acidic), water has a pH of 7.0 (neutral), and a solution of baking soda has a pH of 8.5 (basic).

Phage.—(See *bacteriophage*.)

Phenotype.—The visible properties of an organism that are produced by the interaction of the genotype and the environment.

Plasmid.—Hereditary material that is not part of a chromosome. Plasmids are circular and self-replicating. Because they are generally small and relatively simple, they are used in recombinant DNA experiments as acceptors of foreign DNA.

Plastid.—Any specialized organ of the plant cell other than the nucleus, such as the chloroplast.

Ploidy.—Describes the number of sets of chromosomes present in the organism. For example, humans are diploid, having two homologous sets of 23 chromosomes (one set from each parent) for a total of 46 chromosomes; many plants are haploid, having only one copy of each chromosome.

Polymer.—A long-chain molecule formed from smaller repeating structural units.

Polysaccharide.—A long-chain carbohydrate containing at least three molecules of simple sugars linked together; examples would include cellulose and starch.

Progestogens.—Hormones involved with ovulation.

Prostaglandin.—Refers to a group of naturally occurring, chemically related long-chain fatty acids that have certain physiological effects (stimulate contraction of uterine and other smooth muscles, lower blood pressure, affect action of certain hormones).

Protein.—A linear polymer of amino acids; proteins are the products of gene expression and are the functional and structural components of cells.

Protoplast.—A cell without a wall.

Protoplast fusion.—A means of achieving genetic transformation by joining two protoplasts or joining a protoplast with any of the components of another cell.

Recessive gene.—Any gene whose expression is dependent on the absence of a dominant gene.

Recombinant DNA.—The hybrid DNA produced by joining pieces of DNA from different sources.

Restriction enzyme.—An enzyme within a bacterium that recognizes and degrades DNA from foreign organisms, thereby preserving the genetic integrity of the bacterium. In recombinant DNA experiments, restriction enzymes are used as tiny biological scissors to cut up foreign DNA before it is recombined with a vector.

Reverse transcriptase.—An enzyme that can synthesize a single strand of DNA from a messenger RNA, the reverse of the normal direction of processing genetic information within the cell.

RNA (ribonucleic acid).—In its three forms—messenger RNA, transfer RNA, and ribosomal RNA—it assists in translating the genetic message of DNA into the finished protein.

Somatic cell.—One of the cells composing parts of the body (e.g., tissues, organs) other than a germ cell.

Tissue culture.—An in vitro method of propagating healthy cells from tissues, such as fibroblasts from skin.

Transduction.—The process by which foreign DNA becomes incorporated into the genetic complement of the host cell.

Transformation.—The transfer of genetic information by DNA separated from the cell.

Vector.—A transmission agent; a DNA vector is a self-replicating DNA molecule that transfers a piece of DNA from one host to another.

Virus.—An infectious agent that requires a host cell in order for it to replicate. It is composed of either RNA or DNA wrapped in a protein coat.

Zygote.—A cell formed by the union of two mature reproductive cells.

Acronyms and Abbreviations

AA — amino acids
ACS — American Cancer Society
ACTH — adrenocorticotropic hormone
AI — artificial insemination
AIPL — Animal Improvement Programs Laboratory
APAP — acetaminophen
ASM — American Society for Microbiology
bbl — barrel(s)
bbl/d — barrels per day
BOD5 — 5-day biochemical oxygen demand
BRM — Biological Response Modifier Program
bu — bushel
CaMV — cauliflower mosaic virus
CCPA — The Court of Customs and Patent Appeals
CDC — Center for Disease Control
CERB — Cambridge Experimentation Review Board
DHHS — Department of Health and Human Services (formerly Health, Education, and Welfare)
DHI — Dairy Herd Improvement
DNA — deoxyribonucleic acid
DOC — Department of Commerce
DOD — Department of Defense
DOE — Department of Energy
DPAG — Dangerous Pathogens Advisory Group
EOR — enhanced oil recovery
EPA — Environmental Protection Agency
FDA — Food and Drug Administration
FMDV — foot-and-mouth disease virus
ft^2 — square foot
ft — foot
FTC — Federal Trade Commission
g — gram
gal — gallon
GH — growth hormone
ha — hectares
HEW — Department of Health, Education, and Welfare
hGH — human growth hormone
HYV — high-yielding varieties

IBCs — Institutional Biosafety Committees
ICI — Imperial Chemical Industries
IND — Investigational New Drug Application (FDA)
kg — kilogram
l — liter
lb — pound
mg — milligram
μg — microgram
μm — micrometer (formerly micron)
MUA — Memorandum of Understanding and Agreement
NCI — National Cancer Institute
NDA — new drug application (FDA)
NDAB — National Diabetics Advisory Board
NDCHIP — National Cooperative Dairy Herd Program
NIAID — National Institute of Allergy and Infectious Diseases
NIAMDD — National Institute of Arthritis, Metabolism, and Digestive Diseases
NIH — National Institutes of Health
NIOSH — National Institute of Occupational Safety and Health
NSF — National Science Foundation
OECD — The Organization for Economic Cooperation and Development
ORDA — Office of Recombinant DNA Activities
PD — predicted difference
pH — unit of measure for acidity/basicity
ppm — parts per million
R&D — research and development
RAC — Recombinant DNA Advisory Committee
rDNA — recombinant DNA
SCP — single-cell protein
T-DNA — a smaller segment of the Ti plasmid
Ti — tumor inducing
TSCA — Toxic Substances Control Act
UCSF — University of California at San Francisco
U.S.C. — United States Code
USDA — United States Department of Agriculture

GENETIC TECHNOLOGY

A New Frontier

Chapter 1
Summary: Issues and Options

Chapter 1

Summary: Issues and Options

The genetic alteration of plants, animals, and micro-organisms has been an important part of agriculture for centuries. It has also been an integral part of the alcoholic beverage industry since the invention of beer and wine; and for the past century, a mainstay of segments of the pharmaceutical and chemical industries.

However, only in the last 20 years have powerful new genetic technologies been developed that greatly increase the ability to manipulate the inherited characteristics of plants, animals, and micro-organisms. One consequence is the increasing reliance the pharmaceutical and chemical industries are placing on biotechnology. Micro-organisms are being used to manufacture substances that have previously been extracted from natural sources. Animal and plant breeders are using the new techniques to help clarify basic questions about biological functions, and to improve the speed and efficiency of the technologies they already use. Other industries—from food processing and pollution control to mining and oil recovery—are considering the use of genetic engineering to increase productivity and cut costs.

Genetic technologies will have a broad impact on the future. They may contribute to filling some of the most fundamental needs of mankind—from health care to supplies of food and energy. At the same time, they arouse concerns about their potential effects on the environment and the risks to health involved in basic and applied scientific research and development (R&D). Because genetic technologies are already being applied, it is appropriate to begin considering their potential consequences.

Congressional concern with applied genetics dates back to 1976, when 30 Representatives requested an assessment of recombinant DNA (rDNA) technology. Support for the broader study reported here came in letters to the Office of Technology Assessment from the then Senate Committee on Human Resources and the House Committee on Interstate and Foreign Commerce, Subcommittee on Health and the Environ-

ronment. In addition, specific subtopics are of interest to other committees, notably those having jurisdiction over science and technology and those concerned with patents.

This report describes the potentials and problems of applying the new genetic technologies to a range of major industries. It emphasizes the present state of the art because that is what defines the basis for the future applications. It then makes some estimates of economic, environmental, and institutional impacts—where, when, and how some technologies might be applied and what some of the results might be. The report closes with the possible roles that Government, industry, and the public might play in determining the future of applied genetics.

The term *applied genetics*, as used in this report, refers to two groups of technologies:

- *Classical genetics*—natural mating methods for the selective breeding of organisms for desired characteristics—e.g., breeding cows for increased milk production. The pool of genes available for selection is comprised of those that cause natural differences among individuals in a population and those obtained by mutation.

- *Molecular genetics* includes the technologies of genetic engineering that involve the directed manipulation of the genetic material itself. These technologies—such as rDNA and the chemical synthesis of genes—can increase the size of the gene pool for any one organism by making available genetic traits from many different populations. Molecular genetics also includes technologies in which manipulation occurs at a level higher than that of the gene—at the cellular level, e.g., cell fusion and in vitro fertilization.

Significant applications of molecular genetics to micro-organisms, such as the efforts to manufacture human insulin, are already underway in several industries. Most of these applications depend on fermentation—a technology in which substances produced by micro-organisms can be obtained in large quantities. Applications to plants and animals, which are biologically more complex and more difficult to manipulate successfully, will take longer to develop.

Biotechnology

Biotechnology—the use of living organisms or their components in industrial processes—is possible because micro-organisms naturally produce countless substances during their lives. Some of these substances have proved commercially valuable. A number of different industries have learned to use micro-organisms as natural factories, cultivating populations of the best producers under conditions designed to enhance their abilities.

Applied genetics can play a major role in improving the speed, efficiency, and productivity of these biological systems. It permits the manipulation, or engineering, of the micro-organisms' genetic material to produce the desired characteristics. Genetic engineering is not in itself an industry, but a technique used at the laboratory level that allows the researcher to modify the hereditary apparatus of the cell. The population of altered identical cells that grows from the first changed micro-organism is, in turn, used for various industrial processes. (See figure 1.)

The first major commercial effects of the application of genetic engineering will be in the pharmaceutical, chemical, and food processing industries. Potential commercial applications of value to the mining, oil recovery, and pollution control industries—which may desire to use manipulated micro-organisms in the open environment—are still somewhat speculative.

The pharmaceutical industry

FINDINGS

The pharmaceutical industry has been the first to take advantage of the potentials of applied molecular genetics. Ultimately, it will probably benefit more than any other, with the largest percentage of its products depending on advances in genetic technologies. Already,

micro-organisms have been engineered to produce human insulin, interferon, growth hormone, urokinase (for the treatment of blood clots), thymosin-α 1 (for controlling the immune response), and somatostatin (a brain hormone). (See figure 2.)

The products most likely to be affected by genetic engineering in the next 10 to 20 years are nonprotein compounds like most antibiotics, and protein compounds such as enzymes and antibodies, and many hormones and vaccines. Improvements can be made both in the products and in the processes by which they are produced. Process costs may be lowered and even entirely new products developed.

The most advanced applications today are in the field of hormones. While certain hormones have already proved useful, the testing of others has been hindered by their scarcity and high cost. Of 48 human hormones that have been identified so far as possible candidates for production by genetically engineered micro-organisms, only 10 are used in current medical practice. The other 38 are not, partly because they have been available in such limited quantities that tests of their therapeutic value have not been possible.

Genetic technologies also open up new approaches for vaccine development for such intractable parasitic and viral diseases as amebic dysentery, trachoma, hepatitis, and malaria. At present, the vaccine most likely to be produced is for foot-and-mouth disease in animals. However, should any one of the vaccines for human diseases become available, the social, economic, and political consequences of a decrease in morbidity and mortality would be significant. Many of these diseases are particularly prevalent in less industrialized countries; the developments of vaccines for them may profoundly affect the lives of tens of millions of people.

**Figure 1.—Recombinant DNA: The Technique of Recombining Genes
From One Species With Those From Another**

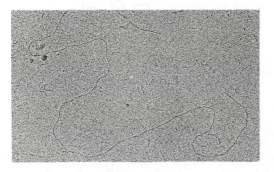

Electron micrograph of the DNA, which is the plasmid SP01 from *Bacillus subtilis.* This plasmid which has been sliced open is used for recombinant DNA research in this bacterial host

Photo credits: Professor F. A. Eiserling, UCLA Molecular Biology Institute

Electron micrograph of *Bacillus subtilis* in the process of cell division. The twisted mass in the center of each daughter cell is the genetic material, DNA

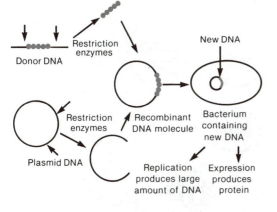

Restriction enzymes recognize certain sites along the DNA and can chemically cut the DNA at those sites. This makes it possible to remove selected genes from donor DNA molecules and insert them into plasmid DNA molecules to form the recombinant DNA. This recombinant DNA can then be cloned in its bacterial host and large amounts of a desired protein can be produced.

SOURCE: Office of Technology Assessment.

For some pharmaceutical products, biotechnology will compete with chemical synthesis and extraction from human and animal organs. Assessing the relative worth of each method must be done on a case-by-case basis. But for other products, genetic engineering offers the only method known that can ensure a plentiful supply; in some instances, it has no competition.

By making a pharmaceutical available, genetic engineering may have two types of effects:

• Drugs that already have medical promise will be available in ample amounts for clinical testing. Interferon, for example, can be tested for its efficacy in cancer and viral therapy, and human growth hormone can be evaluated for its ability to heal wounds.

• Other pharmacologically active substances for which no apparent use now exists will be available in sufficient quantities and at low enough cost to enable researchers to explore new uses. As a result, the potential for totally new therapies exists. Regulatory proteins, for example, which are an entire

Figure 2.—The Product Development Process

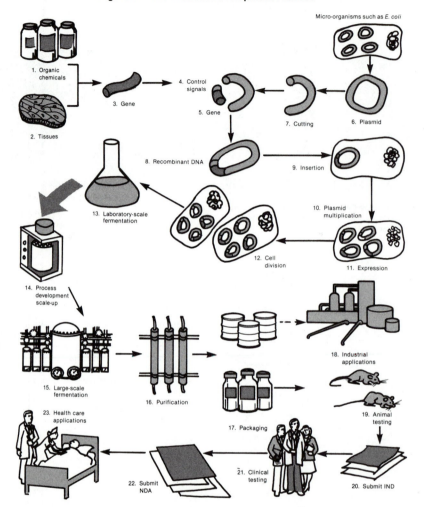

Micro-organisms such as *E. coli*

1. Organic chemicals
2. Tissues
3. Gene
4. Control signals
5. Gene
6. Plasmid
7. Cutting
8. Recombinant DNA
9. Insertion
10. Plasmid multiplication
11. Expression
12. Cell division
13. Laboratory-scale fermentation
14. Process development scale-up
15. Large-scale fermentation
16. Purification
17. Packaging
18. Industrial applications
19. Animal testing
20. Submit IND
21. Clinical testing
22. Submit NDA
23. Health care applications

The development process begins by obtaining DNA either through organic synthesis (1) or derived from biological sources such as tissues (2). The DNA obtained from one or both sources is tailored to form the basic "gene" (3) which contains the genetic information to "code" for a desired product, such as human interferon or human insulin. Control signals (4) containing instructions are added to this gene (5). Circular DNA molecules called plasmids (6) are isolated from micro-organisms such as *E. coli*; cut open (7) and spliced back (8) together with genes and control signals to form "recombinant DNA" molecules. These molecules are then introduced into a host cell (9).

Each plasmid is copied many times in a cell (10). Each cell then translates the information contained in these plasmids into the desired product, a process called "expression" (11). Cells divide (12) and pass on to their offspring the same genetic information contained in the parent cell.

Fermentation of large populations of genetically engineered micro-organisms is first done in shaker flasks (13), and then in small fermenters (14) to determine growth conditions, and eventually in larger fermentation tanks (15). Cellular extract obtained from the fermentation process is then separated, purified (16), and packaged (17) either for industrial use (18) or health care applications.

Health care products are first tested in animal studies (19) to demonstrate a product's pharmacological activity and safety, In the United States, an investigational new drug application (20) is submitted to begin human clinical trials to establish safety and efficacy. Following clinical testing (21), a new drug application (NDA) (22) is filed with the Food and Drug Administration (FDA). When the NDA has been reviewed and approved by the FDA the product may be marketed in the United States (23).

SOURCE: Genentech, Inc.

class of molecules that control gene activity, are present in the body in only minute quantities. Now, for the first time, they can be recognized, isolated, characterized, and produced in quantity.

The mere availability of a pharmacologically active substance does not ensure its adoption in medical practice. Even if it is shown to have therapeutic usefulness, it may not succeed in the marketplace.

The difficulty in predicting the economic impact is exemplified by interferon. If it is found to be broadly effective against both viral diseases and cancers, sales would be in the tens of billions of dollars annually. If its clinical effectiveness is found to be only against one or two viruses, sales would be significantly lower.

At the very least, even if there are no immediate medical uses for compounds produced by genetic engineering, their indirect impact on medical research is assured. For the first time, almost any biological phenomenon of medical interest can be explored *at the cellular level*. These molecules are valuable tools for understanding the anatomy and functions of cells. The knowledge gained may lead to the development of new therapies or preventive measures for diseases.

The chemical industry

FINDINGS

The chemical industry's primary raw material, petroleum, is now in limited supply. Coal is one appealing alternative; another is biomass, a renewable resource composed of plant and animal material.

Biomass has been transformed by fermentation into organic chemicals like citric acid, ethanol, and amino acids for decades. Other organic chemicals such as acetone, butanol, and fumaric acid were at one time made by fermentation until chemical production methods, combined with cheap oil and gas, proved to be more economical. In theory, most any industrial organic chemical can be produced by a biological process.

Commercial fermentation using genetically engineered micro-organisms offers several ad-

vantages over current chemical production techniques.

- *The use of renewable resources:* starches, sugars, cellulose, and other components of biomass can serve as the raw material for synthesizing organic chemicals. With proper agricultural management, biomass can assure a continuous renewable supply for the industry.
- *The use of physically milder conditions:* chemical processes often require high temperatures and extreme pressures. These conditions are energy intensive and pose a hazard in case of accidents. Biological processes operate under milder conditions, which are compatible with living systems.
- *One-step production methods:* micro-organisms can carry out several steps in a synthetic process, eliminating the need for intermediate steps of separation and purification.
- *Decreased pollution:* because biological processes are highly specific in the reactions they catalyze, they offer control over the products formed and decrease undesirable side-products. As a result, they produce fewer pollutants that require management and disposal.

The impact of this technology will cut across the entire spectrum of chemical groups: plastics and resin materials, flavors and perfumes materials, synthetic rubber, medicinal chemicals, pesticides, and the primary products from petroleum that serve as the raw materials for the synthesis of organic chemicals. Nevertheless, the specific products that will be affected in each group can only be chosen on a case-by-case basis, with the applicability of genetics depending on a variety of factors. Crude estimates of the expected economic impacts are in the billions of dollars per year for dozens of chemicals within 20 years.

INDUSTRY AND MANPOWER IMPACTS

Although genetic engineering will develop new techniques for synthesizing many substances, the direct displacement of any current industry seems doubtful. Genetic engineering should be considered simply another industrial tool. Industries will probably use genetic engineering to maintain their positions in their respective markets. This is already illustrated by the variety of companies in the pharmaceutical, chemical, and energy industries that have invested in or contracted with genetic engineering firms. Some large companies are already developing inhouse genetic engineering research capabilities.

Any predictions of the number of workers that will be required in the production phase of biotechnology will depend on the expected volume of chemicals that will be produced. At present, this figure is unknown. An estimated $15 billion worth of chemicals may be manufactured by biological processes. This will employ approximately 30,000 to 75,000 workers for supervision, services, and production. Whether this will represent a net loss or gain in the number of jobs is difficult to predict since new jobs in biotechnology will probably displace some of those in traditional chemical production.

Food processing industry

FINDINGS

Genetics in the food processing industry can be used in two ways: to design micro-organisms that transform inedible biomass into food for human consumption or into feed for animals; and to design organisms that aid in food processing, either by acting directly on the food itself or by providing materials which can be added to food.

The use of genetics to design organisms with desired properties for food processing is an established practice. Fermented foods and beverages have been made by selected strains of mutant organisms (e.g., yeasts) for centuries. Only recently, however, have molecular technologies opened up new possibilities. In particular, large-scale availability of enzymes will play an increasing role in food processing.

The applications of molecular genetics are likely to appear in the food processing industry in piecemeal fashion:

- Inedible biomass, human and animal wastes, and even various industrial effluents are now being transformed into edible micro-organisms high in protein content (called single-cell protein or SCP). Its present cost of production in the United States is relatively high, and it must compete with cheaper sources of protein such as soybeans and fishmeal, among others.
- Isolated successes can be anticipated for the production of such food additives as fructose (a sugar) and the synthetic sweetener aspartame, and for improvements in SCP production.

An industrywide impact is not expected in the near future because of several major conflicting factors:

- The basic knowledge of the genetic characteristics that could improve food has not been adequately developed.
- The food processing industry is conservative in its expenditures for R&D to improve processes. Generally, only one-third to one-half as much is allocated for this purpose as in technologically intensive industries.
- Products made by new microbial sources must satisfy the Food and Drug Administration's (FDA) safety regulations, which include undergoing tests to prove lack of harmful effects. It may be possible to reduce the amount of required testing by transferring the desired gene into micro-organisms that already meet FDA standards.

The use of genetically engineered micro-organisms in the environment

FINDINGS

Genetically engineered micro-organisms are being designed now to perform in three areas (aside from agricultural uses) that require their large-scale release into the environment:

- mineral leaching and recovery,
- enhanced oil recovery, and
- pollution control.

All of these are characterized by:

- the use of large volumes of micro-organisms,
- decreased control over the behavior and fate of the micro-organisms,

- the possibility of ecological disruption, and
- less development in basic R&D (and more speculation) than in the industries in which micro-organisms are used in a controlled environment.

MINERAL LEACHING AND RECOVERY

Bacteria have been used to leach metals, such as uranium and copper, from low-grade ores. Although there is reason to believe leaching ability is under genetic control in these organisms, practically nothing is known about the precise mechanisms involved. Therefore, the application of genetic technologies in this area remains speculative. Progress has been slow in obtaining more information, partly because very little research has been conducted.

In addition to leaching, micro-organisms can be used to recover valuable metals or eliminate polluting metals from dilute solutions such as industrial waste streams. The process makes use of the ability of micro-organisms to bind metals to their surfaces and then concentrate them internally.

The economic competitiveness of biological methods is still unproved, but genetic modifications have been attempted only recently. The cost of producing the micro-organisms has been a major consideration. If it can be reduced, the approach might be useful.

ENHANCED OIL RECOVERY

Many methods have been tried in efforts to remove oil from the ground when natural expulsive forces alone are no longer effective. Injecting chemicals into a reservoir has, in many cases, aided recovery by changing the oil's flow characteristics.

Micro-organisms can produce the necessary chemicals that help to increase flow. Theoretically, they can also be grown in the wells themselves, producing those same chemicals in situ. The currently favored chemical, xanthan, is far from ideal for increasing flow. Genetic engineering should be able to produce chemicals with more useful characteristics.

The current research approach, funded by the Department of Energy (DOE) and independently by various oil companies, is a two-phase process to find micro-organisms that can function in an oil reservoir environment, and then to improve their characteristics genetically.

The genetic alteration of micro-organisms to produce chemicals useful for enhanced oil recovery has been more successful than the alteration of micro-organisms that may be used in situ. However, rDNA technology has not been applied to either case. All attempts have employed artificially induced or naturally occurring mutations.

POLLUTION CONTROL

Many micro-organisms can consume various kinds of pollutants, changing them into relatively harmless materials before they die. These micro-organisms always have had a role in "natural" pollution control: nevertheless, cities have resisted adding microbes to their sewerage systems. Although the Environmental Protection Agency (EPA) has not recommended addition of bacteria to municipal sewerage systems, it suggests that they might be useful in smaller installations and for specific problems in large systems. In major marine spills, the bacteria, yeast, and fungi already present in the water participate in degradation. The usefulness of added microbes has not been demonstrated.

Nevertheless, in 1978, the estimated market of biological products for pollution control was $2 million to $4 million/year, divided among some 20 companies; the potential market was estimated to be as much as $200 million/year.

To date, genetically engineered strains have not been applied to pollution problems. Restricting factors include the problems of liability in the event of health, economic, or environmental damage; the contention that added organisms are not likely to be a significant improvement; and the assumption that selling microbes rather than products or processes is not likely to be profitable.

Convincing evidence that microbes could remove or degrade an intractable pollutant would encourage their application. In the meantime, however, these restrictions have acted to inhibit the research necessary to produce marked improvements.

CONSTRAINTS IN USING GENETIC ENGINEERING TECHNOLOGIES IN OPEN ENVIRONMENTS

The genetic data base for the potentially useful micro-organisms is lacking. Only the simplest methods of mutation and selection for desirable properties have been used thus far. These are the only avenues for improvement until more is learned about the genetic mechanisms.

Even when the scientific knowledge is available, two other obstacles to the use of genetically engineered micro-organisms will remain. The first is the need to develop engineered systems on a scale large enough to exploit their biological activity. This will necessitate a continual dialog among microbial geneticists, geologists, chemists, and engineers; an interdisciplinary approach is required that recognizes the needs and limitations of each discipline.

The second obstacle is ecological. Introducing large numbers of genetically engineered micro-organisms into the environment might lead to ecological disruption or detrimental effects on human health, and raise questions of legal liability.

Issue and Options—Biotechnology

ISSUE: How can the Federal Government promote advances in biotechnology and genetic engineering?

The United States is a leader in applying genetic engineering and biotechnology to industry. One reason is the long-standing commitment by the Federal Government to the funding of basic biological research; several decades of support for some of the most esoteric basic research has unexpectedly provided the foundation for a highly useful technology. A second is the availability of venture capital, which has allowed the formation of small, innovative companies that can build on the basic research.

The chief argument *for* Government subsidization for R&D in biotechnology and genetic engineering is that Federal help is needed in areas such as general (generic) research or highly speculative investigations *not* now being developed by industry. The argument *against* the need for this support is that industry will develop everything of commercial value on its own.

A look at what industry is now attempting indicates that sufficient investment capital is available to pursue specific manufacturing objectives. Some high-risk areas, however, that might be of interest to society, such as pollution control, may justify promotion by the Government, while other, such as enhanced oil recovery might might not be profitable soon enough to attract investment by industry.

OPTIONS:

A. *Congress could allocate funds specifically for genetic engineering and biotechnology R&D in the budget of appropriate agencies.*

Congress could promote two types of programs in biotechnology: those with long-range payoffs (basic research), and those that industry is not willing to undertake but that might be in the national interest.

B. *Congress could establish a separate Institute of Biotechnology as a funding agency.*

The merits of a separate institute lie in the possibility of coordinating a wide range of efforts, all related to biotechnology. On the other hand, biotechnology and genetic engineering cover such a broad range of disciplines that a new funding agency would overlap the mandates of existing agencies. Furthermore, the creation of yet another agency carries with it all the disadvantages of increased bureaucracy and competition for funds at the agency level.

C. *Congress could establish research centers in universities to foster interdisciplinary approaches to biotechnology. In addition, a program of grants could be offered to train scientists in biological engineering.*

The successful use of biological techniques in industry depends on a multidisciplinary approach involving biochemists, geneticists, microbiologists, process engineers, and chemists.

Little is now being done publicly or privately to develop the expertise necessary.

D. *Congress could use tax incentives to stimulate biotechnology.*

The tax laws could be used to stimulate biotechnology by expanding the supply of capital for small, high-risk firms, which are generally considered more innovative than established firms because of their willingness to undertake the risks of innovation. In addition to focusing on the supply of capital, tax policy could attempt to directly increase the profitability of potential growth companies.

A tax incentive could also be directed at increasing R&D expenditures. It has been suggested that companies be permitted to take tax credits: 1) on a certain percentage of their R&D expenses; and 2) on contributions to universities for research.

E. *Congress could improve the conditions under which U.S. companies collaborate with academic scientists and make use of the technology developed in universities, which has been wholly or partly supported by tax funds.*

Developments in genetic engineering have kindled interest in this option. Under legislation that has recently passed both Houses of Congress, small businesses and universities may retain title to inventions developed under federally funded research. Currently, some Federal agencies award contractors these exclusive rights, while others insist on the nonexclusive licensing of inventions.

F. *Congress could mandate support for specific research tasks such as pollution control using microbes.*

Microbes may be useful in degrading intractable wastes and pollutants. Current research, however, is limited to isolating organisms from natural sources or from mutated cultures. More elaborate efforts, involving rDNA techniques or other forms of microbial genetic exchange, will require additional funding.

G. *Most efforts could be left to industry and each Government agency allowed to develop programs in the fields of genetic engineering and biotechnology as it sees fit.*

Generic research will probably not be undertaken by any one company. Leaving all R&D in industry's hands would still produce major commercial successes, but does not ensure the development of needed basic general knowledge or the undertaking of high-risk projects.

Agriculture

The complexity of plants and animals presents a greater challenge to advances in applied genetics than that posed by micro-organisms. Nevertheless, the successful genetic manipulation of microbes has encouraged researchers in the agricultural sciences. The new tools will be used to complement, but not replace, the well-established practices of plant and animal breeding.

The applications of genetics to plants

FINDINGS

It is impossible to exactly determine the extent to which applied genetics has directly contributed to increases in plant yield because of simultaneous improvements in farm management, pest control, and cropping techniques using herbicides, irrigation, and fertilizers. Nevertheless, the impacts of breeding technologies have been extensive.

The plant breeder's approach is determined for the most part by the particular biological factors of the crop being bred. The new genetic technologies potentially offer *additional* tools to allow development of new varieties and even species of plants by circumventing current biological barriers to the exchange of genetic material.

Technologies developed for classical plant breeding and those of the new genetics should not be viewed as being competitive; they are both tools for effectively manipulating genetic information. One new technology—e.g., protoplast fusion, or the artificial fusion of two cells—allows breeders to overcome incompatibility between plants. But the plant that may result still must be selected, regenerated, and evaluated under field conditions to ensure that the genetic change is stable and that the attributes of the new variety meet commercial requirements.

In theory, the new technologies will expand the capability of breeders to exchange genetic information by overcoming natural breeding barriers. To date, however, they have not had a widespread impact on the agricultural industry.

As a note of caution, it must be emphasized that no plant can possess every desirable trait. There will always have to be some tradeoff; often quality for quantity, such as increased protein content but decreased yield.

NEW GENETIC TECHNOLOGIES FOR PLANT BREEDING

The new technologies fall into two categories: those involving genetic transformations through cell fusion and those involving the insertion or modification of genetic information through the cloning of DNA and its vectors. Techniques are available for manipulating organs, tissues, cells, or protoplasts in culture; for regenerating plants; and for testing the genetic basis of novel traits. So far these techniques are routine only in a few species.

The approach to exploiting molecular biology for plant breeding is similar in some respects to the genetic manipulation of micro-organisms. However, there is one major conceptual dif-

A plantlet of loblolly pine grown in Weyerhaeuser Co.'s tissue culture laboratory. The next step in this procedure is to transfer the plantlet from its sterile and humid environment to the soil

Photo credits: Weyerhaeuser Co.

A young Douglas fir tree propagated 4 years ago from a small piece of seedling leaf tissue. Three years ago this was at the test-tube stage seen in the loblolly pine photograph

ference. In micro-organisms, the changes made on the cellular level are the goals of the manipulation. With crops, changes made on the cellular level are meaningless unless they can be reproduced in the entire plant as well. Therefore, unless single cells in culture can be selected and grown into mature plants and the desired traits expressed in the mature plant—procedures which at this time have had limited success—the benefits of genetic engineering will not be widely felt in plant breeding.

Moderate success has been achieved for growing cells in tissue culture into mature plants. Tissue culture programs of commercial significance in the United States include the asparagus, citrus fruits, pineapples, and strawberries. Breeders have had little success, however, in regenerating mature plants of wide agronomic importance, such as corn and wheat.

Some success can be claimed for engineering changes to alter genetic makeup. Both the stable integration of genetic material into a cell and the fusion of genetically different cells are still largely experimental techniques. Technical breakthroughs have come on a species-by-species basis, but key discoveries are not often applicable to all plants. Initial results suggest that agronomically important traits, such as disease resistance, can be transferred from one species to another. Limited success has also been shown in attempts to create totally new species by fusing cells from different genera. Attempts to find both suitable vectors and genes for transferring one plant's genes to another are only now beginning to show promise.

CONSTRAINTS ON USING MOLECULAR GENETICS FOR PLANT IMPROVEMENTS

Molecular engineering has been impeded by a lack of answers to basic questions in molecular biology and plant physiology owing to insufficient research. Federal funding for plant molecular genetics in agriculture has come primarily from the U.S. Department of Agriculture (USDA) and the National Science Foundation (NSF). In USDA, research support is channeled primarily through the flexible competitive grants program (fiscal year 1980 budget of $15 million) for the support of new research directions in plant biology. The total support for the

plant sciences from NSF is approximately $25 million, only $1 million of which is specifically designated for plant genetics.

The shortage of a trained workforce is a significant constraint. Only a few universities have expertise in both plants and molecular biology. In addition, there are only a few people who have the ability to work with modern molecular techniques related to whole plant problems. As a result, a business firm could easily develop a capability in this area exceeding that at any individual U.S. university. However, the building of industrial laboratories and subsequent hiring from the universities could easily deplete the expertise at the university level. With the recent investment activity by many bioengineering firms, this trend has already begun; in the long-run it could have serious consequences for the quality and quantity of university research.

GENETIC VARIABILITY, CROP VULNERABILITY, AND THE STORAGE OF GERMPLASM

Successful plant breeding is based on the availability of genetically diverse plants for the insertion of new genes into plants. The number of these plants has been diminishing for a variety of reasons. However, the rate and extent of this trend is unknown; the data simply do not exist. Therefore, it is essential to have an adequate scientific understanding of how much genetic loss has taken place and how much germplasm (the total genetic variability available to a species) is needed. Neither of these questions can be answered completely at this time.

Even if genetic needs can be adequately identified, there is disagreement about the quantity of germplasm to collect. Furthermore, the extent to which the new genetic technologies will affect genetic variability, vulnerability, or the storage technologies of germplasm has not been determined. As a result, it is currently difficult, if not impossible, to state how much effort should be expended by the National Germplasm System to collect, maintain, and test new gene resources (in this case as seed).

Finally, even if an adequate level of genetic variability can be assessed, the real problem of vulnerability—the practice of planting only a single variety—must be dealt with at an institutional or social level. Even if no genetic tech- nologies existed, farmers would still select only one or a few "best" varieties for planting.

Issues and Options—Plants

ISSUE: Should an assessment be conducted to determine how much diversity in plant germplasm needs to be maintained?

An understanding of how much germplasm should be protected and maintained would make the management of genetic resources simpler.

OPTIONS:

A. *Congress could commission a study of how much genetic variability is necessary or desirable to meet present and future needs.*

A comprehensive evaluation of the National Germplasm System's requirements for collecting, evaluating, maintaining, and distributing genetic resources for plant breeding and research could serve as a baseline for a further assessment.

B. *Congress could commission a study on the need for international cooperation to manage and preserve genetic resources both in natural ecosystems and in repositories.*

This investigation could include an evaluation of the rate at which genetic diversity is being lost from natural and agricultural systems along with an estimate of the effects this loss will have.

C. *Congress could commission a study on how to develop an early warning system to recognize the potential vulnerability of crops.*

Where high genetic uniformity still exists, proposals could be suggested to reduce any risks due to uniformity. Alternatively, the avenues by which private seed companies could be encouraged to increase the levels of genetic diversity could be investigated.

ISSUE: What are the most appropriate approaches in overcoming the various technical constraints that limit the success of molecular genetics for plant improvement?

Although genetic information has been transferred by vectors and protoplast fusion, DNA transformations of commercial value have not yet been performed. Molecular engineering has been impeded by the lack of vectors that can transfer novel genetic material into plants, by insufficient knowledge about which genes would be useful for breeding purposes, and by a lack of understanding of the incompatibility of chromosomes from diverse sources. Another impediment has been the lack of researchers from a variety of disciplines.

OPTIONS:

A. *The level of funding could be increased for plant molecular genetics research supported by NSF and USDA.*

B. *Research units devoted to plant molecular genetics could be established under the auspices of the National Institutes of Health (NIH), with emphasis on potential pharmaceuticals derived from plants.*

C. *An institute for plant molecular genetics could be established under the Science and Education Administration at USDA that would include multidisciplinary teams to consider both basic research questions and direct applications of the technology to commercial needs and practices.*

The discoveries of molecular plant genetics will be used in conjunction with traditional breeding programs. Hence, each of the three options could require additional appropriations for agricultural research.

Advances in reproductive biology and their effects on animal improvement

FINDINGS

Much improvement can be made in the germplasm of all major farm animal species using existing technology. The expanded use of artificial insemination (AI) with stored frozen sperm, especially in beef cattle, would benefit both producers and consumers. New techniques for synchronizing estrus should encourage the wider use of AI. Various manipulations of embryos will find limited use in producing breeding stocks, and sex selection and twinning techniques should be available for limited applications within the next 10 to 20 years.

The most important technology in reproductive physiology will continue to be AI. Due in part to genetic improvement, the average milk yield of cows in the United States has more than doubled in the past 30 years, while the total number of milk cows has been reduced by more than half. AI along with improved management and the availability and use of accurate progeny records on breeding stock have caused this great increase. (See figure 3.)

The improvement lags behind what is theoretically possible. In practice, the observed increase is about 100 lb of milk per cow per year, while a hypothetical breeding program using AI would result in a yearly gain of 220 lb of milk per cow. The biological limits to this rate of gain are not known.

In comparison with dairy cattle, the beef cattle industry has not applied AI technology widely. Only 3 to 5 percent of U.S. beef is artificially inseminated, compared to 60 percent of the dairy herd. This low rate for beef cattle can be explained by several factors, including management techniques (range v. confined housing) and the conflicting objectives of individual breeders, ranchers, breed associations, and commercial farmers.

The national calf crop—calves alive at weaning as a fraction of the total number of cows exposed to breeding each year—is only 65 to 81 percent. An improvement of only a few percentage points through AI would result in savings of

hundreds of millions of dollars to producers and consumers.

Coupled with a technology for estrus-cycle regulation, the use of AI could be expanded for both dairy and beef breeding. Embryo transfer technology, already well-developed but still costly, can be used to produce valuable breeding stock. Sexing technology, which is not yet perfected, would be of enormous benefit to the beef industry because bulls grow faster than heifers.

In the case of animals other than cows:

- Expanded use of AI for swine production will be encouraged by the strong trend to confinement housing, although the poor ability of boar sperm to withstand freezing will continue to be a handicap.
- The benefits of applied genetics have not been realized in sheep production because neither AI nor performance testing has been used. As long as the use of AI continues to be limited by the inability to freeze semen and by a lack of agents on the market for synchronizing estrus, no rapid major gains can be expected.
- Increasing interest in goats in the United States and the demand for goat products throughout the world, should encourage attention to the genetic gains that the use of AI and other technologies make possible.
- Poultry breeders will continue to concentrate on improved egg production, growth rate, feed efficiency, and reduced body fat and diseases. The use of frozen semen should increase as will the use of AI and dwarf broiler breeders.
- Genetics applied to production of fish, mollusks, and crustaceans in either natural environments or manmade culture systems is only at the rudimentary stage.

Breeders must have reliable information about the genetic value of the germplasm they are considering introducing. Since farmers do not have the resources to collect and process data on the performance of animals other than those in their own herds, they must turn to outside sources. The National Cooperative Dairy Herd Improvement Program (NCDHIP) is a mod-

Figure 3.—The Way the Reproductive Technologies Interrelate

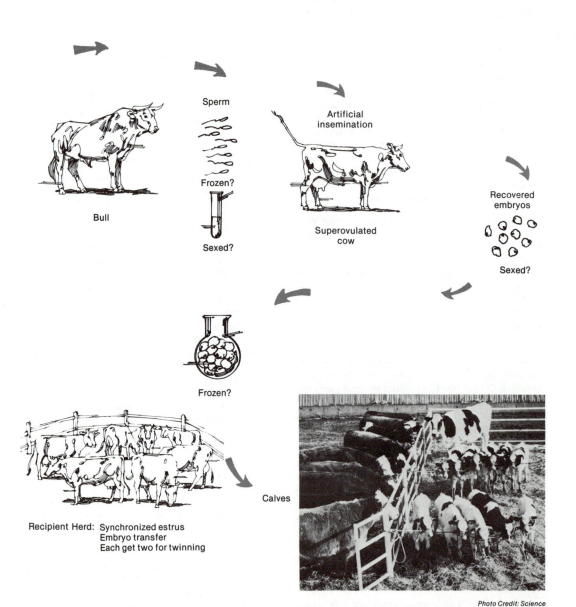

Sperm

Frozen?

Sexed?

Bull

Artificial
insemination

Superovulated
cow

Recovered
embryos

Sexed?

Frozen?

Recipient Herd: Synchronized estrus
Embryo transfer
Each get two for twinning

Calves

Photo Credit: Science

These 10 calves from Colorado State University were the
result of superovulation, in vitro culture, and transfer to
the surrogate mother cows on the left. The genetic
mother of *all* 10 calves is at upper right

SOURCE: Office of Technology Assessment.

el information system and could be adapted to other species.

Selection—deciding which animals to mate —is the breeder's most basic tool. When going outside his herd to purchase new germplasm, the breeder needs impartial information about the quality of the available germplasm. NCDHIP had recorded 2.8 million of the 10.8 million U.S. dairy cattle in 1979. In 1978, cows enrolled in the official plans of NCDHIP outproduced cows not enrolled by 5,000 lb of milk per cow, representing 52 percent more milk per lactation.

No comparable information system exists for other types of livestock. Beef bulls, for example, continue to be sold to a large extent on the basis of pedigrees, but with relatively little objective information on their genetic merit. Data on dairy goats in the United States became available through NCDHIP for the first time in late 1980. No nationwide information systems exist

for other species, although pork production in the United States would greatly benefit from a national swine testing program.

The more esoteric methods of genetic manipulation will probably have little impact on the production of animals or animal products within the next 10 years. Other in vitro manipulations, such as cloning, cell fusion, the production of chimeras, and the use of rDNA techniques, will continue to be of intense interest, especially for research purposes. It is less likely, however, that they will have widespread practical effects on farm production in this century.

Each technique requires more research and refinement. Until specific genes of farm animals can be identified and located, no direct gene manipulation will be practicable. In addition this will be difficult because most traits of importance are due to multiple genes.

Issue and Options—Animals

ISSUE: How can the Federal Government improve the germplasm of major farm animal species?

OPTIONS:

A. *Programs like the NCDHIP could have increased governmental participation and funding. The efforts of the Beef Cattle Improvement Federation to standardize procedures could receive active support, and a similar information system for swine could be established.*

The fastest and least expensive way to upgrade breeding stock in the United States is through effective use of information. Computer technology, along with a network of local representatives for data collecting, can provide the individual farmer or breeder with accurate information on the available germplasm so that he can make his own breeding decisions.

This option implies that the Federal Government would play such a role in new programs, and expand its role in existing ones.

B. *Federal funding could be increased for basic research in total animal improvement.*

This option, in contrast to option A, assumes that it is necessary to maintain or expand basic R&D to generate new knowledge that can be applied to the production of improved animals and animal products.

*   *   *

The wide variety of applications for genetic engineering is summarized in figure 4. Genetics can be used to improve or increase the quality and output of plants and animals for direct use by man. Alternatively, materials can be extracted from plants and animals for use in food, chemical, and pharmaceutical industries.

Biological materials can also be converted to useful products. In this process, genetic engineering can be used to develop micro-organisms that will carry out the conversions. Therefore, genetic manipulation cannot only provide more or better biological raw materials but can also aid in their conversion to useful products.

Figure 4.—Applications of Genetics

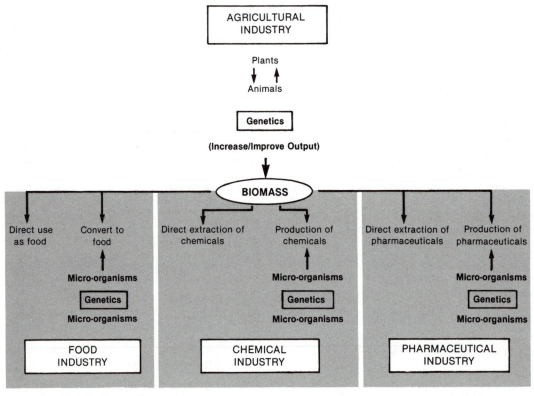

SOURCE: Office of Technology Assessment.

Institutions and society

Regulation of genetic engineering

FINDINGS

No evidence exists that any unexpectedly harmful genetically engineered organism has been created. Yet few experts believe that molecular genetic techniques are totally without risk to health and the environment. Information that has proved useful in assessing the risks from these techniques has come from three sources: experiments designed specifically to test the consequences of working with rDNA, experiments designed for other purposes but

relevant to rDNA, and scientific meetings and workshops.

A program of risk assessment was established at NIH in 1979 to conduct experiments and collate relevant information. It assesses one form of genetic engineering, rDNA. On the basis of these data, conjectured, inadvertant risk is generally regarded as less likely today than originally suspected. Risk due to the manipulation of genes from organisms known to be hazardous is considered to be more realistic. Therefore, microbiological safety precautions that are appropriate to the use of the micro-organisms serving as the source of DNA are required. Nevertheless, it has not been demonstrated that combining those genes in the form of rDNA is any more hazardous than the original source of the DNA.

Perceptions of the nature, magnitude, and acceptability of the risk differ. In addition, public concern has been expressed about possible long-range impacts of genetic engineering. In this context, the problem facing the policymaker is how to address the risk in a way that accommodates the perceptions and values of those who bear it.

The NIH Guidelines for Research Involving Recombinant DNA Molecules and existing Federal laws appear adequate in most cases to deal with the risks to health and the environment presented by genetic engineering. However, the Guidelines are not legally binding on industry, and no single statute or combination will clearly cover all foreseeable commercial applications of genetic engineering.

The Guidelines are a flexible evolving oversight mechanism that combines technical expertise with public participation. They cover the most widely used and possibly risky molecular genetic technique—rDNA—prohibiting experiments using dangerous toxins or pathogens and setting containment standards for other potentially hazardous experiments. Although compliance is mandatory only for those receiving NIH funds, other Federal agencies follow them, and industry has proclaimed voluntary compliance. Rare cases of noncompliance have occurred in universities but have not posed risks to health or the environment. As scientists have learned more about rDNA and molecular genetics, the restrictions have been progressively and substantially relaxed to the point where 85 percent of the experiments can now be done at the lowest containment levels, and virtually all monitoring for compliance now rests with approximately 200 local self-regulatory committees called institutional biosafety committees (IBCs). (See table 1.)

Under the Guidelines, NIH serves an important oversight role by sponsoring risk assess-

Table 1.—Containment Recommended by the National Institutes of Health

Biological—Any combination of vector and host must be chosen to minimize both the survival of the system outside of the laboratory and the transmission of the vector to nonlaboratory hosts. There are three levels of biological containment:

HV1—	Requires the use of *Escherichia coli* K12 or other weakened strains of micro-organisms that are less able to live outside the laboratory.
HV2—	Requires the use of specially engineered strains that are especially sensitive to ultraviolet light, detergents, and the absence of certain uncommon chemical compounds.
HV3—	No organism has yet been developed that can qualify as HV3.

Physical—Special laboratories (P1-P4)

P1—	Good laboratory procedures, trained personnel, wastes decontaminated
P2—	Biohazards sign, no public access, autoclave in building, hand-washing facility
P3—	Negative pressure, filters in vacuum line, class II safety cabinets
P4—	Monolithic construction, air locks, all air decontaminated, autoclave in room, all experiments in class III safety cabinets (glove box), shower room

SOURCE: Office of Technology Assessment.

ment programs, certifying new host-vector systems, serving as an information clearinghouse, and coordinating Federal and local activities. Limitations in NIH's oversight are that: it lacks legal authority over industry; its procedures for advising industry on large-scale projects have not incorporated sufficient expertise on large-scale fermentation technology; its monitoring for either compliance or consistent application of the Guidelines by individuals or institutions is virtually nonexistent; and it has not systematically evaluated other techniques, such as cell fusion, that might present risks.

Federal laws on health and environment will cover most commercial applications of genetic engineering. Products such as drugs, chemicals, and foods can be regulated by existing laws. However, uncertainty exists about the regulation of either production methods using engineered micro-organisms or their intentional release into the environment, when the risk has not been clearly demonstrated. While a broad interpretation of certain statutes, such as the Occupational Safety and Health Act and the Toxic Substances Control Act, might cover these situations, regulatory actions based on such interpretations could be challenged in court. In any event, those agencies that could have substantial regulatory authority over commercial genetic engineering have not yet officially acted to assert that authority.

Issue and Options—Regulation

ISSUE: How could Congress address the risks presented by genetic engineering?

OPTIONS:

A. *Congress could maintain the status quo by letting NIH and the regulatory agencies set the Federal policy.*

Congress might determine that legislation to remedy the limitations in current Federal oversight would result in unnecessary and burdensome regulation. No known harm to health or the environment has occurred under current regulation. Also the agencies generally have the legal authority and expertise to adapt to most new problems posed by genetic engineering.

The disadvantages are the lack of a centralized, uniform Federal response to the problem, and the possibility that risks associated with commercial applications will not be adequately addressed. Conflicting or redundant regulations of different agencies would result in unnecessary burdens on those regulated.

B. *Congress could require that the Federal Interagency Advisory Committee on Recombinant DNA Research prepare a comprehensive report on its members' collective authority to regulate rDNA and on their regulatory intentions.*

The Industrial Practices Subcommittee of this Committee has been studying agency authority over commercial rDNA activities. Presently, there is little official guidance on regulatory requirements for companies that may soon market products made by rDNA methods. A congressionally mandated report would ensure full consideration of these issues by the agencies and expedite the process. On the other hand, the agencies are studying the situation, which must be done before they can act. Also, it is often easier and more efficient to act on each case as it arises, rather than on a hypothetical basis before the fact.

C. *Congress could require that all recombinant DNA activity be monitored for a limited number of years.*

This represents a "wait and see" position by Congress and the middle ground between the status quo and full regulation. It recognizes and balances the following factors: 1) the absence of demonstrated harm to human health or the environment from genetic engineering; 2) the continuing concern that genetic engineering presents risks; 3) the lack of sufficient knowledge and experience from which to make a final judgment; 4) the existence of an oversight mechanism that seems to be working well, but that has clear limitations with respect to commercial activities; 5) the virtual abolition of Federal monitoring of rDNA activities by recent amendments to the Guidelines; and 6) the expected increase in commercial genetic engineering.

This option would provide a data base that could be used for: 1) determining the effectiveness of voluntary compliance with the Guidelines by industry, and mandatory compliance by Federal grantees; 2) determining the quality and consistency of the local self-regulatory actions; 3) continuing a formal risk assessment program; 4) identifying vague or conflicting provisions of the Guidelines for revision; 5) identifying emerging trends or problems; and 6) tracing any long-term adverse impacts on health or the environment to their sources.

The obvious disadvantage of this option would be the required paperwork and effort by scientists, universities, corporations, and the Federal Government.

D. *Congress could make the NIH Guidelines applicable to all rDNA work done in the United States.*

This option would eliminate any concern about the effectiveness of voluntary compliance with the Guidelines, and it has the advantage of using an existing oversight mechanism. The major changes that would have to be made in the area of enforcement. Present penalties for non-compliance—suspension or termination of research funds—are obviously inapplicable to industry. In addition, procedures for monitoring compliance would have to be strengthened.

The main disadvantage of this option is that NIH is not a regulatory agency. Since NIH has traditionally viewed its mission as promoting biomedical research, it would have a conflict of interest between regulation and promotion. One of the regulatory agencies could be given the authority to enforce the Guidelines.

E. *Congress could require an environmental impact statement and agency approval before any genetically engineered organism is intentionally released into the environment.*

There have been numerous cases where an animal or plant species has been introduced into a new environment and has spread in an uncontrolled and undesirable fashion. Yet in pollution control, mineral leaching, and enhanced oil recovery, it might be desirable to release large numbers of engineered micro-organisms into the environment.

The Guidelines currently prohibit deliberate release of any organism containing rDNA without approval of NIH. One disadvantage of this prohibition is that it lacks the force of law. Another is that approval may be granted on a finding that the release would present "no significant risk to health or the environment;" a tougher or more specific standard may be desirable.

A required study of the possible consequences of releasing a genetically engineered organism would be an important step in ensuring safety. An impact statement could be filed before permission is granted to release the organism. However, companies and individuals might be discouraged from developing useful organisms if this process became too burdensome and costly.

F. *Congress could pass legislation regulating all types and phases of genetic engineering from research through commercial production.*

This option would deal comprehensively and directly with the risks of novel molecular genetic techniques. A specific statute would eliminate the uncertainties over the extent to which present law covers particular applications of genetic engineering and any concerns about the effectiveness of voluntary compliance with the Guidelines. Alternatively, the legislation could take the form of amending existing laws to clarify their applicability to genetic engineering.

Other molecular genetic techniques, while not as widely used and effective as rDNA, raise similar concerns. Of the current techniques, cell fusion is the prime candidate for being treated like rDNA in any regulatory framework. No risk assessment of this technique has been done, and no Federal oversight exists.

The principal argument against this option is that the current system appears to be working fairly well, and the limited risks of the techniques may not warrant the significantly increased regulatory burden that would result from such legislation.

G. *Congress could require NIH to rescind the Guidelines.*

Deregulation would have the advantage of allowing money and personnel currently involved in implementing the Guidelines at the Federal and local levels to be used for other purposes.

There are several reasons for retaining the Guidelines. Sufficient scientific concern exists for the Guidelines to prohibit certain experiments and to require containment for others. Most experiments can be done at the lowest, least burdensome containment levels. NIH is serving an important role as a centralized oversight and information coordinating body, and the system has been flexible enough in the past to liberalize the restrictions as evidence indicated lower risk than originally thought.

H. *Congress could consider the need for regulating work with all hazardous micro-organisms and viruses, whether or not they are genetically engineered.*

It was not within the scope of this study to examine this issue, but it is an emerging one that Congress may wish to consider.

Patenting living organisms

On June 16, 1980, in a 5-to-4 decision, the Supreme Court ruled that a human-made micro-organism was patentable under Federal patent statutes. The decision while hailed by some as assuring this country's technological future was at the same time denounced by others as creating Aldous Huxley's Brave New World. It will do neither.

FINDINGS

1. *Meaning and Scope of the Decision.*—The decision held that a patent could not be denied on a genetically engineered micro-organism that otherwise met the legal requirements for patentability solely because it was alive. It was based on the Court's interpretation of a provision of the patent law which states that a patent may be granted on ". . . any new and useful . . . manufacture, or composition of matter. . . ." (35 U.S.C. § 101)

It is uncertain whether the case will serve as a legal precedent for patenting more complex organisms. Such organisms, however, will probably not meet other legal prerequisites to patentability that were not at issue here. In any event, fears that the case would be legal precedent sometime in the distant future for patenting human beings are unfounded because the 13th amendment to the Constitution absolutely prohibits ownership of humans.

2. *Impact on the Biotechnology Industry.*—The decision is not crucial to the development of the industry. It will stimulate innovation by encouraging the dissemination of technical information that otherwise would have been maintained as trade secrets because patents are public documents that fully describe the inventions. In addition, the ability to patent genetically engineered micro-organisms will reduce the risks and uncertainties facing individual companies in the commercial development of those organisms and their products, but only to a limited degree because reasonably effective alternatives exist. These are: 1) maintaining the orga-

nisms as trade secrets; 2) patenting microbiological processes and their products; and 3) patenting inanimate components of micro-organisms, such as genetically engineered plasmids.

3. *Impact on the Patent Law and the Patent and Trademark Office.*—Because of the complexity, reproducibility, and mutability of living organisms, the decision may cause some problems for a body of law designed more for inanimate objects than for living organisms. It raises questions about the proper interpretation and application of the patent law requirements of novelty, nonobviousness, and enablement. In addition, it raises questions about how broad the scope of patent coverage on important micro-organisms should be, and about the continuing need for two statutes, the Plant Patent Act of 1930 and the Plant Variety Protection Act of 1970. These uncertainties could result in increased litigation, making it more difficult and costly for owners of patents on living organisms to enforce their rights.

The impact on the Patent and Trademark Office is not expected to be significant in the next few years. Although the number of patent applications on micro-organisms have almost doubled during 1980, the approximately 200 pending applications represent less than 0.2 percent of those processed each year by the Office. While the number of such applications is expected to increase in the next few years because of of the decision and developments in the field, the Office should be able to accommodate the increase. A few additional examiners may be needed.

4. *Impact on Academic Research.*—Because the decision may encourage academic scientists to commercialize the results of their research, it may inhibit the free exchange of information, but only if scientists rely on trade secrecy rather than patents to protect their inventions from competitors in the marketplace. In this respect, it is not clear how molecular biology differs from other research fields with commercial potential.

Issue and Options—Patenting Living Organisms

ISSUE: **To what extent could Congress provide for or prohibit the patenting of living organisms?**

OPTIONS:

The Supreme Court stated that it was undertaking only the narrow task of determining whether or not Congress, in enacting the patent statutes, had intended a manmade micro-organism to be excluded from patentability solely because it was alive. Moreover, the opinion specifically invited Congress to overrule the decision if it disagreed with the Court's interpretation. Congress can act to resolve the questions left unanswered by the Court, overrule the decision, or develop a comprehensive statutory approach. Most importantly, Congress can draw lines; it can decide which organisms, if any, should be patentable.

A. Congress could maintain the status quo.

Congress could choose not to address the issue of patentability and allow the law to be developed by the courts. The advantage of this option is that issues will be addressed as they arise, in the context of a tangible, nonhypothetical case.

There are two disadvantages to this option: a uniform body of law may take time to develop; and the Federal judiciary is not designed to take sufficient account of the broader political and social interests involved.

B. Congress could pass legislation dealing with the specific legal issues raised by the Court's decision.

Many of the legal questions are so broad and varied that they do not readily lend themselves to statutory resolution. The precise meaning of the requirements for novelty, nonobviousness, and enablement as applied to biological inventions will be most readily developed on a case-by-case basis by the Patent Office and the Federal courts. On the other hand, some questions are fairly narrow and well-defined; thus, they could be better resolved by statute. The most important question is whether there is a continuing need for the two plant protection

acts that grant ownership rights to plant breeders who develop new and distinct varieties of plants.

C. Congress could mandate a study of the Plant Patent Act of 1930 and the Plant Variety Protection Act of 1970.

These Acts could serve as models for studying the broader, long-term potential impacts of patenting living organisms. Such a study would be timely not only because of the Court's decision, but also because of allegations that the Acts have encouraged the planting of uniform varieties, loss of genetic diversity, and increased concentration in the plant breeding industry.

D. Congress could prohibit patents either on any living organism or on organisms other than those already subject to the plant protection Acts.

By prohibiting patents on any living organisms, Congress would be accepting the arguments of those who consider ownership rights in living organisms to be immoral, or who are concerned about other potentially adverse impacts of such patents. A total prohibition would slow but not stop the development of molecular genetic techniques and the biotechnology industry because there are several good alternatives for maintaining exclusive control of biological inventions. Development would be slowed primarily because information that might otherwise become public would be withheld as trade secrets. A major consequence would be that desirable products would take longer to reach the market.

Alternatively, Congress could overrule the Supreme Court's decision by amending the patent law to prohibit patents on organisms other than the plants covered by the two statutes mentioned in option C. This would demonstrate congressional intent that living organisms could be patented only by specific statute.

E. Congress could pass a comprehensive law covering any or all organisms (except humans).

This option recognizes that Congress can draw lines where it sees fit in this area. It could specifically limit patenting to micro-organisms, or it could encourage the breeding of agriculturally important animals by granting patent rights to breeders of new and distinct breeds. In the interest of comprehensiveness and uniformity, one statute could cover plants and all other organisms that Congress desires to be patentable.

Genetics and society

FINDINGS

Continued advances in science and technology are beginning to provide choices that strain human value systems in areas where previously no choice was possible. Existing ethical and moral systems do not provide clear guidelines and directions for those choices. New programs, both in public institutions and in the popular media, have been established to explore the relationships among science, technology, society, and value systems, but more work needs to be done.

Genetics—and other areas of the biological sciences—have in common a much closer relationship to certain ethical questions than do most advances in the physical sciences or engineering. The increasing control over the characteristics of organisms and the potential for altering inheritance in a directed fashion raise again questions about the relationship of humans to each other and to other living things. People respond in different ways to this potential; some see it (like many predecessor developments in science) as a challenging opportunity, others as a threat, and still others respond with vague unease. Although many people cannot articulate fully the basis for their concern, ethical, moral, and religious reasons are often cited.

The public's increasing concern about the advance of science and impacts of technology has led to demands for greater participation in decisions concerned with scientific and technological issues, not only in the United States but throughout the world. The demands imply new challenges to systems of representative government. In every Western country, new mechanisms have been devised for increasing citizen participation.

The public has already become involved in decisionmaking with regard to genetics. As the science develops, additional issues in which the public will demand involvement can be anticipated for the years ahead. The question then becomes one of how best to involve the public in decisionmaking.

Issues and Options—Genetics and Society

ISSUE: How should the public be involved in determining policy related to new applications of genetics?

Because public demands for involvement are unlikely to diminish, ways to accommodate these demands must be considered.

OPTIONS:

A. *Congress could specify that public opinion must be sought in formulating all major policies concerning new applications of genetics, including decisions on the funding of specific research projects. A "Public Participation Statement" could be mandated for all such decisions.*

B. *Congress could maintain the status quo, allowing the public to participate only when it decides to do so on its own initiative.*

If option A were followed, there would be no cause for claiming that public involvement was inadequate (as occurred after the first set of Guidelines for Recombinant DNA Research was promulgated). Option A poses certain problems: How to identify a major policy and at what stage public involvement would be required. Should it take place only when technological development and application are imminent, or at the basic research stage?

Option B would be less cumbersome to effect. It would permit the establishment of ad hoc mechanisms when necessary.

ISSUE: How can the level of public knowledge concerning genetics and its potential be raised?

There are some educators who believe that too little time is spent on genetics within the traditional educational system. Outside the traditional school system, a number of sources may contribute to increased public understanding of science and the relationship between science and society.

Efforts to increase public understanding should, of course, be combined with carefully designed evaluation programs so that the effectiveness of a program can be assessed.

OPTIONS:

A. *Programs could be developed to increase public understanding of science and the relationship between science, technology, and society.*

Public understanding of science in today's world is essential, and there is concern about the adequacy of the public's knowledge.

B. *Programs could be established to monitor the level of public understanding of genetics and of science in general, and to determine whether public concern with decisionmaking in science and technology is increasing.*

Selecting this option would indicate that there is need for additional information, and that Congress is interested in involving the public indeveloping science policy.

C. *The copyright laws could be amended to permit schools to videotape television programs for educational purposes.*

Under current copyright law, videotaping television programs as they are being broadcast may infringe on the rights of the program's owner, generally its producer. The legal status of such tapes is presently the subject of litigation.

In favor of this option, it should be noted that many of the programs are made at least in part with public funds. Removing the copyright constraint on schools would make these programs more available for another public good, educa-

tion. On the other hand, this option could have significant economic consequences to the copyright owner, whose market is often limited to educational institutions.

ISSUE: Should Congress begin preparing now to resolve issues that have not yet aroused much public debate but which may in the future?

As scientific understanding of genetics and the ability to manipulate inherited characteristics develops, society may face some difficult questions that could involve tradeoffs between individual freedom and the needs of society. This will be increasingly the case as genetic technologies are applied to humans. Developments are occurring rapidly. Recombinant DNA technology was developed in the 1970's. In the spring of 1980, investigators succeeded in the first gene replacement in mammals; in the fall of 1980, the first gene substitution in humans was attempted.

Although this study was restricted to nonhuman applications, many people assume from these and other examples that what can be done with lower animals can be done with humans and will be. Therefore, some action might be taken to better prepare society for decisions on the application of genetic technologies to humans.

OPTIONS:

A. *A commission could be established to identify central issues, the probable time frame for application of various genetic technologies to humans, and the probable effects on society, and to suggest courses of action. The commission might also consider the related area of how participatory democracy might be combined with representative democracy in decisionmaking.*

B. *The life of the President's Commission could be extended for the study of Ethical Problems in Medicine and Biomedical and Behavioral Research, for the purpose of addressing these issues.*

This 11-member Commission was established in November 1978 and terminates on December 31, 1982. It could be asked to broaden its coverage to additional areas. This would require that the life span of the commission be extended and additional funds be appropriated.

A potential disadvantage to using the existing commission to address societal issues associated with genetic engineering is that a number of issues already exist, and more are likely to arise in the years ahead. Yet there are also other issues in medicine and biomedical and behavioral research not associated with genetic engineering that also need review. Whether all these issues can be addressed by one commission should be considered. Comments from the existing commission would assist in deciding the most appropriate course of action.

Chapter 2
Introduction

Chapter 2

Figures

Introduction

Humankind is gaining an increasing understanding of heredity and variation among living things—the science of genetics. This report examines both the critical issues arising from the science and technologies that spring from genetics, and the potential impacts of these advances on society. They are the most rapidly progressing areas of human knowledge in the world today.

Genetic technologies exist only within the larger context of a maturing science. The key to planning for their potential is understanding not simply a particular technology, or breeding program, or new opportunity for investment, but how the field of genetics works and how it interacts with society as a whole.

The technologies that this report assesses can be expected to have pervasive effects on life in the future. They touch on the most fundamental and intimate needs of mankind: health care, supplies of food and energy, and reproduction. At the same time, they trigger concerns in areas equally as important: the dwindling supplies of natural resources, the risks involved in basic and applied scientific research and development, and the nature of innovation itself.

As always, some decisions concerning the use of the new technologies will be made by the marketplace, while others will be made by various institutions, both public and private. In the coming years, the public and its representatives in Congress and other governmental bodies will be called on to make difficult decisions because of society's knowledge about genetics and its capabilities.

This report does not make recommendations nor does it attempt to resolve conflicts. Rather, it clarifies the bases for making judgments by defining the likely impacts of a group of technologies and tracing their economic, societal, legal, and ethical implications. The new genetics will be influential for a long time to come. Although it will continue to change, it is not too early to begin to monitor its course.

The origins of genetics

For the past 10,000 years, a period encompassing less than one-half of 1 percent of man's time on Earth, the human race has developed under the impetus of applied genetics. As techniques for planning, cultivating, and storing crops replaced subsistence hunting and foraging, the character of humanity changed as well. From the domestication of animals to the development of permanent settlements, from the rise of modern science to the dawn of biotechnology, the genetic changes that mankind has directed have, in turn, affected the nature of his society.

Applied genetics depends on a fundamental principle—that organisms both resemble and differ from their parents. It must have required great faith on the part of Neolithic man to bury perfectly good grain during one season in the hope of growing a new crop several months later—faith not only that the seed would indeed return, but that it would do so in the form of the same grain-producing crop from which it had sprung. This permanence of form from one generation to the next has been scientifically understood only within the past century, but the understanding has transformed vague beliefs in the inheritance of traits into the science of genetics, and rule-of-thumb animal and plant breeding into the modern manipulations of genetic engineering.

The major conceptual boost for the science of genetics required a shift in perspective, from the simple observation that characteristics passed from parents to offspring, to a study of the underlying agent by which this transmission is accomplished. That shift began in the garden of Gregor Mendel, an obscure monk in mid-19th century Austria. By analyzing generations of controlled crosses between sweet pea plants, Mendel was able to identify the rudimentary characteristics of what was later termed the gene.

Mendel reasoned that genes were the vehicle and repository of the hereditary mechanism, and that each inherited trait or function of an organism had a specific gene directing its development and appearance. An organism's observable characteristics, functions, and measurable properties taken together had to be based somehow on the total assemblage of its genes.

Mendel's analysis showed that the genes of his pea plants remained constant from one generation to the next, but more importantly, he found that genes and observable traits were not simply matched one-for-one. There were, in fact, two genes involved in each trait, with a single gene contributed by each parent. When the genes controlling a particular trait are identical, the organism is homozygous for that trait; if they are not, it is heterozygous.

In the Mendelian crosses, homozygous plants always retained the expected characteristics. But heterozygous plants did not simply display a mixture of their different genes; one of the two tended to predominate. Thus, when homozygous yellow-seed peas were crossed with homozygous green-seed plants, all the offspring were now heterozygous for seed color, possessing a "green" gene from one parent and a "yellow" from the other. Yet all of them turned out to be indistinguishable from the yellow-seed parent: Yellow-seed color in peas was dominant to green.

But even though the offspring resembled their dominant parent, they could be shown to contain a genetic difference. For when the heterozygotes were now crossed with each other, a certain number of recessive green-seed plants again appeared among the offspring. This occurred whenever an offspring was endowed with a pair of genes that was homozygous for

the green-seed trait—and it occurred at a rate consistent with the random selection of one of two genes from each parent for passage to the new generation. (See figure 5.)

Genes were real—Mendel's work made that clear. But where were they located, and what were they? The answer, lay within the nucleus of the cell. Unfortunately, most of the contents of the nucleus were unobtainable by biologists in Mendel's time, so his published findings were ignored. Only during the last decades of the 19th century did improved microscopes and new dyes permit cells to be observed with an acuity never before possible. And only by the

Figure 5.—The Inheritance Pattern of Pea Color

Y = yellow gene   g = green gene

Homozygous yellow-seed peas have the genetic composition: YY.
Homozygous green-seed peas have the genetic composition: gg.

Each parent contributes only one seed-color gene to the offspring. When the two YY and gg homozygotes are crossed, the genetic composition of all offspring is Yg:

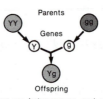

All Yg offspring are heterozygous, and all have yellow seeds, indicating that the Y yellow gene is dominant over the g green gene.

When these Yg heterozygotes are crossed with each other:

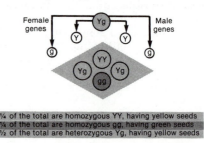

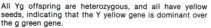

¼ of the total are homozygous YY, having yellow seeds
¼ of the total are homozygous gg, having green seeds
½ of the total are heterozygous Yg, having yellow seeds

Thus, ¾ of these offspring will have yellow seeds, but their individual genetic composition, YY of Yg, may be different.

SOURCE: Office of Technology Assessment.

beginning of the 20th century did scientists *rediscover* Mendel's work and begin to appreciate fully the significance of the cell nucleus and its contents.

Even in the earliest microscopic studies, however, certain cellular components stood out; they were deeply stained by added dye. As a result, they were dubbed "colored bodies," or chromosomes. Chromosomes were seen relatively rarely in cells, with most cells showing just a central dark nucleus surrounded by an extensive light grainy cytoplasm. But periodically the nucleus seemed to disappear, leaving in its place long thready material that consolidated to form the chromosomal bodies. (See figure 6a.) Once formed, the chromosomes assembled along the middle of the cell, copied themselves, and then moved apart while the cell pinched itself in half, trapping one set of chromosomes in each of the two halves. Then the chromosomes themselves seemed to dissolve as two new nuclei appeared, one in each of the two newly formed cells. (See figure 6b.)

Thus, the same number of chromosomes appeared in precisely the same form in every cell of an organism except the germ, or sex, cells. Furthermore, the chromosomes not only remained constant in form and number from one generation to the next, but were inherited in pairs. They were, in short, manifesting all the traits that Mendel had prescribed for genes almost three decades earlier. By the beginning of the 20th century, it was clear that chromosomes were of central importance to the life history of the cell, acting in some unspecified manner as the vehicle for the Mendelian gene.

If this conclusion was strongly implied by the events of cell division, it became obvious when reproduction in whole organisms was analyzed. It had been established by the latter part of the 19th century that the germ cells of plants and animals—pollen and ovum, sperm and egg—actually fuse in the process of fertilizaton. Germ cells differ from other body cells in one important respect—they contain only half the usual number of chromosomes. This chromosome halving within the cell was apparently done very precisely, for every sperm and egg contained exactly one representative from each chromosome pair. When the two germ cells then fused during fertilization, the offspring were supplied with a fully reconstituted chromosome complement, half from each parent. Clearly, chromosomes were the material link from one generation to the next. Somewhere locked within them was the substance of both heredity—the fidelity of traits between generations; and diversity—the potential for genetic variation and change.

Figure 6.—Chromosomes

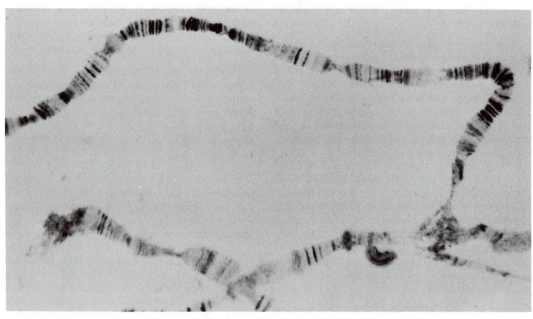

Photo credit: Professor Judith Lengyel, Molecular Biology Institute, UCLA

Optical micrograph of chromosomal material from the salivary gland of the larva of the
common fruit fly, *Drosophila melanogaster*

6a. An example of a chromosome body from a higher organism.

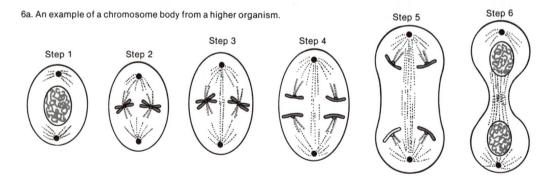

Step 1 Step 2 Step 3 Step 4 Step 5 Step 6

6b. In Step 1, the chromosome bodies are still uncondensed.

In Steps 2 and 3, the chromosomes condense into thread-like bodies and align themselves near the center of the cell.

In Steps 4 and 5, the chromosomes begin to separate and are pulled to the opposite poles of the cell.

In Step 6, the chromosomes return to an uncondensed state and the cell begins to constrict about the middle to form two new cells.

SOURCE: Office of Technology Assessment.

Genetics in the 20th century

During the first few decades of the 20th century, scientists searched for progressively simpler experimental organisms to clarify progressively more complex genetic concepts. First was Thomas Hunt Morgan's *Drosophila*—gnat-sized fruit flies with bulbous eyes. These insects have a simple array of four easily distinguishable chromosome pairs per cell. They reproduce rapidly and in large numbers under the simplest of laboratory conditions, supplying a new generation every month or so. Thus, researchers could carry out an enormous number of crosses employing a whole catalog of different fruit fly traits in a relatively brief time.

It became obvious from the extensive *Drosophila* data that certain traits were more likely to be inherited together than others. Yellow bodies and ruby eyes, for instance, almost always went together, with both in turn, appearing more frequently than expected with the trait known as "forked bristles." All three traits, however, showed up only randomly with curved wings. Certain genes thus seemed to be linked to one another. The entire *Drosophila* genome, in fact, fell into four distinct linkage groups. The physical basis for these groups, not surprisingly, consisted of the four fruit fly chromosomes. Linked genes behaved as they did because they were located on the same chromosome.

Soon, scientists learned that they could not only assign particular genes to particular *Drosophila* chromosomes but could identify the relative locations of different genes on a given chromosome. This gene mapping was possible because linkage itself was not permanent, linked genes sometimes separated. For instance, while yellow bodies, ruby eyes, and forked bristles were all linked traits, the first two stayed together far more frequently than either did with the third.

The degree of linkage between two genes was hypothesized to be directly proportional to the distance between them on the chromosome, mainly because of a unique event that occurs during the development of germ cells. Before the normal chromosome number is halved, the chromosomes crowd together in the center of the cell, coiling tightly around each other, practically fusing along their entire length. It is in this state that crossing-over (or natural recombination)—the actual physical exchange of parts between chromosomes—occurs. No chromosome emerges from the exchange in the same condition as before; the lengths of chromosomes are reshuffled before being transferred to the next generation.

The idea of linkage meant that Mendel's formulations had to be modified. Clearly, genes were not completely independent units. Further work with *Drosophila* in the 1920's showed that genes were also not permanent and could change over time. Although natural mutations occurred at a very slow rate, exposing fruit flies to X-rays accelerated their frequency enormously. Exposure of a parental fly population led to an array of new traits among their offspring—traits which, if they were neither lethal nor sterilizing, could be passed from one generation to the next.

The riddle of the gene

With all this research, nobody yet knew what the gene was made of. The first evidence that it consisted of deoxyribonucleic acid (DNA) emerged from the work of Oswald Avery, Colin MacLeod, and Maclyn McCarty at the Rockefeller Institute in New York in the early 1940's. Avery's group took as its starting point some intriguing observations made a decade earlier by a British physician, Fred Griffith. He had worked with two types of pneumococcus (the bacteria responsible for pneumonia) and with two different bacteria within each type. One bacterium in each type was coated in a polysaccharide capsule; the other was bare. Bare bacteria gave rise only to bare progeny, while those with capsules produced only encapsulated forms. Only the encapsulated forms of both types II and III could cause disease; bare bacteria were benign. (See figure 7a.) But when Griffith took some encapsulated type III bacteria that had been killed and rendered harmless and mixed them with bare bacteria of type II, the presumably safe mixture became virulent: Mice injected with it died of a massive pneumonia infection. Bacteria recovered from these animals were found to be of type II—the only living bacteria the mice had received—now wrapped in type III capsules. (See figure 7b.)

Avery's group recognized Griffith's finding as a genetic phenomenon; the dead type III bacteria must have delivered the gene for making capsules into the genetic complement of the living type II recipients. By meticulous research, Avery's group found that the substance which caused the genetic transformation was DNA.

It had been in 1868, just 3 years after Mendel had published his findings, that DNA was discovered by Friedrich Miescher. It is an extremely simple molecule composed of a small sugar molecule, a phosphate group (a phosphorous atom surrounded by four oxygen atoms), and four kinds of simple organic chemicals known as nitrogenous (nitrogen-containing) bases. Together, one sugar, one phosphate, and one base form a nucleotide—the basic structural unit of the large DNA molecule. Because it is so simple, DNA had appeared to be little more than a monotonous conglomeration of simple nucleotides to scientists in the early 20th century. It seemed unlikely that such a prosaic molecule could direct the appearance of genetic traits while faithfully reproducing itself so that information could be transferred between generations. Although Avery's results seemed clear enough, many were reluctant to accept them.

Those doubts were finally laid to rest in a brief report published in 1953 by James Watson and Francis Crick. By using X-ray crystallographic techniques and building complex models—and without ever having actually seen the molecule itself—Watson and Crick reported that they had discovered a consistent scientifically sound structure for DNA.

Figure 7.—The Griffith Experiment

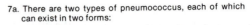

7a. There are two types of pneumococcus, each of which can exist in two forms:

where R represents the rough, nonencapsulated, benign form; and

S represents the smooth, encapsulated, virulent form.

7b. The experiment consists of four steps:

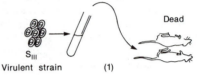

Mice injected with the virulent S_{III} die.

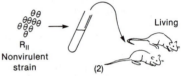

Mice injected with nonvirulent R_{II} do not become infected.

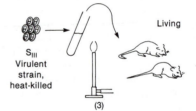

The virulent S_{III} is heat-killed. Mice injected with it do not die.

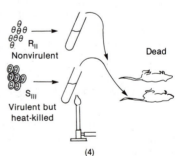

When mice are injected with the nonvirulent R_{II} and the heat-killed S_{III}, they die. Type II bacteria wrapped in type III capsules are recovered from these mice.

SOURCE: Office of Technology Assessment.

The structure that Crick and Watson uncovered solved part of the genetic puzzle. According to them, the phosphates and sugars formed two long chains, or backbones, with one nitrogenous base attached to each sugar. The two backbones were held together like the supports of a ladder by weak attractions between the bases protruding from the sugar molecules. Of the four different nitrogenous bases—adenine, thymine, guanine, and cytosine—attractions existed only between adenine(A) and thymine(T), and between guanine(G) and cytosine(C). (See figure 8a) Thus, if a stretch of nucleotides on one backbone ran:

A-T-G-C-T-T-A-A. . . .

the other backbone had to contain the directly opposite complementary sequence:

T-A-C-G-A-A-T-T. . .

The complementary pairing between bases running down the center of the long molecule was responsible for holding together the two otherwise independent chains. (See figure 8b.) Thus, the DNA molecule was rather like a zipper, with the bases as the teeth and the sugar-phosphate chains as the strands of cloth to which each zipper half was sewn. Crick and Watson also found that in the presence of water, the two polynucleotide chains did not stretch out to full length, but twisted around each other, forming what has undoubtedly become the most glorified structure in the history of biology—the double helix. (See figure 8c.)

The structure was scientifically elegant. But it was received enthusiastically also because it implied how DNA worked. As Crick and Watson themselves noted:

If the actual order of the bases on one of the pair of chains were given, one could write down the exact order of the bases on the other one, because of the specific pairing. Thus one chain is, as it were, the complement of the other, and it is this feature which suggests how the desoxyribonucleic acid molecule might duplicate itself.[1]

When a double-stranded DNA molecule is unzipped, it consists of two separate nucleotide chains, each with a long stretch of unpaired bases. In the presence of a mixture of nucleotides, each base attracts its complementary match in accordance with the inherent affinities of adenine for thymine, thymine for adenine, guanine for cytosine, and cytosine for guanine. The result of this replication is two DNA molecules, both precisely identical to each other and to the original molecule—which explains the faithful duplication of the gene for passage from one generation to the next. (See figure 9.)

Crick and Watson's work solved a major riddle in genetic research. Because George Beadle and Edward Tatum had recently discovered that genes control the appearance of specific proteins, and that one gene is responsible for producing one specific protein, scientists now knew what the genetic material was, how it replicated, and what it produced. But they had yet to determine how genes expressed themselves and produced proteins.

---

[1]James D. Watson and Francis Crick, "Genetic Implications of the Structures of Deoxyribose Nucleic Acid," *Nature* 171, 1953. pp. 737-8.

36

Figure 8.—The Structure of DNA

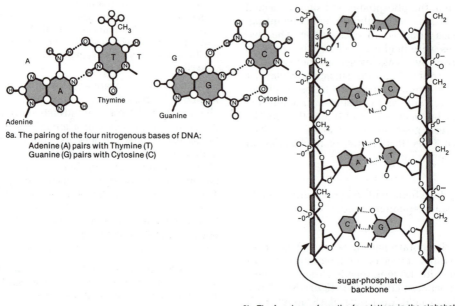

8a. The pairing of the four nitrogenous bases of DNA:
   Adenine (A) pairs with Thymine (T)
   Guanine (G) pairs with Cytosine (C)

sugar-phosphate
backbone

8b. The four bases form the four letters in the alphabet of the genetic code. The *sequence* of the bases along the sugar-phosphate backbone encodes the genetic information.

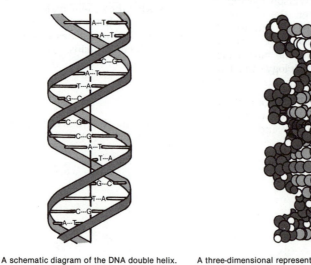

A schematic diagram of the DNA double helix.   A three-dimensional representation of the DNA double helix.

8c. The DNA molecule is a double helix composed of two chains. The sugar-phosphate backbones twist around the outside, with the paired bases on the inside serving to hold the chains together.

SOURCE: Office of Technology Assessment.

Figure 9.—Replication of DNA

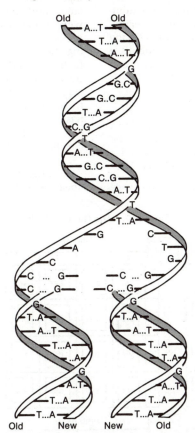

When DNA replicates, the original strands unwind and serve as templates for the building of new complementary strands. The daughter molecules are exact copies of the parent, with each having one of the parent strands.

SOURCE: Office of Technology Assessment.

The genetic code

Proteins are the basic materials of cells. Some proteins are enzymes, which catalyze reactions within a cell. In general, for every chemical reaction in a living organism, a specific enzyme is required to trigger the process. Other proteins are structural, comprising most of the raw material that forms cells.

Ironically, proteins are far more complex and diverse than the four nucleotides that help create them. Proteins, too, are long chains made up of small units strung together. In this case, however, the units are amino acids rather than nucleotides—and there are 20 different kinds of amino acids. Since an average protein is a few hundred amino acids in length, and since any one of 20 amino acids can fill each slot, the number of possible proteins is enormous. Nevertheless, each protein requires the strictest ordering of amino acids in its structure. Changing a single amino acid in the entire sequence can drastically change the protein's character.

It was now possible for scientists to move nearer to an appreciation of how genes functioned. First had come the recognition that DNA determined protein; now it was evident that the sequence of nucleotides in DNA determined a linear sequence of amino acids in proteins.

By the early 1980's, the way proteins were

manufactured, how their synthesis was regulated, and the role of DNA in both processes were understood in considerable detail. The process of transcribing DNA's message—carrying the message to the cell's miniature protein factories and building proteins—took place through a complex set of reactions. Each amino acid in the protein chain was represented by three nucleotides from the DNA. That three-base unit acted as a word in a DNA sentence that spelled out each protein—the genetic code. (See figure 10.)

Through the genetic code, an entire gene—a linear assemblage of nucleotides—could now be

Figure 10.—The Genetic Code

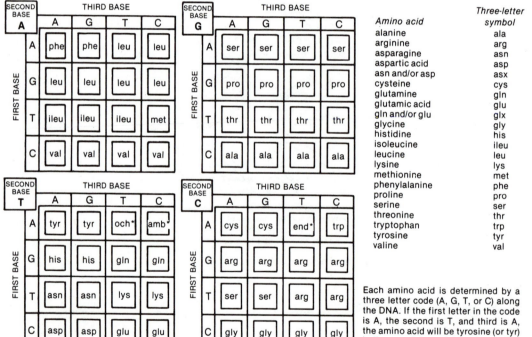

Amino acid	Three-letter symbol
alanine	ala
arginine	arg
asparagine	asn
aspartic acid	asp
asn and/or asp	asx
cysteine	cys
glutamine	gln
glutamic acid	glu
gln and/or glu	glx
glycine	gly
histidine	his
isoleucine	ileu
leucine	leu
lysine	lys
methionine	met
phenylalanine	phe
proline	pro
serine	ser
threonine	thr
tryptophan	trp
tyrosine	tyr
valine	val

Each amino acid is determined by a three letter code (A, G, T, or C) along the DNA. If the first letter in the code is A, the second is T, and third is A, the amino acid will be tyrosine (or tyr) in the complete protein molecule. For leucine (or leu), the code is GAT, and so forth. The dictionary above gives the entire code.

*och (ochre), amb (amber), and end are stop signals for translation, i.e., signal the end of synthesis of the protein chain.

SOURCE: Office of Technology Assessment.

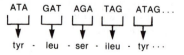

read like a book. By the 1970's, researchers had learned to read the code of certain proteins, synthesize their DNA, and insert the DNA into bacteria so that the protein could be produced. (See figure 11.)

Meanwhile, other scientists were studying the genetics of viruses and bacteria. The combination of these studies with those investigating the genetic code led to the innovations of genetic engineering.

Figure 11.—The Expression of Genetic Information in the Cell

DNA ⟶ mRNA ⟶ Protein
(Transcription)   (Translation)
process   process

The "central dogma" of molecular biology: DNA in the genes is transcribed into messenger RNA (mRNA) which is then translated by reactions in the cell into protein. Each gene contains the information for a specific protein.

SOURCE: Office of Technology Assessment.

Developing genetic technologies

In the early 1960's, scientists discovered exactly how genes move from one bacterium to another. One such mechanism uses bacteriophages—viruses that infect bacteria—as intermediaries. Phages act like hypodermic needles, injecting their DNA into bacterial hosts, where it resides before being passed along from one generation to the next as part of the bacterium's own DNA. Sometimes, however, the injected DNA enters an active phase and produces a crop of new virus particles that can then burst out of their host. Often during this process, the viral DNA inadvertently takes a piece of the bacterial

DNA along with it. Thus, when the new virus particles now infect other bacteria, they bring along several genes from their previous host. This viral transduction—the transfer of genes by an intermediate viral vector or vehicle—could be used to confer new genetic traits on recipient bacteria. (See figure 12.)

Bacteria also transfer genes directly in a process called conjugation, in which one bacterium attaches small projections to the surface of a nearby bacterium. DNA from the donor bacterium is then passed to the recipient through the

Figure 12.—Transduction: The Transfer of Genetic Material in Bacteria by Means of Viruses

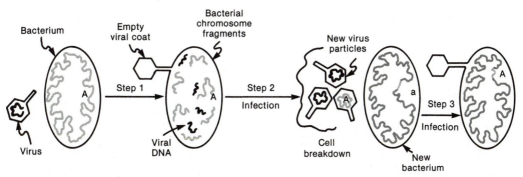

In step 1 of viral transduction, the infecting virus injects its DNA into the cell. In step 2 when the new viral particles are formed, some of the bacterial chromosomal fragments, such as gene A, may be accidently incorporated into these progeny viruses instead of the viral DNA. In step 3 when these particles infect a new cell, the genetic elements incorporated from the first bacterium can recombine with homologous segments in the second, thus exchanging gene A for gene a.

SOURCE: Office of Technology Assessment.

projections. The ability to form projections and donate genes to neighbors is a genetically controlled trait. The genes controlling this trait, however, are not located on the bacterial chromosomes. Instead, they are located on separate genetic elements called plasmids—relatively small molecules of double-stranded DNA, arranged as closed circles and existing autonomously within the bacterial cytoplasm. (See figure 13.)

Plasmids and phages are two vehicles—or vectors—for carrying genes into bacteria. As such, they became tools of genetic engineering; for if a specifically selected DNA could be introduced into these vectors, it would then be possible to transfer into bacteria the blueprints for proteins—the building blocks of genetic characteristics.

But bacteria had been confronting the invasion of foreign DNA for millennia, and they had evolved protective mechanisms that preserved their own DNA while destroying the DNA that did not belong. Bacteria survive by producing restriction enzymes. These cut DNA molecules in places where specific sequences of nucleotides occur—snipping the foreign DNA, yet leaving the bacteria's own genetic complement alone. The first restriction enzyme that was isolated, for instance, would cut DNA only when it located the sequence:

$$G\text{-}A\text{-}A\text{-}T\text{-}T\text{-}C$$
$$C\text{-}T\text{-}T\text{-}A\text{-}A\text{-}G$$

If the sequence occurred once in a circular plasmid, the effect would simply be to open the circle. If the sequence were repeated several times along a length of DNA, the DNA would be chopped into several small pieces.

By the late 1970's, scores of different restriction enzymes had been isolated from a variety of bacteria, with each enzyme having a unique specificity for one specific nucleotide sequence. These enzymes were another key to genetic engineering: they not only allowed plasmids to be opened up so that new DNA could be inserted, but offered a way of obtaining manageable pieces of new DNA as well. (See figure 14.) Using restriction enzymes, almost any DNA molecule could be snipped, shaped, and trimmed with precision.

Cloning DNA—that is obtaining a large quantity of exact copies of any chosen DNA molecule by inserting it into a host bacterium—became technically almost simple. The piece in question was merely snipped from the original molecule, inserted into the vector DNA, and provided with

Figure 13.—Conjugation: The Transfer of Genetic Material in Bacteria by Mating

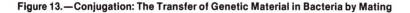

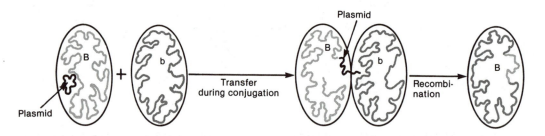

In conjugation, a plasmid inhabiting a bacterium can transfer the bacterial chromosome to a second cell where homologous segments of DNA can recombine, thus exchanging gene B from the first bacterium for gene b from the second.

SOURCE: Office of Technology Assessment.

Figure 14.—Recombinant DNA: The Technique of Recombining Genes From One Species With Those From Another

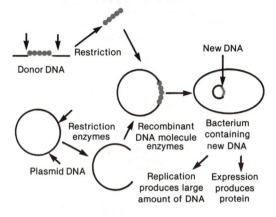

Restriction enzymes recognize certain sites along the DNA and can chemically cut the DNA at those sites. This makes it possible to remove selected genes from donor DNA molecules and insert them into plasmid DNA molecules to form the recombinant DNA. This recombinant DNA can then be cloned in its bacterial host and large amounts of a desired protein can be produced.

SOURCE: Office of Technology Assessment.

a bacterial host as a suitable environment for replication. The desired piece of DNA could be recombined with a plasmid vector, a procedure that gave rise to recombinant DNA (rDNA), also known as gene splicing. Since bacteria can be grown in vast quantities, this process could result in large-scale production of otherwise scarce and expensive proteins.

Although placing genes inside of bacteria is now a relatively straightforward procedure, obtaining precisely the right gene can be difficult. Three techniques are currently available:

• Ribonucleic acid—RNA—is the vehicle through which the message of DNA is read and transcribed to form proteins. The RNA that carries the message for the desired protein is first isolated. An enzyme, called 'reverse transcriptase,' is then added to the RNA. The enzyme triggers the formation of DNA—reversing the normal process of protein production. The DNA is then inserted

into an appropriate vector. This was the procedure used to obtain the gene for human insulin in 1979. (See figure 15.)

• The gene can also be synthesized, or created, directly, since the nucleotide sequence of the gene can be deduced from the amino acid sequence of its protein product. This procedure has worked well for small proteins—like the growth regulatory hormone somatostatin—which have relatively short stretches of DNA coding. But somatostatin is a tiny protein, only 14 amino acids long. With three nucleotides coding for each amino acid, scientists had to synthesize a DNA chain 42 nucleotides long to produce the complete hormone. For larger proteins, the gene-synthesis approach rapidly becomes highly impractical.

• The third method is also the most controversial. In this "shotgun" approach, the entire genetic complement of a cell is chopped up by restriction enzymes. Each of the DNA fragments is attached next to vectors and transferred into a bacterium; the bacteria are then screened to find those making the desired product. Screening thousands of bacterial cultures was part of the technique that enabled the isolation of the human interferon gene.*

At present, these techniques of recombination work mainly with simple micro-organisms. Scientists have only recently learned how to introduce novel genetic material into cells of higher plants and animals. These higher cells are being 'engineered' in totally different ways, by growing plant or animal cells in 'tissue culture' systems, in vitro.

Tissue culture systems work with isolated cells, with entire pieces of tissue, and to a far more limited extent, with whole organs or even early embryos. The techniques make it possible to manipulate cells experimentally and under controlled conditions. Several techniques are available. For example, in one set of experiments, complete plants have been grown from single cells—a breakthrough that may permit

*Strictly speaking, RNA was transcribed using the shotgun approach into DNA, which was then cloned into bacteria and screened.

**Figure 15.—An Example of How the Recombinant DNA Technique May Be Used
To Insert New Genes Into Bacterial Cells**

I. The first part of the technique involves the manipulations necessary to isolate and reconstruct the desired gene from the donor:
   a) The RNA that carries the message (mRNA) for the desired protein product is isolated.
   b) The double-stranded DNA is reconstructed from the mRNA.
   c) In the final step of this sequence, the enzyme terminal transferase acts to extend the ends of the DNA strands with short sequences of identical bases (in this case four guanines).

II. A bacterial plasmid, which is a small piece of circular DNA, serves as the vehicle for introducing the new gene (obtained in part I above) into the bacterium:
   a) The circular plasmid is cleaved by the appropriate restriction enzyme.
   b) The enzyme terminal transferase extends the DNA strands of the broken circle with identical bases (four cytosines in this case, to allow *complementary base pairing* with the guanines added to the gene obtained in part I).

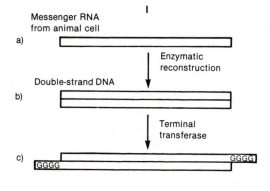

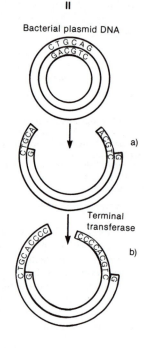

III. The final product, a bacterial plasmid containing the new gene, is obtained. This plasmid can then be inserted into a bacterium where it can be replicated and produce the desired protein product:
   a) The gene obtained in part I and the plasmid DNA from part II are mixed together and anneal because of the complementary base-pairing between them.
   b) Bacterial enzymes fill in any gaps in the circle, sealing the connection between the plasmid DNA and the inserted DNA to generate an intact circular plasmid now containing a new gene.

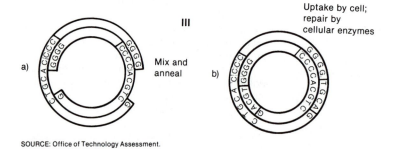

SOURCE: Office of Technology Assessment.

hundreds of plants to be grown asexually from a small sample of plant material. Just as with bacteria, the cells can be induced to take up pieces of DNA in a process called transformation. They can also be exposed to mutation-causing agents so that they produce mutants with desired properties. In another set of experiments, two different cells have been fused to form a new, single-cell "hybrid" that contains the genetic complements of both antecedents. In both cases, the success of tissue culture and cell fusion* can be used to direct efficient, fast genetic changes in plants. (See ch. 8.)

Cell culture techniques, while not strictly genetic manipulation, form a major aspect of modern biotechnology. Combined with genetic approaches, their potential is only on the verge of being realized.

*A related technique is protoplast fusion, or the fusion of cells whose walls have been removed to leave only membrane-bound cells. The cells of bacteria, fungi, and plants must all be freed of their walls before they can be fused.

The basic issues

Applied genetics is like no other technology. By itself, it may enable tremendous advances in conquering diseases, increasing food production, producing new and cheaper industrial substances, cleaning up pollution, and understanding the fundamental processes of life. Because the technology is so powerful, and because it involves the basic roots of life itself, it carries with it potential hazards, some of which might arise from basic research, others of which may stem from its applications.

As the impacts of genetic technologies are discussed, two fundamental questions must be kept in mind:

How will applied genetics be used?

Interest in the industrial use of biological processes stems from a merging of two paths: the revolution in scientific understanding of the nature of genetics; and the accelerated search for a sustainable society in which most industrial processes are based on the use of renewable resources. The new genetic technologies will spur that search in three ways: they will provide a means of doing something biologically—with renewable raw materials—that previously required chemical processes using nonrenewable resources; they will offer more efficient, more economical, less polluting ways for producing both old and new products; and they will increase the yield of the plant and animal resources that are responsible for providing the world's supplies of food, fibers, and some fuels.

What are the dangers?

Even before scientists recognized the potential power of applied genetics, some questioned its consequences; for with its benefits, appeared hypothetical risks. Although most experts today agree that the immediate hazards of the basic research itself appear to be minimal, nobody can be certain about all the consequences of placing genetic characteristics in micro-organisms, plants, and animals that have never carried them before. There are at least three separate areas of concern:

First, genetically engineered micro-organisms might have potentially deleterious effects on human health, other living organisms, or the environment in general. Unlike toxic chemicals, organisms may reproduce and spread of their own accord; if they are released into the environment, they may be impossible to control.

Second, some observers question whether sufficient knowledge exists to allow the extinction of diverse species of "genetically inferior" plants and animals in favor of a few strains of "superior" ones. Evolution thus far has depended, in part, on genetic diversity; replacing in nature diverse inferior strains by genetically engineered superior strains may increase the susceptibility of living things to disease and environmental insults.

Finally, this new knowledge affects the understanding of life itself. It is tied to the ultimate questions of how humans view themselves and what they legitimately control in the world.

Because of the significant and wide-ranging scope of applied genetics, society as whole must begin to debate the issues with a view toward allocating and monitoring its benefits and burdens. That process requires knowledge. The following sections of the report describe the impacts of applied genetics on specific industries, and assess many of their consequences.

Part I
Biotechnology

Chapter 3

Genetic Engineering and the Fermentation Technologies

Chapter 3

Table

Figures

Genetic Engineering and the Fermentation Technologies

Biotechnology—an introduction

Biotechnology involves the use in industry of living organisms or their components (such as enzymes). It includes the introduction of genetically engineered micro-organisms into a variety of industrial processes.

The pharmaceutical, chemical, and food processing industries, in that order, are most likely to take advantage of advances in molecular genetics. Others that might also be affected, although not as immediately, are the mining, crude oil recovery, and pollution control industries.

Because nearly all the products of biotechnology are manufactured by micro-organisms, fermentation is an indispensible element of biotechnology's support system. The pharmaceutical industry, the earliest beneficiary of the new knowledge, is already producing pharmaceuticals derived from genetically engineered micro-organisms. The chemical industry will take longer to make use of biotechnology, but the ultimate impact may be enormous. The food processing industry will probably be affected last.

This report examines many of the pharmaceutical industry's products in detail, as well as some of the secondary impacts that the technologies might have. Because the chemical and food industries will feel the major impact of biotechnology later, specific impacts are less certain and particular products are less identifiable. The mining, oil recovery, and pollution control industries are also candidates for the use of genetic technologies. However, because of technical, scientific, legal, and economic uncertainties, the success of applications in these industries is more speculative.

The generalizations made with respect to each of the industries should be viewed as just that—generalizations. Because a wide array of products can be made biologically, and because different factors influence each instance of production, isolated examples of success may appear throughout the industries at approximately the same time. In almost every case, specific predictions can only be made on a product-by-product basis; for while it may be true that biotechnology's overall impact will be profound, identifying many of the products most likely to be affected remains speculative.

Fermentation

There are several ways that DNA can be cut, spliced, or otherwise altered. But engineered DNA by itself is a static molecule. To be anything more than the end of a laboratory exercise, the molecule must be integrated into a system of production; to have an impact on society at large, it must become a component of an industrial or otherwise useful process.

The process that is central to the economic success of biotechnology has been around for centuries. It is fermentation, essentially the process used to make wine and beer. It can also produce organic chemical compounds using micro-organisms or their enzymes.

Over the years, the scope and efficiency of the fermentation process has been gradually improved and refined. Two processes now exist, both of which will beneft from genetic engineering. In fermentation technology, living organisms serve as miniature factories, converting raw materials into end products. In enzyme technology, biological catalysts extracted from those living organisms are used to make the products.

Fermentation industries

The food processing, chemical, and pharmaceutical industries are the three major users of fermentation today. The food industry was the first to exploit micro-organisms to produce alcoholic beverages and fermented foods. Mid-16th century records describe highly sophisticated methods of fermentation technology. Heat processing techniques, for example, anticipated pasteurization by several centuries.

In the early 20th century, the chemical industry began to use the technology to produce organic solvents like ethanol, and enzymes like amylase, used at the time to treat textiles. The chemical industry's interest in fermentation arose as the field of biochemistry took shape around the turn of the century. But it was not until World War I that wartime needs for the organic solvent acetone—to produce the cordite used in explosives—substantially increased research into the potential of fermentation. Thirty years later after World War II, the pharmaceutical industry followed the chemical industry's lead, applying fermentation to the production of vitamins and new antibiotics.

Today, approximately 200 companies in the United States and over 500 worldwide use fermentation technologies to produce a wide variety of products. Most use them as part of production processes, usually in food processing. But others manufacture either proteins, which can be considered primary products, or a host of secondary products, which these proteins help produce. For genes can make enzymes, which are proteins; and the enzymes can help make alcohol, methane, antibiotics, and many other substances.

Proteins, the primary products, function as:

- enzymes such as asparaginase which are used in the treatment of leukemia;

- structural components, such as collagen, used in skin transplants following burn trauma;
- certain hormones, such as insulin and human growth hormone;
- substances in the immune system, such as antibodies and interferon; and
- specialized functional components, such as hemoglobin.

Fermentation technologies are so useful for producing proteins partly because these are the direct products of genes. But proteins (as enzymes) can also be used in thousands of additional conversions to produce practically any organic chemical and many inorganic ones as well: (See figure 16.)

Figure 16.—Diagram of Products Available From Cells

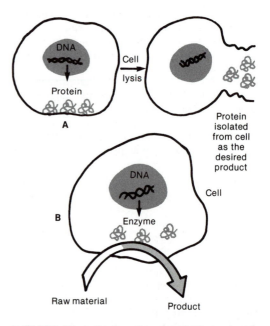

In (**A**) DNA directs the formation of a protein, such as insulin, which is itself the desired product. In (**B**), DNA directs the formation of an enzyme which, in turn, converts some raw material, such as sugar, to a product, such as ethanol.

SOURCE: Office of Technology Assessment.

- carbohydrates, such as fructose sweeteners;
- lipids, such as vitamins A, E, and K;
- alcohols, such as ethanol;
- other organic compounds, such as acetone; and
- inorganic chemicals, such as ammonia, for use in fertilizers.

Fermentation is not the only way to manufacture or isolate these products. Some are traditionally produced by other methods. If a change from one process to another is to occur, both economic and societal pressures will help determine whether an innovative approach will be used to produce a particular product. Alan Bull has identified four stimuli for change and innovation:[1]

1. abundance of a potentially useful raw material;
2. scarcity of an established product;
3. discovery of a new product; and
4. environmental concerns.

And conditions existing today have added a fifth stimulus:

5. scarcity of a currently used raw material.

Each of these factors has tended to accelerate the application of fermentation.

1. *Abundance of a potentially useful raw material.*—The use of a raw material can be the driving force in developing a process. When straight chain hydrocarbons (n-alkanes) were produced on a large scale as petroleum refinery byproducts, fermentation processes were developed to convert them to single-cell proteins for use in animal feed.
2. *Scarcity of an established product.*—The new-found potential for producing human hormones through fermentation technology is a major impetus to the industry today. Similarly, many organic compounds once obtained by other processes—like citric acid, which was extracted directly

from citrus fruits—are now made by fermentation. As a result of more efficient technology, products from vitamin B_{12} to steroids have come into wider use.
3. *Discovery of a new product.*—The discovery that antibiotics were produced by microorganisms sparked searches for an entirely new group of products. Several thousand antibiotics have been discovered to date, of which over a hundred have proved to be clinically useful.
4. *Environmental concerns.*—The problems of sewage treatment and the need for new sources of energy have triggered a search for methods to convert sewage and municipal wastes to methane, the principal component of natural gas. Because micro-organisms play a major role in the natural cycling of organic compounds, fermentation has been one method used for the conversion.
5. *Scarcity of a currently used raw material.*—Because the Earth's supplies of fossil fuels are rapidly dwindling, there is intense interest in finding methods for converting other raw materials to fuel. Fermentation offers a major approach to such conversions.

Fermentation technologies can be effective in each of these situations because of their outstanding versatility and relative simplicity. The processes of fermentation are basically identical, no matter what organism is selected, what medium used, or what product formed. The same apparatus, with minor modifications, can be used to produce a drug, an agricultural product, a chemical, or an animal feed supplement.

Fermentation using whole living cells

Originally, fermentation used some of the most primitive forms of plant life as cell factories. Bacteria were used to make yogurt and antibiotics, yeasts to ferment wine, and the filamentous fungi or molds to produce organic acids. More recently, fermentation technology has begun to use cells derived from higher plants and animals under growth conditions known as cell or tissue culture. In all cases, large quantities of cells with uniform character-

[1]A. T. Bull, D. C. Ellwood, and C. Ratledge, *Microbial Technology: Current State, Future Prospects*, 29th Symposium of the Society for General Microbiology at University of Cambridge, April 1979 (Cambridge, England: Cambridge University Press, 1979), pp. 4-8.

istics are grown under defined, controlled conditions.

In its simplest form, fermentation consists of mixing a micro-organism with a liquid broth and allowing the components to react. More sophisticated large-scale processes require control of the entire environment so that fermentation proceeds efficiently and, more importantly, so that it can be repeated exactly, with the same amounts of raw materials, broth, and micro-organisms producing the same amount of product. Strict control is maintained of such variables as pH (acidity/alkalinity), temperature, and oxygen supply. (See figure 17.) The newest models are regulated by sensors that are monitored by computers. The capacity of industrial-sized fermenters can reach 50,000 gal or more. The one-shot system of fermentation is called batch fermentation—i.e., fermentation in which a single batch of material is processed from start to finish.

In continuous fermentation, an improvement on the batch process, fermentation goes on without interruption, with a constant input of raw materials and other nutrients and an attendant output of fermented material. The most recent approaches use micro-organisms that have been immobilized in a supporting structure. (See figure 18.) As the solution containing the raw material passes over the cells, the micro-organisms process the material and release the products into the solution flowing out of the fermenter.

In general, products obtained by fermentation also can be produced by chemical synthesis, and to a lesser extent can be isolated by extraction from whole organs or organisms. A fermentation process is usually most competitive when the chemical process requires several

Figure 18.—Immobilized Cell System

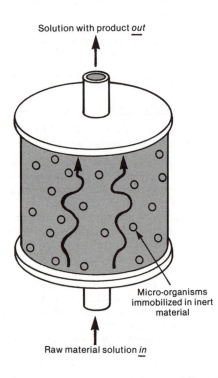

Solution with product *out*

Micro-organisms immobilized in inert material

Raw material solution *in*

Typically, a solution of raw materials is pumped through a bed of immobilized micro-organisms which convert the materials to the desired product.

SOURCE: Office of Technology Assessment.

Figure 17.—Features of a Standard Fermenter

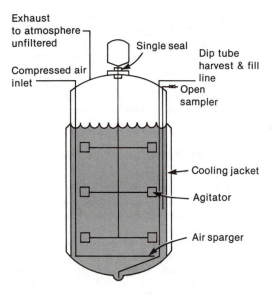

Exhaust to atmosphere unfiltered

Single seal

Compressed air inlet

Dip tube harvest & fill line

Open sampler

Cooling jacket

Agitator

Air sparger

SOURCE: Eli Lilly & Co.

individual steps to complete the conversion. In a chemical synthesis, the raw material (shown in figure 19 as *a*) might have to be transformed to an intermediate *b*, which, in turn, might have to be converted to intermediates *c* and *d* before final conversion to the product *e*—each step necessitating the recovery of its products before the next conversion. In fermentation technology, all steps take place within those miniature chemical factories, the micro-organisms; the microbial chemist merely adds the raw material *a* and recovers the product *e*.

A wide variety of carbohydrate raw materials can be used in fermentation. These can be pure substances (sucrose or table sugar, glucose, or fructose) or complex mixtures still in their original form (cornstalks, potato mash, sugarcane, sugar beets, orcellulose). They can be of recent biological origin (biomass) or derived from fossil fuels (methane or oil). The availability of raw materials varies from country to coun-

try and even from region to region within a country; the economics of the production process varies accordingly.

The cost of the raw material can contribute significantly to the cost of production. Usually, the most useful micro-organisms are those that consume readily available inexpensive raw materials. For large volume, low-priced products (such as commodity chemicals), the relationship between the cost of the raw material and the cost of the end product is significant. For low volume, high-priced products (such as certain pharmaceuticals), the relationship is negligible.

The process of enzyme technology

Although live yeast had been used for several thousand years in the production of fermented foods and beverages, it was not until 1878 that the active agents of the fermentation process were given the name "enzymes" (from the Greek, meaning "in yeast"). The inanimate nature of enzymes was demonstrated less than two decades later when it was shown that extracts from yeast cells could effect the conversion of glucose to ethanol. Finally, their actual chemical nature was established in 1926 with the purification and crystallization of the enzyme urease.

Fermentation carried out by live cells provided the conceptual basis for designing fermentation processes based on isolated enzymes. A single enzyme situated within a living cell is needed to convert a raw material into a product. A lactose-fermenting organism, e.g., can be used to convert the sugar lactose, which is found in milk, to glucose (and galactose). But if the actual enzyme responsible for the conversion is identified, it can be extracted from the cell and used in place of a living cell. The purified enzyme carries out the same conversion as the cell, breaking down the raw material in the absence of any viable micro-organism. An enzyme that acts inside a cell to convert a raw material to a product can also do this outside of the cell.

Both batch and continuous methods are used in enzyme technology. However, in the batch method, the enzymes cannot be recovered eco-

Figure 19.—Diagram of Conversion of Raw Material to Product

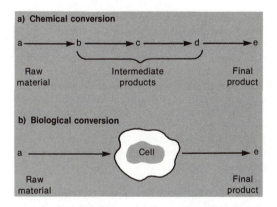

a) In the chemical conversion of raw material *a* to final product *e*, intermediates *b, c,* and *d* must be synthesized. Each intermediate must be recovered and purified before it can be used in the next step of the conversion.

b) A cell can perform the same conversion of *a* to *e*, but with the advantage that the chemist does not have to deal with the intermediates: the raw material *a* is simply added and the final product *e*, recovered.

SOURCE: Office of Technology Assessment.

nomically, and new enzymes must be added for each production cycle. Furthermore, the enzymes are difficult to separate from the end product and constitute a potential contaminant. Because enzymes used in the continuous method are reusable and tend not to be found in the product, the continuous method is the method of choice for most processes. Depending on the desired conversion, the immobilized micro-organisms of figure 18 could be replaced by an appropriate immobilized enzyme.

Although more than 2,000 enzymes have been discovered, fewer than 50 are currently of industrial importance. Nevertheless, two major features of enzymes make them so desirable: their specificity and their ability to operate under relatively mild conditions of temperature and pressure. (The most frequently used enzymes are listed in table 2.)

Comparative advantages of fermentations using whole cells and isolated enzymes

At present, it is still uncertain whether the use of whole cells or isolated enzymes will be more useful in the long run. There are advantages and disadvantages to each. The role of genetic engineering in the future of the industry, however, will be partly determined by which method is chosen. With isolated enzymes, genetic manipulation can readily increase the supply of enzymes, while with whole organisms, a wide variety of manipulations is possible in constructing more productive strains.

The relationship of genetics to fermentation

Applied genetics is intimately tied to fermentation technology, since finding a suitable species of micro-organism is usually the first step in developing a fermentation technique. Until recently, geneticists have had to search for an organism that already produced the needed product. However, through genetic manipulation a *totally new capability* can be engineered; micro-organisms can be made to produce substances beyond their natural capacities. The most striking successes have been in the pharmaceutical industry, where human genes have been transferred to bacteria to produce insulin, growth hormone, interferon, thymosin α-1, and somatostatin. (See ch. 4.)

In general, once a species is found, conventional methods have been used to induce mutations that can produce even more of the desired compound. The geneticist searches from among hundreds of mutants for the one micro-organism that produces most efficiently. Most of the many methods at the microbiologist's disposal involve trial-and-error. Newer genetic technologies, such as the use of recombinant DNA (rDNA), allow approaches in which useful genetic traits can be inserted directly into the micro-organism.

The current industrial approach to fermentation technologies therefore considers two problems: First, whether a biological process can produce a particular product; and second, what micro-organism has the greatest potential for production and how the desired characteristics can be engineered for it. Finding the desired micro-organism and improving its capability is so fundamental to the fermentation industry, that geneticists have become important members of fermentation research teams.

Table 2.—Enzyme Products

Source/name	Commercially available before: 1900	Commercially available before: 1950	Commercially available before: 1980	Current production tons/yr
Animal				
Rennet	X			2
Trypsin		X		15
Pepsin		X		5
Plant				
Malt amylase	X			10,000
Papain		X		100
Microbial				
Koji	X			?
Fungal protease	X			10
Bacillus protease		X		500
Amyloglucosidase			X	300
Fungal amylase	X			10
Bacterial amylase		X		300
Pectinase		X		10
Glucose isomerase			X	50
Microbial rennet			X	10

SOURCE: Office of Technology Assessment.

Genetic engineering can increase an organism's productive capability (a change that can make a process economically competitive); but it can also be used to construct strains with characteristics other than higher productivity. Properties such as objectionable color, odor, or slime can be removed. The formation of spores that could lead to airborne spread of the microorganism can be suppressed. The formation of harmful byproducts can be eliminated or reduced. Other properties, such as resistance to bacterial viruses and increased genetic stability, can be given to micro-organisms that lack them.

Applying recent genetic engineering techniques to the production of industrially valuable enzymes may also prove useful in the future. For example, a strain of micro-organism that carries the genes for a desired enzyme may be pathogenic. If the genes that express (produce) the enzyme can be transferred to an innocuous micro-organism, the enzyme can be produced safely.

CURRENT TECHNICAL LIMITS ON GENETIC ENGINEERING

Despite the many genetic manipulations that are theoretically possible, there are several notable technical limitations:

* Genetic maps—the identification of the location of desired genes on various chromosomes have not been constructed for most industrially useful micro-organisms.
* Genetic systems for industrially useful micro-organisms, such as the availability of useful vectors, are at an early stage of development.
* Physiological pathways—the sequence of enzymatic steps leading from a raw material to the desired product, are not known for many chemicals. Much basic research will be necessary to identify all the steps. The *number* of genes necessary for the conversion is a major limitation. Currently, rDNA is most useful when only a *single* easily identifiable gene is needed. It is more difficult to use when several genes must be transferred. Finally, the problems are formidable, if not impossible, when the genes have not yet been identified. This is the

case with many traits of agronomic importance, such as plant height.
* Even if the genes are identified and successfully transferred, methods must be developed to *recognize* the bacteria that received them. Therefore, the need to develop appropriate selection methods has impeded the application of molecular genetics.

As a consequence of these limitations, genetic engineering will be applied to the development of capabilities that require the transfer of only one or a few identified genes.

Fermentation and industry

Genetic engineering is not in itself an industry, but a technology used at the laboratory level. It allows the researcher to alter the hereditary apparatus of a living cell so that the cell can produce more or different chemicals, or perform completely new functions. The altered cell, or more appropriately the population of altered identical cells is, in turn, used in industrial production. It is within this framework that the impacts of applied genetics in the various industries is examined.

Regardless of the industry, the same three criteria must be met before genetic technologies can become commercially feasible. These criteria represent major constraints that industry must overcome before genetic engineering can play a part in bringing a product to market. They include the need for:

1. a useful biochemical product;
2. a useful biological fermentation approach to commercial production; and
3. a useful genetic approach to increase the efficiency of production.

The three criteria interrelate and can be met in any order; the demonstration of usefulness can begin with any of the three. Insulin, e.g., was first found to have value in therapy; fermentation was then shown to be useful in its production; and, now genetic engineering promises to make the fermentation process economically competitive. In contrast, the value of thymosin α-1, has not yet been proved, although the usefulness of genetic engineering and fermentation in its production have been demonstrated.

As these examples indicate, the limits on a product's commercial potential vary with the product. In some cases, the usefulness of the product has already been shown, and the usefulness of genetic technologies must be proved. In others, the genetic technologies make production at the industrial level possible, but their market has not yet been established. In still others, the feasibility of fermentation is the major problem.

The Pharmaceutical Industry

Chapter 4

The Pharmaceutical Industry

Background

The domestic sales of prescription drugs by U.S. pharmaceutical companies exceeded $7.5 billion in 1979. Of these, approximately 20 percent were products for which fermentation processes played a significant role. They included anti-infective agents, vitamins, and biologicals, such as vaccines and hormones. Genetics is expected to be particularly useful in the production of these pharmaceuticals and biologicals, which can only be obtained by extraction from human or animal tissues and fluids.

Although the pharmaceutical industry was the last to adopt traditional fermentation technologies, it has been the first industry to make widespread use of such advanced genetic technologies as recombinant DNA (rDNA) and cell fusion. Two major factors triggered the use of genetics in the pharmaceutical industry:

- The biological sources of many pharmacologically active products are micro-organisms, which are readily amenable to genetic engineering.
- The major advances in molecular genetic engineering have been made under an institutional structure that allocates funds largely to biomedical research. Hence, the Federal support system has tended to foster studies that have as their ostensible goal the improvement of health.

Two factors, however, have tended to discourage the application of genetics in the chemical and food industries. In the former, economic considerations have not allowed biological production systems to be competitive with the existing forms of chemical conversion, with rare exceptions. And in the latter, social and institutional considerations have not favored the development of foods to which genetic engineering might make a contribution.

Past uses of genetics

Genetic manipulation of biological systems for the production of pharmaceuticals has two general goals:

1. to increase the level or efficiency of the production of pharmaceuticals with proven or potential value; and
2. to produce totally new pharmaceuticals and compounds not found in nature.

The first goal has had the strongest influence on the industry. It has been almost axiomatic that if a naturally occurring organism can produce a pharmacologically valuable substance, genetic manipulation can increase the output. The following are three classic examples.

- The genetic improvement of penicillin production is an example of the elaborate long-term efforts that can lead to dramatic increases. The original strains of *Penicillium chrysogenum*, NRRL-1951, were treated with chemicals and irradiation through successive stages, as shown in figure 20, until the strain E-15.1 was developed. This strain had a 55-fold improvement in productivity over the fungus in which penicillin was originally recognized—the Fleming strain.
- Chemically induced mutations improved a strain of *Escherichia coli* to the point where it produced over 100 times more L-asparaginase (which is used to fight leukemia) than the original strain. This increase made the task of isolating and purifying the pharmaceutical much easier, and resulted in lowering the cost of a course of therapy from nearly $15,000 to approximately $300.
- Genetic manipulation sufficiently improved the production of the antibiotic, gentami-

Figure 20.—The Development of a High Penicillin-Producing Strain via Genetic Manipulation

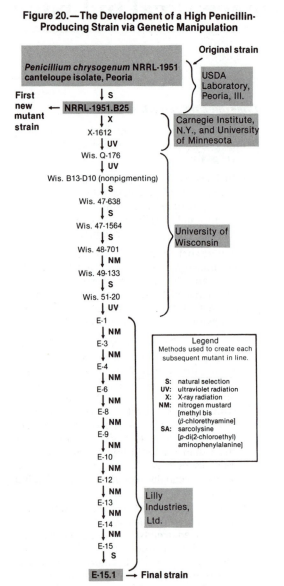

An illustration of the extensive use of genetics to increase the yield of a commercially valuable substance. A variety of laboratories and methods were responsible for the successful outcome.

SOURCE: Adapted by Office of Technology Assessment from R. P. Elander in *Genetics of Industrial Microorganisms*, O. K. Sebek and A. I. Laskin (eds.) (Washington, D.C.: American Society for Microbiology, 1979), p. 23.

cin, so that Schering-Plough, its producer, did not have to build a scheduled manufacturing plant, thereby saving $50 million.

Most industry analysts agree that, overall, genetic manipulation has been highly significant in increasing the availability of many pharmaceuticals or in reducing their production costs.

The second major goal of genetic manipulation, the production of new compounds, has been achieved to a lesser degree. A recent new antibiotic, deoxygentamicin, was obtained by mutation and will soon be clinically tested in man. Earlier, an important new antibiotic, amikacin, was produced through classical molecular genetic techniques. And before that, the well-known antibiotic, tetracycline, which is normally not found in nature, was produced by a strain of the bacterium, *Streptomyces*, after appropriate genetic changes had been carried out in that bacterium.

Potential uses of molecular genetic technologies ___

Polypeptides—proteins—are the first abundant end products of genes. They include peptide hormones, enzymes, antibodies, and certain vaccines. Producing them is the goal of most current efforts to harness genetically directed processes. However, it is just a matter of time and the evolution of technology before complex nonproteins like antibiotics can also be manufactured through rDNA techniques.

Hormones

The most advanced applications of genetics today, in terms of technological sophistication and commercial development, are in the field of hormones, the potent messenger molecules that help the body coordinate the actions of various tissues. (See Tech. Note 1, p. 80.) The capacity to synthesize proteins through genetic engineering has stemmed in large part from attempts to prepare human peptide hormones (like insulin and growth hormone). The diseases caused by their deficiencies are presently treated with extracts made from animal or human glands.

The merits of engineering other peptide hormones depend on understanding their actions and those of their derivatives and analogs. Evidence that they might be used to improve the treatment of diabetes, to promote wound healing, or to stimulate the regrowth of nerves will stimulate new scientific investigations. Other relatively small polypeptides that influence the sensation of pain, appetite suppression, and cognition and memory enhancement are also being tested. If they prove useful, they will unquestionably be evaluated for production via fermentation.

While certain hormones have already attained a place in pharmacology, their testing and use has been hindered to some extent by their scarcity and high cost. Until recently, animal glands, human-cadaver glands, and urine were the only sources from which they could be drawn. Their use is also limited because polypeptide hormones must be administered by injection. They are digested if they are taken orally, a process that curtails their usefulness and causes side-effects.

There are four technologies for producing polypeptide hormones and polypeptides:

- extraction from human or animal organs, serum, or urine;
- chemical synthesis;
- production by cells in tissue culture; and
- production by microbial fermentation after genetic engineering.

One major factor in deciding which technology is best for which hormone is the length of the hormone's amino acid chains. (See table 3.) Modern methods of chemical synthesis have made the preparation of low-molecular weight polypeptides a fairly straightforward task, and chemically synthesized hormones up to at least 32 amino acids (AA) in length—like calcitonin

Table 3.—Large Human Polypeptides Potentially Attractive for Biosynthesis

	Amino acid residues	Molecular weight
Prolactin	198	
Placental lactogen	192	
*Growth hormone.	191	22,005
Nerve growth factor	118	13,000
Parathyroid hormone (PTH)	84	9,562 bovine
Proinsulin	82	
Insulin-like growth factors		
(IGF-I & IGF-2)	70, 67	7,649, 7471
Epidermal growth factor		6,100
*Insulin	51	5,734
Thymopoietin	49	
Gastric inhibitory polypeptide		
(GIP)	43	5,104 porcine
*Corticotropin (ACTH)	39	4,567 porcine
Cholecystokinin (CCK-39)	39	
Big gastrin (BG).	34	
Active fragment of PTH	34	4,109 bovine
Cholecystokinin (CCK-33)	33	3,918 porcine
*Calcitonin	32	3,421 human
		3,435 salmon
Endorphins	31	3,465
*Glucagon.	29	3,483 porcine
Thymosin-α_1	28	3,108
Vasoactive intestinal peptide (VIP)	28	3,326 porcine
*Secretin	27	
*Active fragment of ACTH	24	
Motilin	22	2,698

*Currently used in medical practice.

SOURCE: Office of Technology Assessment.

—have become competitive with those derived from current biological sources. Since fragments of peptide hormones often express activities comparable or sometimes superior to the intact hormone, a significant advantage of chemical synthesis for research purposes is that analogs having slight pharmacological differences from natural hormones can be prepared by incorporating different amino acids into their structures. In principle however, genetically engineered biosynthetic schemes can be devised for most desirable peptide hormones and their analogs, although the practicality of doing so must be assessed on a case-by-case basis. Ultimately, the principal factors bearing on the practicality of the competing alternatives are:

- The cost of raw materials. For genetically engineered biosynthesis, this includes the cost of the nutrient broth plus some amortization of the cost of developing the synthetic organism. In the case of chemical synthesis, it includes the cost of the pure amino acid subunits plus the chemicals used as activating, protecting, coupling, liberating, and supporting agents in the process.
- The different costs of separating the desired product from the cellular debris and the culture medium in biological production, and from the supporting resin, byproducts, and excess reagents in chemical synthesis.
- The cost of purification and freedom from toxic contaminants. The process is more expensive for biologically produced material than for materials produced by conventional chemistry, although hormones from any source can be contaminated.
- Differences in the costs of labor and equipment. Chemical synthesis involves a sequence of similar (but different) operations during a time period roughly proportional to the length of the amino acid chain (three AA per day) in an apparatus large enough to produce 100 grams (g) to 1 kilogram (kg) per batch; biological fermentations use vats —with capacities of several thousand gallons—for a few days, regardless of the length of the amino acid chain.

- The cost and suitability of comparable materials gathered from organs or fluids obtained from animals or people.

In the past decade, some simpler hormones have been chemically synthesized and a few are being marketed. However, synthesizing glycoproteins—proteins bound to carbohydrates—is still beyond the capabilities of chemists. Data obtained from companies directly involved in the production of peptides by chemical synthesis indicate that the cost of chemically preparing polypeptides of up to 50 AA in length is extremely sensitive to volume (see Tech. Note 2, p. 80.); although the costs are high, the production of large quantities by chemical synthesis offers a competitive production method.

Nevertheless, rDNA production, also known as molecular cloning, has already been used to produce low-molecular weight polypeptides. In 1977, researchers at Genentech, Inc., a small biotechnology company in California, inserted a totally synthetic DNA sequence into an *E. coli* plasmid and demonstrated that it led to the production of the 14 AA polypeptide sequence corresponding to somatostatin, a hormone found in the brain. The knowledge of somatostatin's amino acid sequence made the experiment possible, and the existence of sensitive assays allowed the hormone's expression to be detected. Although the primary motive for using this particular hormone for the first demonstration was simply to show that it could be done, Genentech has announced that it plans to market its genetically engineered molecule for research purposes. (See figure 21.)

Somatostatin is one of about 20 recognized small human polypeptides that can be made without difficulty by chemical synthesis. (See table 4.) Unless a sizable market is found for one of them, it is unlikely that fermentation methods will be developed in the foreseeable future. Some small peptides that may justify the development of a biosynthetic process of production are:

- The seven AA sequence known as MSH/ ACTH 4-10, which is reputed to influence memory, concentration, and other psychological-behavioral effects: should such

Figure 21.—The Product Development Process for Genetically Engineered Pharmaceuticals

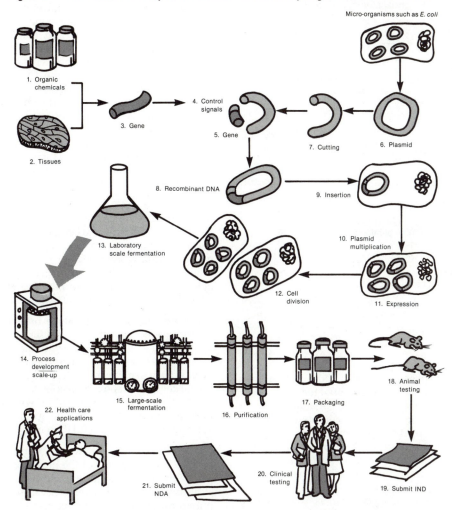

The development process begins by obtaining DNA either through organic synthesis (1) or derived from biological sources such as tissues (2). The DNA obtained from one or both sources is tailored to form the basic "gene" (3) which contains the genetic information to "code" for a desired product, such as human interferon or human insulin. Control signals (4) containing plasmids (6) are isolated from micro-organisms such as *E. coli*; cut open (7) and spliced back (8) together with genes and control signals to form "recombinant DNA" molecules. These molecules are then introduced into a host cell (9).

Each plasmid is copied many times in a cell (10). Each cell then translates the information contained in these plasmids into the desired product, a process called "expression" (11). Cells divide (12) and pass on to their offspring the same genetic information contained in the parent cell.

Fermentation of large populations of genetically engineered micro-organisms is first done in shaker flasks (13), and then in small fermenters (14) to determine growth conditions, and eventually in larger fermentation tanks (15). Cellular extract obtained from the fermentation process is then separated, purified (16), and packaged (17) for health care applications.

Health care products are first tested in animal studies (18) to demonstrate a product's pharmacological activity and safety, In the United States, an investigational new drug application (19) is submitted to begin human clinical trials to establish safety and efficacy. Following clinical testing (20), a new drug application (NDA) (21) is filed with the Food and Drug Administration (FDA). When the NDA has been reviewed and approved by the FDA the product may be marketed in the United States (22).

SOURCE: Genentech, Inc.

**Table 4.—Naturally Occurring Small Peptides of
Potential Medical Interest**

	Number of amino acids	Molecular weight
Dynorphin	17	
Little gastrin (LG)	17	2,178
Somatostatin	14	1,639
Bombesin	14	1,620
Melanocyte stimulating hormone	13	1,655
Active dynorphin fragment	13	
Neurotensin	13	
Mini-gastrin (G13)	13	
Substance?	11	1,347 bovine
Luteinizing hormone-releasing hormone (LNRH)	10	1,183
Active fragment of CCK	10	
Angiotensin I	10	1,297
Caerulein	10	1,252 porcine
Bradykinin	9	1,060
*Vasopressin (ADH)	9	
*Oxytocin	9	1,007
Facteur thymique serique (FTH)	9	
Substance P (4-11) octapeptide	8	966
Angiotensin II	8	1,046
Angiotensin III	7	931
MSH/ACTH 4-10	7	
Enkephalins	5	575
Active fragment of thymopoietin (TP5)	5	
*Thyrotropin releasing hormone (TRH)	3	362

*Currently used in medical practice.

SOURCE: Office of Technology Assessment.

agents prove of value in wider testing, they
have an enormous potential for use.

- Both cholecystokinin (33 AA) and bombesin
(10 AA), which have been shown to suppress appetite, presumably as a satiety
signal from stomach to brain: there is a
large market for antiobesity agents—approximately $85 million per year at the
manufacturer's level.

- Several hormones, such as somatostatin,
which are released by nerves in the hypo-
thalamus of the brain to stimulate or inhibit release of hormones by the pituitary
gland: hormones produced by these glands
are crucial in human fertility; analogs of
some are being investigated as possible
contraceptives.

- Calcitonin (32 AA), which is currently the
largest polypeptide produced by chemical
synthesis for commercial pharmaceutical
use: it is useful for pathologic bone disorders, such as Paget's disease, that affect

up to 3 percent of the population over 40
years of age, in Western Europe.

- Adrenocorticotropic hormone (ACTH) (39
AA), which promotes and maintains the
normal growth and development of the
adrenal glands and stimulates the secretion
of other hormones: in the United States,
ACTH is used primarily as a diagnostic
agent for adrenal insufficiency, but in
principle, ACTH might be used for at least
one-third of the medical indications—like
rheumatic disorders, allergic states, and
eye inflammation—for which about 5 million Americans annually receive corticosteroids.

Within the last 5 years, other small polypeptides have been identified in many tissues and
have been linked to a variety of activities. Some
certainly bind to the same receptor sites as the
pain-relieving opiates related to the morphine
family. These peptides are called endogenous
opiates: the smaller (5 AA) peptides are called
enkephalins and the larger (31 AA), endorphins.

Certain enkephalins produce brief analgesia
when injected directly into the brains of mice.
Synthetic analogs that are less susceptible to enzymatic inactivation produce longer analgesia
even if they are injected intravenously, as does
the larger β-endorphin molecule. Very recently,
a 17 AA polypeptide, dynorphin, was reported
to be the most potent pain killer yet found—it is
1,200 times more powerful than morphine.

The preparation of new analgesic agents appears a likely outcome of the new research, but
problems similar to those associated with classical opiates must be overcome. Consequently,
unnatural analogs—including some made with
amino acids not found in micro-organisms—
might prove more useful. The value of microbial biosynthesis for these substances is questionable at this time. However, the importance
of genetic technologies in clarifying the
underlying mechanisms should not be under-
estimated.

Higher molecular weight polypeptides cannot
be made practically by chemical synthesis, and
must be extracted from human or animal tissues or produced in cells growing in culture.

Now they can also be manufactured by fermentation using genetically designed bacteria, as has been demonstrated by the production of insulin and human growth hormone.

INSULIN

Insulin, is composed of two chains—A and B—of amino acids. It is initially produced as a single, long chain called pre-proinsulin, which is cut into a shorter chain, proinsulin. Proinsulin, in turn, is cut into the A and B chains when a piece is cleaved from the middle. (See figure 22.) Work on the genetic engineering of insulin has

proceeded quickly. A year after one group reported that the insulin gene had been incorporated into *E. coli* without expression, a second group managed to grow colonies of *E. coli* that actually excreted rat proinsulin. Then, within a couple of months, workers at Genentech, in collaboration with a group at City of Hope Medical Center, announced the separate synthesis of the A (21 AA) and B (30 AA) chains of human insulin. The synthesis of the DNA sequences depended on advances in organic chemistry as well as in genetics. Six months were required simply to synthesize the necessary building blocks.

Figure 22.—The Amino Acid Sequence of Proinsulin

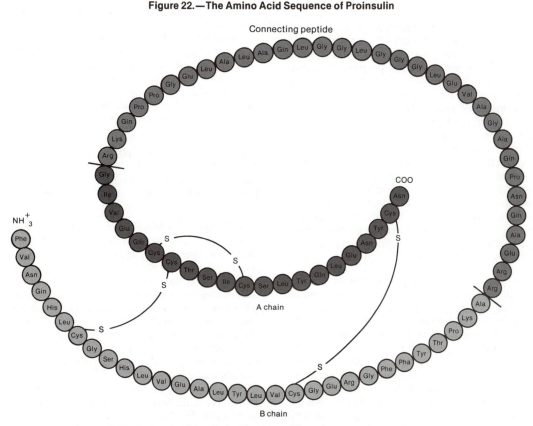

Proinsulin is composed of 84 amino acid residues. When the connecting peptide is removed, the remaining A and B chains form the insulin molecule. The A chain contains 21 amino acids; the B chain contains 30 amino acids.

SOURCE: Office of Technology Assessment.

A comparison with the traditional source of animal insulin is interesting. If 0.5 milligram (mg) of pure insulin can be obtained from a liter of fermentation brew, 2,000 liters (l) (roughly 500 gal) would yield 1 g of purified insulin—the amount produced by about 16 lb of animal pancreas. If, on the other hand, the efficiency of production could be increased to that achieved for asparaginase (which is produced commercially by the same organism, *E. coli*), 2,000 l would yield 100 g of purified insulin—the amount extracted from 1,600 lb of pancreas. (The average diabetic uses the equivalent of about 2 mg of animal insulin per day.)

The extent of the actual demand for insulin is a controversial issue. Eli Lilly & Co. estimates that there are 60 million diabetics in the world (35 million in underdeveloped countries, where few are diagnosed or treated). Of the 25 million in the developed countries, perhaps 15 million have been diagnosed; according to Lilly's estimate, 5 million are treated with insulin. Only one-fourth of those diabetics treated with insulin live in the United States, but they use 40 to 50 percent of the insulin consumed in the world. A number of studies indicate that while the emphasis on diet (alone) and oral antidiabetic drugs varies, approximately 40 percent of American patients in large diabetes clinics or practices take insulin injections. In the United States, diabetes ranks as the fifth most common cause of death and second most common cause of blindness. Roughly 2 million persons require daily injections of insulin.

Today, at least, there is no real shortage of glands from slaughter houses for the production of animal (principally bovine and porcine) insulin. A study conducted by the National Diabetes Advisory Board (NDAB) concluded that a maximum demand and a minimum supply would lead to shortages in the 1990's. Eli Lilly's projection, presented in that report, also anticipates these shortages. But, Novo Industri, a major world supplier of insulin, told the NDAB that it estimates that the 1976 free-world consumption of insulin of 51×10^9 units constituted only 23 percent of the potential supply, and the 87×10^9 units projected for 1996 would only equal 40 percent of the supply, assuming that the animal population stays constant.

For insulin, therefore, the limitation on bringing the fruits of genetic engineering to the marketplace is not technological but institutional. The drug must first be approved by the Food and Drug Administration (FDA) and then marketed as a product as good as or better than the insulin extracted by conventional means. Lilly has stated that it anticipates a 6-month testing period in humans. Undoubtedly, FDA will examine the evidence presented in the investigational new drug application (INDA) with special care. Its review will establish criteria that may influence the review of subsequent applications in at least the following requirements:

- evidence that the amino acid sequence of the material is identical to that of the normal human hormone;
- freedom from bacterial endotoxins that may cause fever at extremely low concentrations—an inherent hazard associated with any process using *E. coli;* and
- freedom from byproducts, including substances of very similar structure that may give rise to rare acute or chronic reactions of the immune system.

Furthermore, as development continues, FDA might require strict assurances that the molecules produced from batch to batch are not subject to subtle variations resulting from their genetic origin.

If the insulin obtained from rDNA techniques manages to pass FDA requirements, it must overcome a second obstacle—competition in the marketplace. The clinical rationale for using human rather than animal insulin rests on the differences in structure among insulins produced by different species. Human and porcine insulins for example, differ in a single amino acid, while human and cattle insulins differ with respect to three. As far as is known, these variations do not impair the effectivenes of the insulin, but no one has ever been in a position to conduct a significant test of the use of human insulin in a diabetic population. Many consequences of the disease, such as retinopathy (retinal disease) and nephropathy (kidney disease), are not prevented by routine injection of animal insulin. Patients also occasionally respond adversely or produce antibodies to animal insulin, with subsequent allergic or resistant reaction.

It remains to be seen how many patients will be better off with human insulin. The proof that it improves therapy will take years. Progress on the etiology of the disease—especially in identifying it in those at risk or in improving the dosage form and administration of insulin—may have far more significant effects than new developments in insulin production. Nevertheless, as long as private enterprise sees fit to invest in such developments, and as long as the cost of treating diabetics who respond properly to animal insulin is not increased, biological production of human insulin may become a kind of insurance for diabetics within the next few decades.

GROWTH HORMONE

The second polypeptide hormone currently a candidate for FDA approval is growth hormone (GH). It is one of a family of closely related, relatively large pituitary peptide hormones—single-chain polypeptides 191- to 198-AA in length. It is best known for the growth it induces in many soft tissues, cartilage, and bone, and it is a requirement for postnatal growth in man.

The growth of an organism is a highly complex process that depends on the correct balance of many variables: The action of GH in the body for example, depends on the presence of insulin, whose secretion is stimulated by GH. Under some circumstances, one or more intermediary polypeptides produced under the influence of GH by the liver (and possibly the kidneys) may actually be the proximate causes of some of the effects attributed to GH. In any case, the biological significance of GH is most clearly illustrated by the growth retardation that characterizes its absence before puberty, and by the benefits of replacement therapy.

In the United States, most of the demand for human growth hormone (hGH) is met by the National Pituitary Agency, which was created in the early 1960's by the College of Pathologists and the National Institute of Arthritis, Metabolism, and Digestive Diseases (NIAMDD) to collect pituitary glands from coroners and private donors. Under the programs of the NIAMDD, hGH is provided without charge to treat children with hypopituitarism, or dwarfism (about

1,600 patients, each of whom receives therapy for several years), and for research.

While the National Pituitary Agency feels that it can satisfy the current demand for hGH (see Tech. Note 3, p. 80.), it welcomes the promise of additional hGH at relatively low cost to satisfy areas of research that are handicapped more by a scarcity of funds than by a scarcity of the hormone. However, if hGH is shown to be therapeutically valuable in these areas, widespread use could severely strain the present supply. At present, the potential seems greatest for patients with:

- senile osteoporosis (bone decalcification);
- other nonpituitary growth deficiences such as Turner's syndrome (1 in 3,000 live female births);
- intrauterine growth retardation;
- bleeding ulcers that cannot be controlled by other means; and
- burn, wound, and bone-fracture healing

Two groups have already announced the preparation of micro-organisms with the capacity for synthesizing GH. (See Tech. Note 4, p. 80.) In December 1979, one of these groups—Genentech —requested and received permission from the National Institutes of Health (NIH), on the recommendation of the Recombinant DNA Advisory Committee (RAC), to scale-up its process. Its formation of a joint-venture with Kabi Gen AB is typical of the kind of alliance that develops as a result of the different expertise of groups in the multidisciplinary biomedical field. Kabi has been granted a New Drug Application (NDA) under which to market pituitary GH imported from abroad.

OTHER HORMONES

Additional polypeptide hormones targeted for molecular cloning (rDNA production) include:

- Parathyroid hormone (84 AA), which may be useful alone or in combination with calcitonin for bone disorders such as osteoporosis.
- Nerve growth factor (118 AA), which influences the development, maintenance, and

repair of nerve cells and thus could be significant for nerve restoration in surgery.

- Erythropoietin, a glycopeptide that is largely responsible for the regulation of blood cell development. Its therapeutic applications may range from hemorrhages and burns to anemias and other hematologic conditions. (See Tech. Note 5, p. 80.)

Immunoproteins

Immunoproteins include all the proteins that are part of the immune system—antigens, interferons, cytokines, and antibodies. Since polypeptides, the primary products of every molecular cloning scheme, are at the heart of immunology, developments made possible by recent breakthroughs will presumably affect the entire field. There is little doubt that applied genetics will play a critical role in developing a pharmacology for controlling immunologic functions, since it provides the only apparent means of synthesizing many of the agents that will comprise immunopharmacology.

ANTIGENS (VACCINES)

One early dramatic benefit should be in the area of vaccination, where genetic technologies may lead to the production of harmless substances capable of eliciting specific defenses against various stubborn infectious diseases.

Vaccination provides effective immunity by introducing relatively harmless antigens into the immune system thereby allowing the body to establish, *in advance*, adequate levels of antibody and a primed population of cells that can grow when the antigen reappears in its virulent form. Obviously, however, the vaccination itself should not be dangerous. As a result, several methods have been developed over the past two centuries to modify the virulence of micro-organisms used in vaccines without destroying their ability to trigger the production of antibodies. (See Tech. Note 6, p. 80.)

Novel pure vaccines based on antigens synthesized by rDNA have been proposed to fight communicable diseases like malaria, which have resisted classical preventive efforts. Pure vaccines have always been scarce; if they were available, they might reduce the adverse effects of conventional vaccines and change the methods and the dosages in which vaccines are administered.

Some vaccines are directed against toxic proteins (like the diphtheria toxin produced by some organisms), preparing the body to neutralize them. Molecular cloning might make it possible to produce inactivated toxins, or better nonvirulent fragments of toxins, by means of micro-organisms that are incapable of serving as disease-causing organisms.

Immunity conferred by live vaccines invariably exceeds that conferred by nonliving antigenic material—possibly because a living micro-organism creates more antigen over a longer period of time, providing continuous "booster shots." Engineered micro-organisms might become productive sources of high-potency antigen, offering far larger, more sustained, doses of vaccine without the side-effects from the contaminants found in those vaccines that consist of killed micro-organisms.

However, it is clear that formidable Federal regulatory requirements would have to be met before permission is granted for a novel living organism to be injected into human subjects. Because of problems encountered with live vaccines, the most likely application will lie in the area of killed vaccines (often using only *parts* of micro-organisms).

It is impossible in the scope of this report to discuss the pros, cons, and consequences of developing a vaccine for each viral disease. However, the most commercially important are the influenza vaccines, with an average of 20.8 million doses given per year from 1973 to 1975—a smaller number than the 25.0 million doses per year of polio vaccine, but more profitable.

Influenza is caused by a virus that has remained uncontrolled largely because of the frequency with which it can mutate and change its antigenic structures. It has been suggested that antigenic protein genes for influenza could be kept in a "gene bank" and used when needed. In addition, the genetic code for several antigens could be introduced into an organism such as *E.*

coli, so that a vaccine with several antigens might be produced in one fermentation.[1]

Two more viral diseases deserve at least brief comment. Approximately 800 million doses of foot-and-mouth disease virus (FMDV) vaccine are annually used worldwide, making it the largest volume vaccine produced. This vaccine must be given frequently to livestock in areas where the disease is endemic, which includes most of the world outside of North America. The present methods of producing the vaccine require that enormous quantities of hazardous virus be contained. Many outbreaks are attributed to incompletely inactivated vaccine or to the escape of the virus from factories. (See figure 23.)

Molecular cloning of the antigen could produce a stable vaccine at considerably less expense, without the risk of the virus escaping. On the basis of that potential, RAC has approved a joint program between the U.S. Department of Agriculture (USDA) and Genentech to clone pieces of the FMDV genome to produce pure antigen. The RAC decision marked the first exception to the NIH prohibition against cloning DNA that is derived from a virulent pathogen.[2] FMDV vaccine made by molecular cloning will probably be distributed commercially by 1985, although not in the United States. It will be the first vaccine to achieve that status, and illustrates the potential veterinary uses of genetic technologies.

Hepatitis has also received significant attention. Vaccines against viral hepatitis, which affects some 300,000 Americans each year, may be produced by molecular cloning. This disease is second only to tuberculosis as a cause of death among reportable infectious diseases. It is extremely difficult to cultivate the causative agents. Hepatitis A has a good chance of being the first human viral disease for which the initial preparation of experimental vaccine will involve molecular cloning. A vaccine against hepatitis B, made from the blood of chronic carriers,

---

[1]For other aspects of vaccine production see: Office of Technology Assessment, U.S. Congress, *Working Papers, The Impacts of Genetics,* vol. 2. (Springfield, Va.: National Technical Information Service, 1981).
[2]Ibid.

Figure 23.—Recombinant DNA Strategy for Making Foot-and-Mouth Disease Vaccine

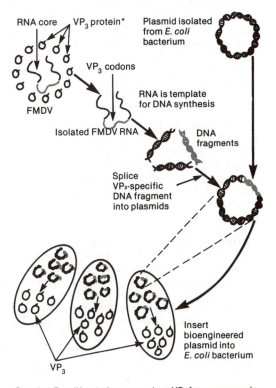

Growing *E. coli* bacteria may produce VP₃ for use as vaccine for foot-and-mouth disease. No virus or infectious RNA is produced by the harmless bacteria strain.

*VP₃ is the protein from the shell of the virus, which can act as a vaccine for immunizing livestock against foot-and-mouth disease. The idea outlined above is to make this VP₃ protein without making any virus or infectious RNA.

SOURCE: Office of Technology Assessment.

is in the testing stage, but cloning is being investigated as a better source of an appropriate antigen. The causative agent for a third form of hepatitis has not even been identified. Since at least 16 million U.S. citizens are estimated to be at high risk of contracting hepatitis, there is keen interest in the development of vaccines among academic and industrial researchers.[3]

---

[3]Ibid.

More hypothetically, molecular cloning may lead to three other uses of antigens as well: vaccination against parasites, such as malaria and hookworm (see Tech. Note 7, p. 80.); immunization in connection with cancer treatment; and counteracting abnormal antibodies, which are made against normal tissues in the so-called "autoimmune diseases," such as multiple sclerosis. (See Tech. Note 8, p. 81.)

INTERFERONS

Interferons are glycoproteins normally made by a variety of cells in response to viral infection. All interferons (see Tech. Note 9, p. 81) can induce an antiviral state in susceptible cells. In addition, interferon has been found to have at least 15 other biochemical effects, most of which involve other elements of the immune system.

Promising preliminary studies have supported the use of interferon in the treatment of such viral diseases as rabies, hepatitis, varicella-zoster (shingles), and various herpes infections. To date, the effect of interferon has been far more impressive as a prophylactic than as a therapeutic agent. The interferon produced by Genentech, for example, has been shown to protect squirrel monkeys from infection by the lethal myocarditis virus. Once interferon is available in quantity, large-scale tests on human populations can be conducted to confirm its efficacy in man.

Several production techniques are being explored. (See Tech. Note 10, p. 81.) Extraction of interferon from leukocytes (white blood cells), the current method of choice, may have to compete with tissue culture production as well as rDNA. (See table 5.)

Recombinant DNA is widely regarded as the key to mass production of interferons, and important initial successes have already been achieved. Each of the four major biotechnology companies is working on improved production methods, and all have reported some success.

An enormous amount remains to be learned about the interferon system. It now appears that the interferons are simply one of many families of molecules involved in physiological regulation of response to disease. Only now have molecular biology and genetics made their study—and perhaps their use—possible.

Table 5.—Summary of Potential Methods for Interferon Production

Means of production	Types of interferon produced	Potential for scale-up	Present projected ($/10^6 units)		Problems	Potential for improvement
"Buffy coat" leukocytes	leukocyte, 95% fibroblast, 5%	No	50	—	—lack of scale-up —pathogen contamination	—minimal
Lymphoblastoid cells	leukocyte, 80% fibroblast, 20%	Yes	—	≅25	—poor yields —cells derived from tumor	—improved yields —expression of fibroblast interferon
Fibroblasts	fibroblast	Yes	43-200	≅1-10	—cell culture —economic competition with recombinant DNA	—improved yields —improved cell-culture technology —expression of leukocyte-type interferon
Recombinant DNA	leukocyte or fibroblast	Yes	—	≅1-10	—does not produce interferon —in vitro drug stability —poor yields —drug approval —possible economic competition with fibroblast cell production	—improved yields —modified interferons

SOURCE: Office of Technology Assessment.

The interferons are presently receiving attention largely because studies in Sweden and the United States stimulated the appropriation of $5.4 million by the American Cancer Society (ACS) for expanded clinical trials in the treatment of cancer. That commitment by the nonprofit ACS—the greatest by far in its history—was followed by a boost in NIH funding for interferon research from $7.7 million to $19.9 million for fiscal year 1980. Much of the cost of interferon research is allotted to procuring the glycopeptide. Initially, the ACS bought 40 billion units of leukocyte interferon from the Finnish Red Cross for $50 per million units. In March 1980, Warner-Lambert was awarded a contract to supply the National Cancer Institute (NCI) with 50 billion units of leukocyte interferon within the next 2 years at an average price of $18 per million units. NCI is also planning to purchase 50 billion units each of fibroblast and lymphoblastoid interferons.

The bulk of the NIH funding is included in NCI's new Biological Response Modifier (BRM) program—interferon accounts for $13.9 million of the $34.1 million allocated for BRM work in fiscal year 1980. (NCI expenditures on interferon in 1979 were $2.6 million, 19 percent of the amount budgeted for 1980.) Other important elements of that BRM program concern immunoproteins known as lymphokines and thymic hormones, for which molecular genetics has major implications. The program is aimed at identifying and testing molecules that control the activities of different cell types.

LYMPHOKINES AND CYTOKINES

Lymphokines and cytokines are regulatory molecules that have begun to emerge from the obscure fringes of immunology in the past 10 years. (Interferon is generally considered a lymphokine that has been characterized sufficiently to deserve independent status.)

Lymphokines are biologically active soluble factors produced by white blood cells. Studied in depth only within the last 15 years, they are being implicated at virtually every stage in the complex series of events that make up the immune response. They now include about 100 different compounds. Cytokines, which have effects similar to lymphokines, include several compounds associated with the thymus gland, referred to as thymic hormones.[4]

In 1979, the BRM subcommittee concluded that several of these agents probably have great potential for cancer treatment. Nevertheless, adequate quantities for laboratory and clinical testing of many of them will probably not be available until the problems of producing glycoproteins by molecular cloning are overcome. No system is currently available for the industrial production of glycoproteins, although yeasts may prove to be the most useful micro-organisms.

ANTIBODIES

Antibodies are the best known and most exploited protein components of the immune system. Until recently, all antibodies were obtained from the blood of humans or animals; and they were often impure. Within the past 5 years, however, it has become possible to produce antibodies from cells in culture, and to achieve levels of purity previously unattainable. As with previous advances in antibody technology, researchers are examining ways to put this new level of purity to use. There have been hundreds, if not thousands, of examples of new diagnostic and research methods, new methods of purification, and new therapies published within the first 3 years that the technique has been available. (See Tech. Note 11, p. 81.)

This high level of purity was attained by the development of monoclonal antibodies. These antibodies that recognize only *one* kind of antigen were the unanticipated fruit of fundamental immunological research conducted by Drs. Caesar Milstein and Georges Kohler at the Medical Research Council in England in 1975. They fused two types of cells—myeloma and plasma-spleen cells—to form hybridomas that produce the monoclonal antibodies. (See Tech. Note 12, p. 81.) Not only are the antibodies specific, but because the hybridomas can be grown in mass culture, a virtually limitless supply is available.

The most immediate medical application for monoclonal antibodies lies in diagnostic testing.

[4]For 40 of the best characterized cytokines, see footnote 1, p. 69.

Over the past 20 years, large segments of the diagnostic and clinical laboratory industries have sprung up to detect and quantify particular substances in specimens. Because monoclonal antibodies are so specific, hybridomas seem certain to replace animals as the source of antibodies for virtually all diagnosis and monitoring. Their use will not only improve the accuracy of tests and decrease development costs, but should result in a more uniform product.

Today, such assays are used to:

- determine hormone levels in order to assess the proper functioning of an endocrine gland or the inappropriate production of a hormone by a tumor;
- detect certain proteins, the presence of which has been found to correlate with a tumor or with a specific prenatal condition;
- detect the presence of illicit drugs in a person's blood, or monitor the blood or tissue level of a drug to ensure that the dosage achieves a therapeutic level without exceeding the limits that could cause toxic effects; and
- identify microbial pathogens.

The extent of the use of antibodies and the biochemical properties that they can identify is suggested by table 6. No one assay constitutes a major market, and short product lifetime has been characteristic of this business.

Other applications of monoclonal antibodies include:

- the improvement of the acceptance of kidney (and other organ) transplants by injection of the recipient with antibodies against certain antigens;
- passive immunization against an antigen involved in reproduction, as a reversible immunological approach to contraception.
- localizing tumors with tumor-specific antibodies (see Tech. Note 13, p. 81); and
- targeting cancer cells with antibodies that have anticancer chemicals attached to them.

Enzymes and other proteins

ENZYMES

Enzymes are involved in virtually every biological process and are well-understood. Nevertheless, despite their potency, versatility, and diversity, they play a small role in the practice of medicine today. Therapeutic enzymes accounted for American sales of about $70 million (wholesale) in 1978, but one-half of those sales involved the blood-plasma-derived coagulation factors used to treat hemophilia. Although the figure is difficult to estimate, the total number of patients receiving any type of enzyme therapy in 1980 probably does not exceed 50,000.

Enzymes cannot be synthesized by conventional chemistry. Almost all those presently employed in medicine are extracted from human blood, urine, or organs, or are produced by micro-organisms. Already the possibility of using rDNA clones as the source of enzymes—primarily to reduce the cost of production—is being explored.

However, problems associated with the use of nonhuman enzymes (such as immune and febrile responses) and the scarcity of human enzymes, have hindered research, development, and clinical exploitation of enzymes for therapeutic purposes. Today, the experimental genetic technologies of rDNA and somatic cell fusion and culture open the only conceivable routes to relatively inexpensive production of compatible human enzymes.

The genetic engineering of enzymes is probably the best example of a dilemma that hampers the exploitation of rDNA: Without a clinical need large enough to justify the investment, there is no incentive to produce a product; yet without adequate supplies, the therapeutic possibilities cannot be investigated. The substances that break this cycle will probably be those that are already produced in quantity from natural tissue.

The only enzymes administered today are given to hemophiliacs—and they are actually

Table 6.—Immunoassays

Analgesics and narcotics	Deslanoside	Malathion	Methaqualone
Anileridine	Digitoxin	Narcotic antagonists	Steroid hormones
Antipyrine	Digoxin	Cyclazoncine	Skeletal muscle relaxants
Codeine	Gitoxin	Naloxone	d-Tubocurarine
Etorphine	Methyl digoxin	Nucleosides and	Synthetic peptides
Fentanyl	Ouabain	nucleotides	DDAVP
Meperidine	Proscillaridin	Cyclic AMP	Saralasin
Methadone	Dihydroergotamine	Cyclic GMP	Synthetic steroids
Morphine	Propranolol	N^2-Dimethylguanosine	Anabolic steroids
Pentazocine	Quinidine	7-Methylguanosine	Trienbolone acetate
Antibiotics	Catecholamines	Pseudouridine	Androgens
Amikacin	Epinephrine	Thymidine	Fluoxymesterone
Chloramphenicol	Norepinephrine	Peptide hormones	Estrogens
Clindamycin	Tyramine	Angiotensin	Diethylstilbestrol
Gentamicin	CNS stimulants	Anterior pituitary	Ethinylestradiol
Isoniazid	Amphetamine	Bradykinin	Mestranol
Penicillin	Benzoyl ecgonine	Gastric	Glucocorticoids
Sisomycin	(cocaine metabolite)	Hypothalamic	Dexamethasone
Tobramycin	Methamphetamine	Intestinal	Methylprednisolone
Anticonvulsants	Pimozide	Pancreatic	Prednisolone
Clonazepam	Diuretics	Parathyroid	Prednisone
Phenytoin	Bumetanide	Posterior pituitary	Metyrapone
Primidone	Fibrinopeptides	Thyroid (calcitonin)	Progestins
Anti-inflammatory agents	Fibrinopeptide A	Plant hormones	Medroxyprogesterone
Colchicine	Fibrinopeptide B	Indole-3-acetic acid	acetate
Indomethacin	Hallucinogenic drugs	Gibberelilic acid	Norethindrone
Phenyibutazone	Mescaline	Polyamines	Norethisterone
Antineoplastic agents	Tetrahydrocannabinol	Spermine	Norgestrel
Adriamycin	Hypoglycemic agents	Prostaglandis	Toxins
Bleomycin	Butylbiguanide	Sedatives and	Aflatoxin B$_1$
Daunomycin	Glibenclamid	tranquilizers	Genistein
Methotrexate	Indolealkylamines	Barbituarates	Nicotine and metabolites
Bile acid conjugates	Melatonin	Barbital	Ochratoxin A
Cholylglycine	Serotonin	Pentobarbital	Paralytic shellfish poison
Cholyitaurine	Insect hormones	Phenobarbital	Thyroid hormones
Bronchodilators	Ecdysone	Chlordiazepoxide	Thyroxine
Theophylline	Insecticides	Chlorpromazine	Triodothyronine
Cardiovascular drugs	Aldrin	Desmethylimipramine	Vitamins
Cardiac glycosides	DDT	Diazepam and	Vitamin B$_{12}$
Acetylstrophanthidin	Dieldrin	N-desmethyldiazepam	Vitamin D
Cedilanid		Glutethimide	

SOURCE: "Immunoassays of Drugs—Comprehensive Immunology," *Immunal Pharmacology*, Hadden Caffey (ed.) (New York: Plenum Press, 1977), p. 325.

proenzymes, which are converted to active enzymes in the body when needed. The most common agents are called Factor VIII and Factor IX, which are found in serum albumin and are currently extracted from human blood plasma. Hemophilia A and Hemophilia B—accounting for over 90 percent of all major bleeding disorders—are characterized by a deficiency of these factors. Supplies of the proenzymes will exceed demand well beyond 1980 if the harvesting and processing of plasma continues as it has. Nevertheless, the risk of hepatitis associated with the

use of human plasma-derived products is extremely high. One recent study found chronic hepatitis in a significant percentage of asymptomatic patients treated with Factor VIII and Factor IX.

The plasma fractionation industry, which produces the proenzymes, is currently faced with excess capacity, intense competition, high plasma costs, and tight profit margins.[5] The cost and availability of any one plasma protein is

[5]For details of the factors governing the industry. see footnote 1, p. 69.

coupled to the production of the others. Hence, the industry would still have to orchestrate the production of the other proteins even if just one of them, such as Factor VIII, becomes a target for biological production.

Another enzyme, urokinase, has been targeted for use in removing unwanted blood clots, which lead to strokes, myocardial infarctions, and pulmonary emboli. Currently, the drug is either isolated from urine or produced in tissue culture. (See Tech. Note 14, p. 81.)

Urokinase is thus far the only commercial therapeutic product derived from mammalian cell culture. Nevertheless, some calculations suggest that production by *E. coli* fermentation would have economic advantages. The costs implicit in having to grow cells for 30 days on fetal calf serum (or its equivalent) or in having to collect and fractionate urine—as reflected in urokinase's market price ($150/mg at the manufacturer's level)—should be enough incentive to encourage research into its production. In fact, in April 1980, Abbott Laboratories disclosed that *E. coli* had been induced to produce urokinase through plasmid-borne DNA.

The *availability* of urokinase might be guaranteed by the new genetic technologies, but its *use* is not. For a variety of reasons, the American medical community has not accepted the drug as readily as have the European and Japanese communities. Studies to establish the use of urokinase for deep vein thrombosis, for example, are now being conducted almost exclusively in Europe.[6]

OTHER PROTEINS

In addition to the proteins and polypeptides already mentioned, the structural proteins, such as the collagens (the most abundant proteins in the body), elastins and keratins (the compounds of extracellular structures like hair and connective tissue), albumins, globulins, and a wide variety of others, may also be susceptible to genetic engineering. Structural proteins are less likely to be suitable for molecular genetic manipulations: On the one hand, their size and complexity exceed the synthetic and analytic capabilities that will be available in the next few years; on the other, either their use in medicine has yet to be established or material derived from animals appears adequate, as is the case with collagen, for which uses are emerging.

Plasma, the fluid portion of the blood, contains about 10 percent solids, most of which are proteins. During World War II, a simple procedure was developed to separate the various components. It is still used today.

Serum albumin is the smallest of the main plasma proteins but it constitutes about half of plasma's total mass. Its major therapeutic use is to reverse the effects of shock.[7] It is a reasonable candidate for molecular cloning, although its relatively high molecular weight complicates purification, and its commercial value is relatively low. The market value of normal serum albumin is approximately $3/g, but the volume is such that domestic sales exceed $150 million. Including exports, annual production is in the range of 100,000 kg.

Normal serum albumin for treating shock is already regarded as too expensive compared with alternative treatments, to expand its use would require a lower price. On the other hand, the Federal Government—and especially the Department of Defense—might disregard the immediate economic prospects and conclude that having a source of human serum albumin that does not depend on payments to blood donors might be in the national interest. Since many nations import serum albumin, products derived from molecular cloning could be exported.

Serum albumin is presently the principal product of blood plasma fractionation, a change in the way it is manufactured would significantly affect that industry. Because a number of other products (such as clotting factors) are also derived from fractionation, a growth in the need for plasma-derived albumin could have a significant impact on the availability and the cost of these byproducts.

---

[6]For additional information about how urokinase came to play a role in therapy, see footnote 1, p. 69.

[7]For a detailed discussion of the costs and benefits of using albumin and the structure of the industry, see footnote 1, p. 69.

Antibiotics

Antimicrobial agents for the treatment of infectious diseases have been the largest selling prescription pharmaceuticals in the world for the past three decades. Most of these agents are antibiotics—antimicrobials naturally produced by micro-organisms rather than by chemical synthesis or by isolation from higher organisms. However, one major antibiotic, chloramphenicol—originally produced by a micro-organism, is now synthesized by chemical methods. The field of antibiotics, in fact, provides most of the precedent for employing microbial fermentation to produce useful medical substances. The United States has been prominent in their development, production, and marketing, with the result that American companies account for about half of the roughly $5 billion worth of antimicrobial agents sold worldwide each year. The American market share has been growing as new antibiotics are developed and introduced every year.

For 30 years, high-yielding, antibiotic-producing micro-organisms have been identified by selection from among mutant strains. Initially, organisms producing new antibiotics are isolated by soil sampling and other broad screening efforts. They are then cultured in the laboratory, and efforts are made to improve their productivity.

Antibiotics are complex, usually nonprotein, substances, which are generally the end products of a series of biological steps. While knowledge of molecular details in metabolism has made some difference, not a single antibiotic has had its complete biosynthetic pathway elucidated. This is partly because there is no *single* gene that can be isolated to produce an antibiotic. However, mutations can be induced within the original micro-organism so that the *level* of production can be increased.

Other methods can also increase production, and possibly create new antibiotics. Microbial mating, for example, which leads to natural recombination, has been widely investigated as a way of developing vigorous, high-yielding antibiotic producers. However, its use has been limited by the mating incompatibility of many

industrially important higher fungi, the presence of chromosomal aberrations in micro-organisms improved by mutation, and a number of other problems. Furthermore, natural recombination is most advantageous when strains of extremely diverse origins are mated; the proprietary secrets protecting commercial strains usually prevent the sort of divergent "competitor" strains most likely to produce vigorous hybrids from being brought together.

The technique of protoplast or cell fusion provides a convenient method for establishing a recombinant system in strains, species, and genera that lack an efficient natural means for mating. For example, as many as four strains of the antibiotic-producing bacterium *Streptomyces* have been fused together in a single step to yield recombinants that inherit genes from four parents. The technique is applicable to nearly all antibiotic producers. It will help combine the benefits developed in divergent lines by mutation and selection.

In addition, researchers have compared the quality of an antibiotic-producing fungus, *Cephalosporium acremonium,* produced by mating to one produced by protoplast fusion. (See Tech. Note 15, p. 82.) They concluded that protoplast fusion was far superior for that purpose. What is more, protoplast fusion can give rise to hundreds of recombinants—including one isolate that consistently produced the antibiotic cephalosporin C in 40 percent greater yield than the best producer among its parents—without losing that parent strain's rare capacity to use inorganic sulfate, rather than expensive methionine, as a source of sulfur. It also acquired the rapid growth and sporulation characteristics of its less-productive parent. Thus, desirable attributes from different parents were combined in an important industrial organism that had proved resistant to conventional crossing.

Even more significant are the possibilities for preparation by protoplast fusion between different species or genera of hybrid strains, which could have unique biosynthetic capacities. One group is reported to have isolated a novel antibiotic, clearly not produced by either parent, in an organism created through fusion of actinomycete protoplasts. (See Tech. Note 16,

p. 82.) The value of protoplast fusion, therefore, lies in potentially broadening the gene pool.

Protoplast fusion is genetic recombination on a large scale. Instead of one or a few genes being transferred across genus and species barriers, entire sets of genes can be moved. Success is not assured, however; a weakness today is the inherited instability of the "fused" clones. The preservation of traits and long-range stability has yet to be resolved. Furthermore, it seems that one of the most daunting problems is screening—determining what to look for and how to recognize it. (See Tech. Note 17, p. 82.)

Recombinant DNA techniques are also being examined for their ability to improve strains. Many potentially useful antibiotics do not reach their commercial potential because the micro-organisms cannot be induced to produce sufficient quantities by traditional methods. The synthesis of certain antibiotics is controlled by plasmids, and it is believed that some plasmids may nonspecifically enhance antibiotic production and excretion.

It may also be possible to transfer as a group, all the genes needed to produce an antibiotic into a new host. However, increasing the number of copies of critical genes by phage or plasmid transfer has yet to be achieved in antibiotic-producing organisms because little is known of the potential vectors. The genetic systems of commercial strains will have to be understood before the newer genetic engineering approaches can be used. Genetic maps have been published for only 3 of the 24 or more industrially useful bacteria.

Since 2,000 of the 2,400 known antibiotics are produced by *Streptomyces,* that is the genus of greatest interest to the pharmaceutical industry. Probably every company conducting research on *Streptomyces* is developing vectors, but little of the industrial work has been revealed to date.

Nonprotein pharmaceuticals

In both sales and quantity, over 80 percent of the pharmaceuticals produced today are not made of protein. Instead, they consist of a varie-ty of organic chemical entities. These drugs, except for antibiotics, are either extracted from some natural plant or animal source or are synthesized chemically.

Some of the raw materials for pharmaceuticals are also obtained from plants; micro-organisms are then used to convert the material to useful drugs in one or two enzymatic steps. Such conversions are common for steroid hormones.

In 1949, when cortisone was found to be a useful agent in the treatment of arthritis, the demand for the drug could not be met since no practical method for large-scale production existed. The chemical synthesis was complicated and very expensive. In the early and mid-1950's, many investigators reported the microbial transformation of several intermediates to compounds that corresponded to the chemical synthetic scheme. By saving many chemical steps and achieving higher yields, manufacturers managed to reduce the price of steroids to a level where they were a marketable commodity. A conversion of progesterone, for example, dropped the price of cortisone from $200 to $6/g in 1949. Through further improvements, the price dropped to less than $1/g. The 1980 price is $0.46/g.

Developments based on genetic techniques to increase the production and secretion of key enzymes could substantially improve the economics of some presently inefficient processes. Currently, assessments are being carried out by various companies to determine which of the many nonprotein pharmaceuticals can be manufactured more readily or more economically by biological means.

Approximately 90 percent of the pharmaceuticals used in the treatment of hypertension are obtained from plants, as well as are miscellaneous cardiovascular drugs. Morphine and important vasodilators are obtained from the opium poppy, *Papaver somniferum.* All these chemical substances are produced by a series of enzymes that are coded by corresponding genes in the whole plant. The long-term possibility (over 10 years) of using fermentation methods will depend on identifying the important genes.

The genes that are transferred from plant to bacteria must obviously be determined on a case-by-case basis. The case study on acetaminophen (the active ingredient in analgesics such as Tylenol) demonstrates the steps in such a feasibility study. (See app. I-A.)

The first step in such a study is to determine whether and where enzymes exist to carry out the necessary transformation for a given product. Acetaminophen for instance, can be made from aniline, a relatively inexpensive starting material. The two necessary enzymes can be found in several fungi. Either the enzymes can be isolated and used directly in a two-step conversion or the genes for both enzymes can be transferred into an organism that can carry out the entire conversion by itself.

Given the cost assumptions outlined in the case study and the assumptions on the efficiency of converting aniline to acetaminophen, the cost of producing the drug by fermentation could be 20 percent lower than production by chemical synthesis.

Impacts

Genetic technologies can help provide a variety of pharmaceutical products, many of which have been identified in this report. But the technologies cannot guarantee how a product will be used or even whether it will be used at all. The pharmaceuticals discussed have illustrated the kinds of major economic, technical, social, and legal constraints that will play a role in the application of genetic technologies.

Clearly, the major direct impacts of genetic technologies will be felt primarily through the type of products they bring to market. Nevertheless, each new pharmaceutical will offer its own spectrum and magnitude of impacts. Technically, genetic engineering may lead to the production of growth hormone and interferon with equal likelihood; but if the patient population is a thousandfold higher for interferon, and if its therapeutic effect is to alleviate pain and lower the cancer mortality rate, its impact will be significantly greater.

Many hormones and human proteins cannot be extensively studied because they are still either unavailable or too expensive. Until the physiological properties of a hormone are understood, its therapeutic values remain unknown. Recombinant DNA techniques are being used to overcome this circular problem. In one laboratory, somatostatin is being used as a research tool to study the regulation of the hormonal milieu of burn patients. A single experiment may use as much as 25 mg of the hormone, which, as a product of solid state chemical synthesis, costs as much as $12,000. Reducing its cost would allow for more extensive research on its physiological and therapeutic qualities.

By making a pharmaceutical available, genetic engineering can have two types of impacts. First, pharmaceuticals that already have medical promise will be available for testing. For example, interferon can be tested for its efficacy in cancer and viral therapy, and human growth hormone can be evaluated for its ability to heal wounds. For these medical conditions, the indirect, societal impact of applied genetics could be widespread.

Second, other pharmacologically active substances that have no present use will be available in sufficient quantities and at a low enough cost to enable researchers to explore their possibilities, thus creating the potential for totally new therapies. Genetic technologies can make available for example, cell regulatory proteins, a class of molecules that control gene activity and that is found in only minute quantities in the body. The cytokines and lymphokines typify the countless rare molecules involved in regulation, communication, and defense of the body to maintain health. Now, for the first time, genetic technologies make it possible to recognize, isolate, characterize, and produce these proteins.

The potential importance of this class of pharmaceuticals—the new cell regulatory molecules—is underscored by the fact that half of the 22 active INDs for new molecular entities that have been rated by FDA as promising important therapeutic gains are in the Metabolic and Endocrine Division, which oversees such drugs. It is reasonable to anticipate that they will be employed to treat cancer, to prevent or combat infections, to facilitate transplantation of organs and skin, and to treat allergies and other diseases in which the immune system has turned against the organism to which it belongs. (See table 7.)

At the very least, even if immediate medical uses cannot be found for any of these compounds, their indirect impact on medical research is assured. For the first time, almost any biological phenomenon of medical interest can be explored *at the cellular level* by the appli-

cation of available scientific tools. These new molecules are valuable tools for dissecting the structure and function of the cell. The knowledge gained may lead to the development of new therapies or preventive measures for diseases.

The increased availability of new vaccines might also have serious consequences. But the extent to which molecular cloning will provide useful vaccines for intractable diseases is still unknown. For some widespread diseases, such as amebic dysentery, not enough is known about the interaction between the micro-organism and the patient to help researchers design a rational plan of attack. For others, such as trachoma, malaria, hepatitis, and influenza, there is only preliminary experimental evidence that a useful vaccine could be produced. (See table 8.) To date, the vaccine that is most likely to have an immediate impact combats foot-and-mouth disease in veterinary medicine. There is little doubt however, that should any one of the vaccines for human diseases become available, the societal, economic, and political consequences of a decrease in morbidity and mortality would be significant. Many of these diseases are particularly prevalent in less-developed countries. The effects of developing vaccines

Table 7.—Diseases Amenable to Drugs Produced by Genetic Engineering in the Pharmaceutical Industry

Disease or condition	Drug potentially produced by genetically engineered organism
Diabetes[a]	Insulin
Atherosclerosis	Platelet-derived growth factor (PDGF)
Virus diseases Influenza Hepatitis Polio Herpes Common cold	Interferon
Cancer Hodgkin's disease Leukemia Breast cancer	Interferon
Anovulation	Human chorionic gonadatropin
Dwarfism[a]	Human growth hormone
Pain	Enkephalins and endorphins
Wounds and burns	Human growth hormone
Inflammation, rheumatic diseases[a]	Adrenocorticotrophic hormone (ACTH)
Bone disorders, e.g., Paget's disease[a]	Calcitonin and parathyroid hormone
Nerve damage	Nerve growth factor (NGF)
Anemia, hemorrhage	Erythropoietin
Hemophilia[a]	Factor VIII and Factor IX
Blood clots[a]	Urokinase
Shock[a]	Serum albumin
Immune disorders	Cytokines

[a]Indicates diseases currently treated by the drugs listed.

SOURCE: Office of Technology Assessment.

Table 8.—Major Diseases for Which Vaccines Need To Be Developed

Parasitic diseases
Hookworm
Trachoma
Malaria
Schistosomiasis
Sleeping sickness

Viruses
Hepatitis
Influenza
Foot-and-mouth disease (for cloven-hoofed animals)
Newcastle disease virus (for poultry)
Herpes simplex
Mumps
Measles
Common cold rhinoviruses
Varicella-zoster (shingles)

Bacteria
Dysentery
Typhoid fever
Cholera
Traveller's diarrhea

SOURCE: Office of Technology Assessment.

for them will be felt on an international scale and will involve hundreds of millions of people.

The new technologies may also lower the risks of vaccine production. For example, the FMDV vaccine produced by Genentech is constructed out of 17 of the 20 genes in the entire virus—enough to confer resistance, but too few to develop into a viable organism.

The new technology may also supply pharmaceuticals with effects beyond therapy. At least two promise impacts with broad consequences: MSH/ACTH 4-10 can be expected to be used on a wide scale if it is shown to improve memory; and bombesin and cholecystokinin might expand the appetite suppression market. But neither of these compounds has yet been found to be useful. While genetic technologies may provide large supplies of the drugs, they do not guarantee their value.

Antibody-based diagnostic tests, developed through genetic engineering, may eventually include early warning signals for cancer; they should be able to recognize any one of the scores of cancers that cause about a half-million deaths per year in the United States. If antibodies prove successful as diagnostic screening agents to predict disease, large-scale screening of the population can occur, accelerating the trend toward preventive medicine in the United States.

In addition to drugs and diagnostic agents, proteins could be produced for laboratory use. Expensive, complex media such as fetal calf serum are presently required for growing most mammalian tissue cells. Genetic cloning could make it possible to synthesize vital constituents cheaply, and could markedly reduce the costs of cell culture for both research and production. Ironically, genetic cloning could make economically competitive the very technology that offers an alternative production method for many drugs: tissue culture.

Nevertheless, the mere availability of a pharmacologically active substance does not ensure its adoption in medical practice. Even if it is shown to have therapeutic usefulness, it may not succeed in the marketplace. Consumer resistance limits the use of some drugs. The American aversion to therapies that require frequent injection, for instance, is illustrated by the opinion of some that a drug like ACTH offers few, if any, advantages over steroids.

The use of ACTH is somewhat greater abroad than in the United States. This is due in part because physicians in other cultures make far less use of systemic steroids than their American counterparts, and in part because frequent injections are more acceptable hence more common. Sales of ACTH in Great Britain—with its much smaller population—equal American sales.

At present, the need for injection is a far more likely deterrent to the wider use of ACTH than the cost of the drug itself. Reports that it can be applied by nasal spray suggest that its use may grow. Implantable controlled-release dosages may also become available within the next 5 years. This dependence on appropriate drug delivery mechanisms may lead to another line of research—increased attempts to develop technologies for drug-delivery.

As new pharmaceuticals become available, disruption can be expected to occur in the supply of some old ones. Pharmaceuticals whose production is tied to the production of others might become increasingly expensive to produce. Clotting factors, for example, are extracted with other blood components from plasma. Nevertheless, producing any of the 14 currently approved blood plasma products by rDNA would reduce the incidence of hepatitis caused by contamination from natural blood sources.

Whether new pharmaceuticals are produced or new production methods for existing pharmaceuticals are devised, future *sources* for the drugs may change. Currently, the sources are diverse, including many different plants, numerous animal organs, various tissue culture cells, and a wide range of raw materials used for chemical synthesis. A massive shift to fermentation would narrow the selection. The impacts on present sources can only be judged on a case-by-case basis. The new sources—microorganisms and the materials that feed them—offer the guarantee that the raw materials won't dry up. If one disappears, another can be found.

Clearly, there is no simple formula to identify all the impacts of applied genetics on the pharmaceutical industry. Even projections of economic impacts must remain crude estimates. Nevertheless, the degree to which genetic engineering and fermentation technologies might potentially account for drug production in specific categories is projected in appendix I-B.

Given the assumptions described, the immediate direct economic impact of using genetic manipulation in the industry, measured as sales, can be estimated in the billions of dollars, with the indirect impacts (sales for suppliers, savings due to decreased sick days, etc.) reaching several times that value.

Technical notes

1. Many hormones are simply chains of amino acids (polypeptides); some are polypeptides that have been modified by the attachment of carbohydrates (glycopeptides). Hormones usually trigger events in cells remote from the cells that produced them. Some act over relatively short distances—between segments in the brain, or in glands closely linked to the brain, others act on distant sites in tissues throughout the body.

2. For peptides about 30 AA in length, the cost may approach $1 per mg as the volume approaches the kilogram level—a level of demand rarely existing today but likely to be generated by work in progress. Today, the cost of the 32 AA polypeptide, calcitonin, which is synthesized chemically and marketed as a pharmaceutical product by Armour, is probably in the range of $20 per mg, since the wholesale price in vials containing approximately 0.15 mg is about $85/mg. (That price is an educated guess, since such costs are closely guarded secrets and since the price of a pharmaceutical includes so many variables that the cost of the agent itself is a small consideration.)

3. In addition to those helped by the National Pituitary Agency, another 100 to 400 patients are treated with hGH from commercial sources. The commercial price is approximately $15 per unit (roughly $30/mg). The production cost at the National Pituitary Agency is about $0.75/unit ($1.50/mg). The National Pituitary Agency produces 650,000 international units (IU) (about 325 g) of hGH, along with the thyroid-stimulating hormone, prolactin, and other hormones, from about 60,000 human pituitaries collected each year. That is enough hGH both for the current demand and for perhaps another 100 hypopituitary patients.

4. Workers at the Howard Hughes Medical Institute of the University of California, San Francisco, isolated messenger RNA from a human pituitary tumor and converted it into a DNA-sequence that could be put into E. coli. The sequence, however, was a mixture of hGH and non-hGH material. It has been reported that Eli Lilly & Co., which has provided some grant money to the Institute, has obtained a license to the patents relating to this work. Grants from the National Institutes of Health and the National Science Foundation were also acknowledged in the publication.

At practically the same time, researchers at Genentech, in conjunction with their associates at City of Hope National Medical Center disclosed the production of an hGH analog. This was the first time that a human polypeptide was directly expressed in E. coli in functional form. The work was supported by Kabi Gen AB, and Kabi has the marketing rights.

The level of hGH production reported in the scientific account of the Genentech work was on the same order as that reported for the insulin fragments—approximately 186,000 hGH molecules per cell—a level that might be competitive even before efforts are made to increase yield. Genentech stresses the point that design, rather than classical mutation and selection, is the logical way to improve the system, since the hormone's "blueprint" is incorporated in a plasmid that can be moved between strains of E. coli, between species, or even from simple bacteria into more complex organisms, such as yeast.

5. Since erythropoietin is a glycoprotein, it may not be feasible to synthesize the active hormone with presently available rDNA techniques.

6. Antigens are surface components of pathogenic organisms, toxins, or other proteins secreted by pathogenic micro-organisms. They are also the specific counterparts of antibodies: antibodies are formed by the body's immune system in response to their presence. Antibodies are synthesized by white blood cells and are created in such a way that they are uniquely structured to bind to specific antigens.

7. Many of the most devastating infectious diseases involve complex parasites that refuse to grow under laboratory conditions. The first cultivation of the most malignant of the species of protozoa that causes malaria, using human red blood cells, was described in 1976 by a Rockefeller University parasitologist, William Rager. Experimental immunogens were prepared and showed promise in monkeys, but concern about the existence of the red blood cell remnants—which could give rise to autoimmune reactions—curtailed the prospect for making practical vaccines by that route. Several biotechnology firms are currently trying to synthesize malaria antigens by molecular cloning. This effort may produce technical solutions to such scourges as

schistosomiasis (bilharzia), filariasis (onchocerciasis and elephantiasis), leshmaniasis, hookworm, amebic infections, and trypanosomiasis (sleeping sickness and Chagas' disease).

8. Another potential use of antigens is suggested by the experimental treatment of stage I lung cancer patients with vaccines prepared from purified human lung cancer antigens, which appears to substantially prolong survival. And the Salk Institute is expanding clinical trials in which a porcine myelin protein prepared by Eli Lilly & Co. is injected into multiple sclerosis patients to mop up the antimyelin antibodies that those patients are producing. Fifteen to forty-two g of myelin have been injected without adverse effects, suggesting a new therapeutic approach to auto-immune diseases. The protein appears to suppress the symptoms of experimental allergic encephalomyelitis, an animal disease resembling multiple sclerosis. Should this research succeed, the use of molecular clones to produce human protein antigens seems inevitable.

9. There are at least two distinct kinds of "classical" interferons—leukocyte interferon and fibroblast interferon, so-called for the types of cells from which they are obtained. A third kind, called lymphoblastoid because it is produced from cells derived from a Burkitt's lymphoma, appears to be a mixture of the other two interferons. All produce the antiviral state and are induced by viruses. A fourth kind, known as "immune" interferon, is produced by lymphocytes. Some evidence indicates that it may be a more potent antitumor agent than the classical types. Currently, interferon is obtained chiefly from white blood cells (leukocytes) from the blood bank in Helsinki that serves all of Finland, or from fibroblasts grown in cell culture.

10. Recently, G. D. Searle & Co. announced that new technology developed at its R&D facility in England has increased the yield of fibroblast interferon by a factor of 60. On the basis of this process, Searle expects to supply material for the first large-scale clinical trial of fibroblast interferon. Abbott Laboratories also recently announced plans to resume production of limited quantities of fibroblast interferon for clinical studies it plans to sponsor.

Unlike leukocytes and specially treated fibroblasts, which can be used only once, lymphoblasts derived from the tumor Burkitt's lymphoma grow freely in suspension and produce the least costly interferon presently obtainable. However, they also produce a disadvantageous mixture of both leukocyte and fibroblast interferons. The Burroughs-Wellcome Co. produces lymphoblastoid interferon in 1,000-l fermenters and has begun clinical trials in England, but the U.S. FDA has generally resisted efforts to make use of products derived from malignant cells. It is used extensively in research, and FDA is considering evidence from Burroughs-Wellcome that may lead to a relaxation of the prohibition, under pressure from the National Cancer Institute.

11. What may be a landmark patent has been issued to Hilary Koprowski and Carlos Croce of the Wistar Insti-

tute (for work done under the then Department of Health, Education, and Welfare funding) on the production of monoclonal antibodies against tumor cells. In a number of examples, these researchers demonstrated that an animal can be immunized with tumor cells, and that hybridomas derived from that animal will produce antibodies that demonstrate a specificity for the tumor.

The final sentence of the patent text provides the rationale for the use of antibodies in both cancer and infectious disease therapies: "If the (tumor) antigen is present, the patient can be given an injection of an antibody as an aid to react with the antigen." (U.S. 4,172,124.)

12. Myeloma cells grow vigorously in culture and have the unique characteristic of producing large quantities of antibodies. Each spleen cell of the immune type, on the other hand, produces an antibody that recognizes a single antigen, but these do not grow well in culture. When normal immune spleen cells are fused with myeloma cells, the resulting mixture of genetic capacities forms a cell, called a "hybridoma," which displays the desired characteristics of the parent cells: 1) it secretes the antibody specified by the genes of the spleen cell; and 2) it displays the vigorous growth, production, and longevity that is typical of the myeloma cell.

13. The use of high-correlation antibody assays in cancer studies has only just begun. Antibodies that have been treated so they can be seen with X-rays and that are specific for a tumor, can be used early to detect the occurrence or spread of tumor cells in the body. Because some 785,000 new cancer cases will be detected in 1980 with current diagnostic methods, because cancer will cause 405,000 deaths, and because early detection is the major key to improving survival, the implications are indeed enormous.

14. In the late 1950's, Lederle Laboratories marketed a preparation of 95-percent pure streptokinase (a bacterially produced enzyme that dissolves blood clots) for intravenous administration. They withdrew the product from the market around 1960 because it caused allergic reactions, which dampened clinical enthusiasm for its therapeutic potential.

The presence in human urine of urokinase, an enzyme also capable of removing blood clots, was also discovered in the early 1950's. Urokinase was purified, crystallized, and brought into clinical use in the mid-1960's. From the beginning it was apparent that "an intense thrombolytic state could be achieved with a much milder coagulation defect than occurred with streptokinase; no pyrogenic or allergic reactions were noted, and no antibodies resulted from its administration . . . There did not appear to be as great variation in patient responsiveness." In 1967-68 and 1970-73, the National Heart and Lung Institute organized clinical trials that compared urokinase with streptokinase and heparin, an anticoagulant, in the treatment of pulmonary embolism. The trials indicated that streptokinase and urokinase were equivalent and superior to heparin over the short term, although their long-term benefits were not established. Since then, clinical investigation of urokinase has been hampered by domestic regulatory problems, which have raised the cost of production and restricted its availability in the United States.

In January 1978, Abbott Laboratories obtained a new drug application for urokinase and introduced the product Abbokinase; by that time, however, the sales of urokinase in Japan were already pushing $90 million per year. Recently, Sterling Drug has begun marketing a urokinase product (Breokinase) manufactured by Green Cross of Japan: "According to Japanese reports, urokinase is the first Japanese-made drug formulation to receive production and sales approval from FDA. Green Cross estimates that within 3 years of the start of Sterling's marketing activities, the value of urokinase exports will reach Yen 500 million ($2.12 million) per month, and considers that its profits from exporting a finished product will probably be better than those from bulk drug sales or the licensing of technology." The Green Cross product is made from human urine collected throughout Korea and Japan, and takes advantage of technology licensed from Sterling. Abbott's product, on the other hand, is derived from kidney-cell culture.

15. Intergeneric hybrids have extremely interesting possibilities. For example, it would be beneficial to cephalosporin-process technology to combine in one organism the acyltransferase from *Penicillium chrysogenum* and the enzymes of *C. acremonium*, which does not incorporate side chain precursors onto cephalosporin like *P. chrysogenum* does for penicillins.

16. Another example of recombination between species is that reported for two species of fungi, *Aspergillus nidulans* and *A. rugulosus*, subsequent to protoplast fusion.

The only report of a successful cross between genera using protoplast fusion technology has been between the yeasts *Candida tropicalis* and *Saccariomycopsis fibuligera*, which took place at low frequency and gave rise to types intermediate between the parents.

17. An example of screening is provided by the new β-lactam (penicillin-like) antibiotics. Using older screening methods, no new β-lactams were found from 1956 until 1972 when a new method was devised. A new series of these antibiotics was thus found. Within the past year, 6 new β-lactams have been commercialized and at least 12 more are in clinical trials around the world. The sales forecasts for these new agents are estimated to be over $1 billion.

Chapter 5
The Chemical Industry

Chapter 5

Tables

Figures

The Chemical Industry

Background

The organic substances first used by humans to make useful materials such as cotton, linen, silk, leather, adhesives, and dyes were obtained from plants and animals and are natural and renewable resources. In the late 19th century, coal tar, a nonrenewable substance, was found to be an excellent raw material for many organic compounds. When organic chemistry developed as a science, chemical technology improved. At about the same time relatively cheap petroleum became widely available. The industry shifted rapidly to using petroleum as its major raw material.

The chemical industry's constant search for cheap and plentiful raw materials is now about to come full circle. The supply of petroleum, which presently serves more than 90 percent of the industry's needs, is severely threatened by both dwindling resources and increased costs. It has been estimated that at the current rate of consumption, the world's petroleum supplies will be depleted in the middle of the next century. Most chemical industry analysts, therefore, foresee a shift first back to coal and then, once again, to the natural renewable resources referred to as biomass. The shifts will not necessarily occur sequentially for the entire

chemical industry. Rather, both coal and biomass will be examined for their potential roles on a product-by-product basis.[1]

The chemical industry is familiar with the technology of converting coal to organic chemicals, and a readily available supply exists. Coal-based technologies will be used to produce a wide array of organic chemicals in the near future.* Nevertheless, economic, environmental, and technical factors will increase the industry's interest in biomass as an alternative source for raw materials. Applied genetics will probably play a major role in enhancing the possibilities by allowing biomass and carbohydrates from natural sources to be converted into various chemicals. Biology will thereby take on the dual role of providing both raw materials and a process for production.

---

[1]For further details see *Energy From Biological Processes*, vol. I, OTA-E-124 (Washington, D.C.: Office of Technology Assessment, July 1980).

*Most important organic intermediates (chemical compounds used for the industrial synthesis of commercial products such as plastics and fibers) can be obtained from coal as an alternative raw material. Currently, methods are being developed to convert coal into "synthetic gas," which can then be used as raw material for further conversions.

Overview of the industry

The chemical industry is one of the largest and most important in the world today. The U.S. market for synthetic organic chemicals alone, excluding primary products made from petroleum, natural gas, and coal tar, exceeded $35 billion in 1978.

The industry's basic function is to transform low-cost raw materials into end-use products of greater value. The most important raw materials are petroleum, coal, minerals (phosphate, carbonate), and air (oxygen, nitrogen). Roughly two-thirds of the industry is devoted to produc-

ing inorganic chemicals such as lime, salt, ammonia, carbon dioxide, chlorine gas, and hydrocholoric and other acids.

The other third, which is the target for biotechnology, produces organic chemicals. Its output includes plastics, synthetic fibers, organic solvents, and synthetic rubber. (See figure 24.) In general, petroleum and natural gas are first converted into "primary products" or basic organic chemicals such as the hydrocarbons ethylene and benzene. These are then converted into a wide range of industrial chemicals. Ethylene

Figure 24.—Flow of Industrial Organic Chemicals From Raw Materials to Consumption

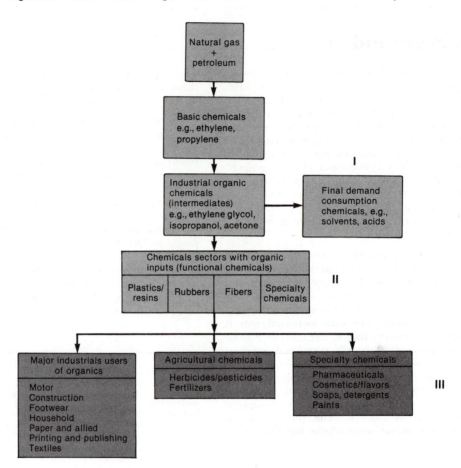

Organic resources

  80% raw material from petroleum/ natural gas

  20% raw material from coal, coke, and renewable resources

SOURCE: *U.S. Industrial Outlook* (Washington, D.C.: Department of Commerce, 1978); *Kline Guide to Chemical Industry*, Fairfield, N.J., adapted from Tong, 1979.

alone serves as the basic chemical for the manufacture of half of the largest volume industrial chemicals. Each of the steps in a chemical conversion process is controlled by a separate reaction, which is often performed by a separate company.

Evaluating the competitiveness both of a process and of the market is critical for the chemical industry, which is intensive for capital, energy, and raw materials. Its plants use large amounts of energy and can cost hundreds of millions of dollars to build, and raw material costs are generally 50 to 80 percent of a product's cost. If a biological process can use the same raw materials and reduce the process cost by even 20 percent, or allow the use of inexpensive raw materials, it could provide the industry with a major price break.

Fermentation and the chemical industry

The production of industrial chemicals by fermentation is not new. Scores of chemicals have been produced by micro-organisms in the past, only to be replaced by chemical production based on petroleum. In 1946, for example, 27 percent of the ethyl alcohol in the United States was produced from grain and grain products, 27 percent from molasses, a few percent each from such materials such as potatoes, pineapple juice, cellulose pulp, and whey, and only 36 percent from petroleum. Ten years later almost 60 percent was derived from petroleum.

Even more dramatically, fumaric acid was at one time produced on a commercial scale through fermentation, but its biological production was stopped when a more economical synthesis from benzene was developed. Frequently, after a fermentation product was discovered, alternative chemical synthetic methods were soon developed that used inexpensive petroleum as the raw material.

Nevertheless, for the few chemical entities still produced by fermentation, applied genetics

has contributed to the economic viability of the process. The production of citric and lactic acids and various amino acids are among the processes that have benefited from genetics. Lactic acid is produced both synthetically and by fermentation. Over the past 10 to 20 years, manufacture by fermentation has experienced competition from chemical processes.

The organisms used for the production of lactic acid are various species of the bacterium *Lactobacillus*. Starting materials may be glucose, sucrose, or lactose (whey). The fermentation per se is efficient, resulting in 90 percent yields, depending on the original carbohydrate. Since most of the problems in the manufacture of lactic acid lie in the recovery procedure and not in fermentation, few attempts have been made to improve the industrial processes through genetics.

Citric acid is the most important acidulant, and historically has held over 55 to 65 percent of the acidulant market for foods.* It is also used in pharmaceuticals and miscellaneous industrial applications. It is produced commercially by the mold *Aspergillus niger*. Surprisingly little work has been published on improving citric acid-producing strains of this micro-organism. Weight yields of 110 percent have recently been reported in *A. niger* mutants obtained by irradiating a strain for which a maximum yield of 29 percent had been reported.

Amino acids are the building blocks of proteins. Twenty of them are incorporated into proteins manufactured in cells, others serve specialized structural roles, are important metabolic intermediates, or are hormones and neurotransmitters. All of the amino acids are used in research and in nutritional preparations, with most being used in the preparation of pharmaceuticals. Three are used in large quantities for two purposes: glutamic acid to manufacture monosodium glutamate, which is a fla-

*The other two important acidulants, or acidifying agents, are phosphoric acid (20 to 25 percent) and malic acid (5 percent).

vor enhancer particularly in oriental cooking;* and lysine and methionine as animal feed additives.

Conventional technology for producing glutamic acid is based on pioneering work that was subsequently applied to other amino acids. The production employed microbial strains to produce amino acids that are not within their normal biosynthetic capabilities. This was accomplished by using two methods: 1) manipulating microbial growth conditions, and 2) isolating naturally occurring mutants.

Although microbial production of all the amino acids has been studied, glutamic acid and L-lysine** are the ones produced in significant quantities by fermentation processes. (See table 9.) The production of L-lysine is an excellent ex-

---

*Monosodium glutamate is the sodium salt of glutamic acid. In 1978, about 18,000 tonnes were manufactured in the United States and about 11,000 tonnes imported. The food industry consumed 97 percent. The fermentation plant of the Stauffer Chemical Co. in San Jose, Calif., is the sole U.S. producer. The microbes used in glutamic acid fermentation (*Corynebacterium glutamicum*, *C. lileum*, and *Brevibacterium flavum*) produce it in 60 percent of theoretical yield. Thus, there is some but not great potential for the use of applied genetics to improve the yield. Many of the genetic approaches have already been thoroughly investigated by industrial scientists.

**The lack of a single amino acid can retard protein synthesis, and therefore growth, in a mammal. The limiting amino acid is a function of the animal and its feed. The major source of animal feed in the United States is soybean meal. The limiting amino acid for feeding swine is methionine, the limiting amino acid for feeding poultry is lysine. Because of increased poultry demand, world demand for lysine is climbing. Eurolysine is spending $27 million to double its production capacity in Amiens, France, to 10 thousand tonnes. The Asian and Mideast markets are estimated to increase to 3 thousand tonnes in 1985. Some bacteria produce lysine at over 90 percent of theoretical yield. Little genetic improvement is likely in this conversion yield, however, significant improvement can be made in the rate and final concentration.

Table 9.—Data for Commercially Produced Amino Acids[a]

Amino acid	Price March 1980 (per kg pure L)	Present source	Production 1978 (tonnes)	Potential for application of biotechnology (de novo synthesis or bioconversion; organisms and enzymes)
Alanine	$ 80	Hydrolysis of protein; chemical synthesis	10 - 50 (J)[b]	—
Arginine	28	Gelatin hydrolysis	200 - 300 (J)	Fermentation in Japan
Asparagine	50	Extraction	10 - 50 (J)	—
Aspartic acid	12	Bioconversion of fumaric acid	500 - 1,000 (J)	Bioconversion
Citrulline	250	—	10 - 90 (J)	Fermentation in Japan
Cysteine	50	Extraction	100 - 200 (J)	—
Cystine	60	Extraction	100 - 200 (J)	—
DOPA (dihydrophenylalanine)	750	Chemical	100 - 200 (J)	—
Glutamic	4	Fermentation	10,000 - 100,000 (J)	De novo: *Micrococcus glutamicus*
Glutamine	55	Extraction	200 - 300 (J)	Fermentation in Japan
Histidine	160	—	100 - 200	Fermentation in Japan
Hydroxyproline	280	Extraction from collagen	10 - 50	—
Isoleucine	350	Extraction	10 - 50 (J)	—
Leucine	55	—	50 - 100 (J)	Fermentation in Japan
Lysine	350	Fermentation (80%) Chemical (20%)	10,000 (J)	(80% by fermentation) De novo: *Corynebacterium glutamicum* and *Brevibacterium flavum*
Methionine	265	Chemical from acrolein	17,000 (D,L)[c] 20,000 (D,L) (J)	—
Ornithine	60	—	10 - 50 (J)	Fermentation in Japan
Phenylalanine	55	Chemical from benzaldehyde	50 - 100 (J)	Fermentation in Japan
Proline	125	Hydrolysis of gelatin	10 - 50 (J)	Fermentation in Japan
Serine	320	—	10 - 50 (J)	Bioconversion in Japan
Threonine	150	—	50 - 10 (J)	Fermentation in Japan
Tryptophan	110	Chemical from indole	55 (J)	—
Tyrosine	13	Extraction	50 - 100 (J)	—
Valine	60	—	50 - 100 (J)	Fermentation in Japan

[a]Production data largely from Japan because of relative small U.S. production.
[b]Japan.
[c]D and L forms.

SOURCE: Massachusetts Institute of Technology.

ample of the competition between chemical and biotechnological methods. Fermentation has been gradually replacing its production by chemical synthesis; in 1980, 80 percent of its worldwide production is expected to be by microbes. It is not produced in the United States, which imported about 7,000 tonnes in 1979, mostly from Japan and South Korea. Recent estimates of primary U.S. cost factors in the competing production methods are summarized in table 10. Fermentation costs are lower for all three components of direct operating costs: labor, material, and utilities.

Table 10.—Summary of Recent Estimates of Primary U.S. Cost Factors in the Production of L-Lysine Monohydrochloride by Fermentation and Chemical Synthesis

	Cost factors in production of 98% L-lysine monohydrochloride					
	By fermentation[a]			By chemical synthesis[b]		
	Requirement (units per unit product)	Estimated 1976 cost per unit product		Requirement (units per unit product)	Estimated 1976 cost per unit product	
		Cents/lb	Cents/kg		Cents/lb	Cents/lb
Total labor[c]	—	8	18	—	9	20
Materials						
Molasses	44	7	16	—	—	—
Soybeanmeal, hydrolized ...	0.462	4	9	—	—	—
Cyclohexanol............	—	—	—	0.595	17	37
Anhydrous ammonia.......	—	—	—	0.645	6	14
Other chemicals[d].........	—	7	15	—	4	10
Nutrients and solvents	—	—	—	—	4	8
Packaging, operating, and maintenance materials ...	—	10	22	—	9	21
Total materials.........	—	28	62	—	45	90
Total utilities[e]	—	6	12	—	7	16
Total direct operating cost	—	42	92	—	56	126
Plant overhead, taxes, and insurance	—	10	21	—	10	21
Total cash cost	—	52	11	—	66	147
Depreciation[f]..........	—	16	35	—	13	28
Interest on working capital	—	1	3	—	1	3
Total cost[g]	—	69	151	—	80	178

[a]Assumes a 23-percent yield on molasses.
[b]Assumes a 65-percent yield on cyclohexanol.
[c]Includes operating, maintenance, and control laboratory labor.
[d]For both the process of fermentation and chemical synthesis, assumed use of hydrochloric acid (36 percent) and ammonia (29 percent). For fermentation includes also potassium diphosphate, urea, ammonium sulfate, calcium carbonate, and magnesium sulfate. For chemical synthesis also includes nitrosyl chloride, sulfuric acid, and a credit for ammonium sulfate byproduct.
[e]Total utilities for both processes include cooling water, steam process water, and electricity. For chemical synthesis, natural gas is also included.
[f]Ten percent per year of fixed capital costs for a new 20 million lb per year U.S. plant built in 1975 at assumed capital cost of $38.6 x 10[s] for fermentation and $32.5 x 10[s] for chemical synthesis exclusive of land costs.
SOURCE: Stanford Research Institute, *Chemical Economics Handbook* 583:3401, May 1979.

New process introduction

The development of biotechnology should be viewed not so much as the creation of a new industry as the revitalization of an old one. Both fermentation and enzyme technologies will have an impact on chemical process development. The first will affect the transition from nonrenewable to renewable raw materials. The second will allow fermentation-derived products to enter the chemical conversion chains, and will compete directly with traditional chemical transformations. (See figure 25.) Fermentation, by replacing various production steps, could act as a complementary technology in the overall manufacture of a chemical.

Figure 25.—Diagram of Alternative Routes to Organic Chemicals

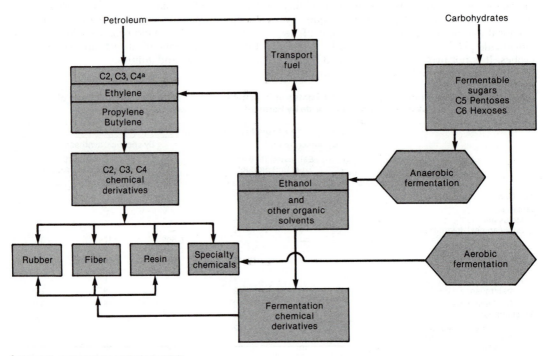

aC followed by number indicates length of carbon chain.

SOURCE: G. E. Tong, "Industrial Chemicals From Fermentation Enzymes," *Microb. Technol.*, vol. 1, 1979, pp. 173-179.

Characteristics of biological production technologies

The major advantages of using commercial fermentation include the use of renewable resources, the need for less extreme conditions during conversion, the use of one-step production processes, and a reduction in pollution. A micro-organism might be constructed, for example, to transform the cellulose in wood directly into ethanol.* (App. I-D, a case study of the impact of genetics on ethanol production, elaborates these points.)

RENEWABLE RESOURCES

Green plants use the energy captured from sunlight to transform carbon dioxide from the

*A request for approval of such an accomplishment by rDNA techniques was submitted to the Recombinant DNA Advisory Committee at the Sept. 25, 1980 meeting.

atmosphere into carbohydrates, some of which are used for their own energy needs. The rest are accumulated in starches, cellulose, lignins, and other materials called the biomass, which is the foundation of all renewable resources.

The technologies of genetic engineering could help ease the chemical industry's dependence on petroleum-based products by making the use of renewable resources attractive. All micro-organisms can metabolize carbohydrates and convert them to various end products. Extensive research and development (R&D) has already been conducted on the possibility of using genetically engineered strains to convert cellulose, the major carbohydrate in plants, to commercial products. The basic building block of cellulose—glucose—can be readily used as a raw material for fermentation.

Other plant carbohydrates include corn-starch, molasses, and lignin. The last, a polymer found in wood, could be used as a precursor for the biosynthesis of aromatic (benzene-like) chemicals, making their production simpler and more economical. Nevertheless, the increase in the price of petroleum is not a sufficient reason for switching raw materials, since the cost of carbohydrates and other biological materials has been increasing at a relative rate.

PHYSICALLY MILDER CONDITIONS

In general, there are two main ways to speed chemical reactions: by increasing the reaction temperature and by adding a catalyst. A catalyst (usually a metal or metal complex) causes one specific reaction to occur at a faster rate than others in a chemical mixture by providing a surface on which that reaction can be promoted. Even using the most effective catalyst, the conditions needed to accelerate industrial organic reactions often require extremely high temperatures and pressures—several hundred degrees Celsius and several hundred pounds per square inch.

Biological catalysts, or enzymes, on the other hand can speed-up reactions without the need for such extreme conditions. Reactions occur in dilute, aqueous solutions at the moderate conditions of temperature, pressure, and pH (a measure of the acidity or alkalinity of a solution) that are compatible with life.

ONE-STEP PRODUCTION METHODS

In the chemical synthesis of compounds, each reaction must take place separately. Because most chemical reactions do not yield pure products, the product of each individual reaction must be purified before it can be used in the next step. This approach is time-consuming and expensive. If, for example, a synthetic scheme that starts with ethylene (a petroleum-based product) requires 10 steps, with each step yielding 90 percent product (very optimistic yields in chemical syntheses), only about one-third of the ethylene is converted into the final end product. Purification may be costly; often, the chemicals involved (such as organic solvents for extractions) and the byproducts of the reaction are toxic and require special disposal.

In biological systems, micro-organisms often complete entire synthetic schemes. The conversion takes place essentially in a single step, although several might occur within the organisms, whose enzymes can transform the precursor through the intermediates to the desired end product. Purification is not necessary.

REDUCED POLLUTION

Metal catalysts are often nonspecific in their action; while they may promote certain reactions, their actions are not ordinarily limited to making only the desired products. Consequently, they have several undesirable features: the formation of side-products or byproducts; the incomplete conversion of the starting material(s); and the mechanical and accidental loss of the product.

The last problem occurs with all types of synthesis. The first two represent inefficiencies in the use of the raw materials. These necessitate the separation and recycling of the side-products formed, which can be difficult and costly because they are often chemically and physically similar to the desired end products. (Most separation techniques are based on differences in physical properties—e.g., density, volatility, and size.)

When byproducts and side-products have no value, or when unconverted raw material cannot be recycled economically, problems of waste disposal and pollution arise. Their solution requires ingenuity, vigilance, energy, and dollars. Many present chemical processes create useless wastes that require elaborate degradation procedures to make them environmentally acceptable. In 1980, the chemical industry is expected to spend $883 million on capital outlays for pollution control, and well over $200 million on R&D for new control techniques and replacement products. These figures do not include the millions of dollars that have been spent in recent years to clean up toxic chemical dumps and to compensate those harmed by poorly disposed wastes, nor do they include the cost of energy and labor required to operate pollution-control systems.

A genetically engineered organism, on the other hand, is designed to be precursor- and product-specific, with each enzyme having essentially 100-percent conversion efficiency. An enzymatic process that carries out the same transformation as a chemical synthesis produces no side-products (because of an enzyme's high specificity to its substrate) or byproducts (because of an enzyme's strong catalytic power). Consequently, biological processes eliminate many conventional waste and disposal problems at the front end of the system—in the fermenter. This high conversion efficiency reduces the costs of recycling. In addition, the efficiency of the biological conversion process generally simplifies product recovery, reducing capital and operating costs. Furthermore, by their nature, biologically based chemical processes, tend to create some waste products that are biodegradable and valuable as sources of nutrients.

Specific comparisons of the environmental hazards produced by conventional and biological systems are difficult. Data detailing the pollution parameters for various current chemical processes exist, but much less information is available for fermentation processes, and few compounds are produced by both methods. However, in most beverage distilling operations, pollution has been reduced to almost zero with the complete recovery of still slops as animal feeds of high nutritional value. Such control procedures are generally applicable to most fermentation processes. (App. I-C describes the pollutants that may be produced by current chemical processes and those expected from biologically based processes.)

The Environmental Protection Agency has estimated that the U.S. Government and industry combined will spend over $360 billion to control air and water pollution in the decade from 1977 through 1986. The share of the chemical and allied industries is about $26 billion. Genetic engineering technology may help alleviate this burden by offering cleaner processes of synthesis and better biological waste treatment systems. The monetary savings could be tremendous. As pure speculation, if just 5 percent of the current chemical industry were affected, spending on pollution could be reduced by about $100 million per year.

Industrial chemicals that may be produced by biological technologies

Despite the benefits of producing industrial chemicals biologically, thus far major fermentation processes have been developed primarily for a few complex compounds such as enzymes. (See table 11.) Biological methods have also been developed for a few of the simpler commodity chemicals: ethanol, butanol, acetone, acetic acid, isopropanol, glycerol, lactic acid, and citric acid.

Two questions are critical to assessing the feasibility or desirability of producing various chemicals biologically:

1. Which compounds can be produced biologically (at least theoretically)?
2. Which compounds may be primarily dependent on genetic technology, given the costs and availability of raw materials?

In principle, virtually all organic compounds can be produced by biological systems. If the necessary enzyme or enzymes are not known to exist, a search of the biological world will probably uncover the appropriate ones. Alternatively, at least in theory, an enzyme can be engineered to carry out the required reaction. Within this framework, the potential appears to be limited only by the imagination of the biotechnologist—even though certain chemicals that are highly toxic to biological systems are probably not amenable to production.

Three variables in particular affect the answer to the second question: the availability of an organism or enzymes for the desired transformation; the cost of the raw material; and the cost of the production process. When specific organisms and production technologies

Table 11.—Some Commercial Enzymes and Their Uses

Enzyme	Source	Industry and application
Amylase	Animal (pancreas)	Pharmaceutical: digestive aids
		Textile: desizing agent
	Plant (barley malt)	Baking: flour supplement
		Brewing, distilling, and industrial alcohol: mashing
		Food: precooked baby foods
		Pharmaceutical: digestive aids
		Textile: desizing agent
	Fungi (Aspergillus niger, A. oryzae)	Baking: flour supplement
		Brewing, distilling, and industrial alcohol: mashing
		Food: precooked baby foods, syrup manufacture
		Pharmaceutical: digestive aids
	Bacteria (Bacillus subtilis)	Paper: starch coatings
		Starch: cold-swelling laundry starch
Bromelin.	Plant (pineapple)	Food: meat tenderizer
		Pharmaceutical: digestive aids
Cellulase and hemicellulase . .	Fungi (Aspergillus niger)	Food: preparation of liquid coffee concentrates
Dextransucrase	Bacteria (Leuconostoc mesenteroides)	Pharmaceutical: preparation of blood-plasma extenders, and dextran for other uses
Ficin	Plant (fig latex)	Pharmaceutical: debriding agent
Glucose oxidase (plus catalase or peroxidase)	Fungi (Aspergillus niger)	Pharmaceutical: test paper for diabetes
		Food: glucose removal from egg solids
Invertase.	Yeast (Saccharomyces cerevisiae)	Candy: prevents granulation of sugars in soft-center candies
		Food: artificial honey
Lactase.	Yeast (Saccharomyces fragilis)	Dairy: prevents crystallization of lactose in ice cream and concentrated milk
Lipase.	Fungi (Aspergillus niger)	Dairy: flavor production in cheese
Papain.	Plant (papaya)	Brewing: stabilizes chill-proof beer
		Food: meat tenderizer
Pectinase	Fungi (Aspergillus niger)	Wine and fruit juice: clarification
Penicillinase	Bacteria (Bacillus cereus)	Medicine: treatment of allergic reaction to penicillin, diagnostic agent
Pepsin.	Animal (hog stomach)	Food: animal feed supplement
Protease.	Animal (pancreas)	Dairy: prevents oxidized flavor
		Food: protein hydrolysates
		Leather: bating
		Pharmaceutical: digestive aids
		Textile: desizing agent
	Animal (pepsin)	Brewing: beer stabilizer
	Animal (rennin, rennet)	Dairy: cheese
	Animal (trypsin)	Pharmaceutical: wound debridement
	Fungi (Aspergillus oryzae)	Baking: bread
		Food: meat tenderizer
	Bacteria (Bacillus subtilis)	Baking: modification of cracker dough
		Brewing: clarifier
Streptodornase	Bacteria (Streptococcus pyrogenes)	Pharmaceutical: wound debridement

SOURCE: David Perlman, "The Fermentation Industries," *American Society for Microbiology News* 39:10, 1973, p. 653.

have been developed, the cost of raw materials becomes the limiting step in production. If a strain of yeast, for example, produces 5 percent ethanol using sugar as a raw material, the process might become economically competitive if the cost of sugar drops or the price of petroleum rises. Even if prices remain stable, the micro-organisms might be genetically improved to increase their yield; genetic manipulation might solve the problem of an inefficient organism. Finally, the production process itself is a factor. After fermentation, the desired product must be separated from the other compounds in the reaction mixture. As an aid to recovery, the production conditions might be altered and improved to generate more of a desired compound.

More than one raw material can be used in a fermentation process. If, in the case of ethanol, the price of sucrose (from sugarcane or sugar beets) is not expected to change, the production technology is being run at optimum efficiency, and the micro-organism is producing as much ethanol as it can, the hurdle to economic competitiveness might be overcome if a less expensive raw material—cellulose, perhaps—were used. But cellulose cannot be used in its natural state: physical, chemical, or biological methods must be devised to break it down to its glucose (also a sugar) components.

The constraints vary from compound to compound. But even though the role of genetics must be examined on a product-by-product basis, certain generalizations can be made. Overall, genetic engineering will probably have an impact on three processes:

- Aerobic fermentation, which produces enzymes, vitamins, pesticides, growth regulators, amino acids, nucleic acids, and other speciality chemicals, is already well-established. Its use should continue to grow. Already, complex biochemicals like antibiotics, growth factors, and enzymes are made by fermentation. Amino acids and nucleotides—somewhat less complicated molecules—are sometimes produced by fermentation. Their production is expected to increase.

- Anaerobic fermentation, which produces organic acids, methane, and solvents, is the industry's area of greatest current growth. Already, 40 percent of the ethanol manufactured in the United States is produced in this way. The main constraint on the production of other organic acids and solvents is the need for cheaper methods for converting cellulose to fermentable sugars.

- Chemical modification of the fermentation products of both aerobic and anaerobic fermentation, which to date has rarely been used on a commercial scale, is of great interest. (See table 12.) Chemical production technologies that employ high temperatures and pressures might be replaced by biological technologies operating at atmospheric pressure and ambient temperature. A patent application has already been filed for the biological production of one of

Table 12.—Expansion of Fermentation Into the Chemical Industry

	Examples
Aerobic fermentation	
Enzymes	Amylases, proteases
Vitamins	Riboflavin B$_{12}$
Pesticides	*Bacillus thuringiensis*
Growth regulators	Gibberellin
Amino acids	Glutamic, lysine
Nucleic acids	
Acids	Malic acid, citric acid
Anaerobic fermentation	
Solvents	Ethanol, acetone, *n*-butanol
Acids	Acetic, propionic, acrylic

SOURCE: Office of Technology Assessment.

these products, ethylene glycol, by the Cetus Corp. in Berkeley, Calif. The process is claimed to be more energy efficient and less polluting. If it proves successful when run at an industrial scale, the technology could become significant to a U.S. market totaling $2½ billion per year.

The chemical industry produces a variety of likely targets for biotechnology. Tables I-B-27 through I-B-32 in appendix I-B present projections of the potential economic impacts of applied genetics on selected compounds that represent large markets, and the time frames for potential implementation. Table I-B-7 lists one large group of organic chemicals that were identified by the Genex Corp. and Massachusetts Institute of Technology (MIT) as amenable to biotechnological production methods. They are in agreement on about 20 percent of the products cited, which underscores the uncertain nature of attempting to predict so far into the future.

Fertilizers, polymers, and pesticides

Gaseous ammonia is used to produce nitrogen fertilizers. About 15 billion tonnes of ammonia were produced chemically for this purpose, in 1978; the process requires large amounts of natural gas. Nitrogen can also be converted, or "fixed," to ammonia by enzymes in micro-organisms; about 175 billion tonnes are fixed per year. For example, one square yard of land planted with certain legumes (such as soybeans) can fix up to 2 ounces of nitrogen, using bacteria associated with their roots. Currently, microbial production of ammonia from nitrogen is not economically competitive. Aside from the difficulties associated with the enzyme's sensitivity to oxygen and the near total lack of understanding of its mechanism, it takes the equivalent of the energy in 4 kilograms (kg) of sugar to make 1 kg of ammonia. Since ammonia costs $0.13/kg and sugar costs $0.22/kg, it is unlikely that the chemical process will be replaced in the near future. On the other hand, the genes for nitrogen fixation have now been transferred into yeast, opening up the possibility that agriculturally useful nitrogen can be made by fermentation.

A large segment of the chemical industry engaged in the manufacture of polymers is shown in table 13. A total of 4.3 million tonnes of fibers, 12 million tonnes of plastics, and 1.1 million tonnes of synthetic rubber were produced in the United States in 1978. All were derived from petroleum, with the exception of the less than 1 percent derived from cellulose fibers. The most likely ones are polyamides (chemically related to proteins), acrylics, isoprene-type rubber, and polystyrene. Because most monomers, the building blocks of polymers, are chemically simple and are presently available in high yield from petroleum, their microbial production in the next decade is unlikely.

While biotechnology is not ready to replace the present technology, its eventual impact on polymer production will probably be large. Biopolymers represent a new way of thinking. Most of the important constituents of cells are polymers: proteins (polypeptides from amino acid monomers), polysaccharides (from sugar monomers), and polynucleotides (from nucleotide monomers). Since cells normally assemble polymers with extreme specificity, the ideal industrial process would imitate the biological production of polymers in all possible respects—using a single biological machine to convert a raw material, e.g., a sugar, into the monomer to polymerize it, then to form the final product. A more likely application is the development of new monomers for specialized applications. Polymer chemistry has largely consisted of the study of how their properties can be modified.

Table 13.—The Potential of Some Major Polymeric Materials for Production Using Biotechnology

Product	Domestic production 1978 (thousand tonnes)
Plastics	
Thermosetting resins	
Epoxy	135
Polyester	544
Urea	504
Melamine	90
Phenolic	727
Thermoplastic resins	
Polyethylene	
Low density	3,200
High density	1,890
Polypropylene	1,380
Polystyrene	2,680
Polyamide, nylon type	124
Polyvinyl alcohol	57
Polyvinyl chloride	2,575
Other vinyl resins	88
Fibers	
Cellulosic fibers	
Acetate	139
Rayon	269
Noncellulosic fibers	
Acrylic	327
Nylon	1,148
Olefins	311
Polyester	1,710
Textile glass	418
Other	7
Rubbers	
Styrene-butadiene	628
Polybutadiene	170
Butyl	69
Nitrile	33
Polychlorophene	72
Ethylene-propylene	78
Polyisoprene	62

SOURCE: Office of Technology Assessment.

Conceivably, biotechnology could enable the modification of their function and form.

Pesticides include fungicides, herbicides, insecticides, rodenticides, and related products such as plant growth regulators, seed disinfectants, soil conditioners, and soil fumigants. The largest market (roughly $500 million annually) involves the chemical and microbial control of insects. Although microbial insecticides have been around for years, they comprise only 5 percent of the market. However, recent successes in developing viruses and bacteria that produce diseases in insects, and the negative publicity given to chemical insecticides, have encouraged the use of microbial insecticides.

Of the 15,000 known species of insects, only 200 are harmful enough to warrant control or destruction. Fortunately for man, most of them are sensitive to certain micro-organisms which, if they are not toxic to man, nontarget animals, and plants, can be used as commercial insecticides.

Approximately 100 known species of bacteria are pathogenic (disease causing) to insects, but only three—*Bacillus popilliae, B. thuringiensis* and *B. moritai*—have been developed into commercial insecticides. *B. popilliae* is found and produced only in the larvae of Japanese beetles.

The other two species can be produced by conventional fermentation techniques. They have been useful because they form spores that can be mass-produced easily and are stable enough to be handled commercially. The actual substances that cause toxicity to the insect are toxins synthesized by the microbes.

Genetic engineering should make it possible to construct more potent bacterial insecticides by increasing the dosage of the genes that code for the synthesis of the toxins involved. Mixtures of genes capable of directing the synthesis of various toxins might also be produced.

Constraints on biological production techniques

The chief impediments to using biological production technology are associated with the need for biomass.[2] They include:

- competition with food needs for starch and sugar;
- cyclic availability;
- biodegradability and associated storage problems;
- high moisture content for cellulosics, and high collection and storage costs;
- mechanical processing for cellulosics;
- the heterogenous nature of cellulosics (mixtures of cellulose, hemicellulose, and lignin); and
- The need for disposal of the nonfermentable portions of the biomass.

For food-related biomass sources, such as sugar, corn, and sorghum, few technological barriers exist for conversion to fermentable sugars; but subsidies are needed to make the fermentation of sugars as profitable as their use as food. For cellulosic biomass sources such as agricultural wastes, municipal wastes, and wood, technological barriers exist in collection, storage, pretreatment, fermentation, and waste disposal. In addition, biomass must always be transformed into sugars by either chemical or enzymatic processes before fermentation can begin.

A second major impediment is associated with the purification stage of production. Most chemical products of fermentation are present in extremely dilute solutions, and concentrating these solutions to recover the desired product is highly energy-intensive. Problems of technology and cost will continue to make this stage an important one to improve.

The developments in genetics show great promise for creating more versatile micro-organisms, but they do not by themselves produce a cheaper fuel or plastic. Associated technologies still require more efficient fermentation facilities and product separation processes; microbes may produce molecules, but they will not isolate, purify, concentrate, mix, or package them for human use.

The interaction between genetic engineering and other technologies is illustrated by the problems of producing ethanol by fermentation. The case study presented in appendix I-D identifies those steps in the biomass-to-ethanol scheme that need technological improvements before the process can become economical.

Genetic engineering is expected to reduce costs in many production steps. For certain ones—such as the pretreatment of the biomass to make it fermentable—genetics will probably not play a role: physical and chemical technologies will be responsible for the greatest advances. For others, such as distillation, genetic

---

[2]*Energy From Biological Processes*, op. cit.

technologies should make it possible to engineer organisms that can ferment at high temperatures (82° to 85° C) so that the fermentation and at least part of the distillation can both take place in the same reactor.* Various technol-

*Thermophilic ethanol producers have already been described in the genus Clostridium (i.e., C. thermocellum). In addition, genetically engineered organisms, described as a cross between yeasts and thermophilic bacteria, can ferment at 70° C.

ogies, such as the immobilization of whole cells in reactor columns, could be developed in parallel with genetic technologies to increase the stability of cells in fermenters.

The advantages of such thermophilic fermentations are significant: fermentation time is considerably reduced; the risk of contamination is nearly eliminated; and cooling requirements are lower due to the higher temperature of the fermenting broth.

An overview of impacts

The cost of raw materials may become cheaper than the petroleum now used—especially if cellulose conversion technologies can be developed. The source of raw materials would also be broader since several kinds of biomass could be interchanged, if necessary. For small quantities of chemicals, the raw material supply would be more dependable, particularly because of the domestic supply of available biomass. For substances produced in large quantities, such as ethanol, the supply of biomass could limit the usefulness of biotechnology.

Raw materials, such as organic wastes, could be processed both to produce products and reduce pollution. Nevertheless, the impact on total imported petroleum will be low. Estimates of the current consumption of petroleum as a raw material for industrial chemicals is approximately 5 to 8 percent of the total imported.

Impacts on the process include relatively cheaper production costs for selected compounds. For these, lower temperatures and pressures can be used, suggesting that the processes might be safer. Chemical pollution from biotechnology may be lower, although methods of disposal or new uses must be found for the micro-organisms used in fermentation. Finally, the biological processes will demand the development of new technologies for the separation and purification of the products.

Impacts on the products include both cheaper existing chemicals as well as entirely new products. Since biotechnology is the method of choice for producing enzymes, new uses for en-

zymes may expand and drive this sector of the industry.

Impacts on other industries

Although genetic engineering will develop new techniques for synthesizing many substances, the direct displacement of any present industry appears to be doubtful: Genetic engineering should be considered simply another industrial tool. As such, any industry's response should be to use this technique to maintain its positions in its respective markets. The point is illustrated by the variety of companies in the pharmaceutical, chemical, and energy industries that have invested in or contracted with genetic engineering firms. Some large companies are already developing inhouse genetic engineering research capabilities.

The frequent, popular reference to the small, innovative "genetic engineering companies" as a major new industry is somewhat misleading. The companies (see table 14) arose primarily to convert micro-organisms with little commercial use into micro-organisms with commercial potential. A company such as the Cetus Corp. initially used mutation and selection to improve strains, whereas other pioneers such as Genentech, Inc., Biogen, S. A., and Genex Corp. were founded to exploit recombinant DNA (rDNA) technology. Part of their marketing strategy includes the sale or licensing of genetically engineered organisms to large established commercial producers in the chemical, pharmaceutical, food, energy, and mining industries. Each engi-

Table 14.—Some Private Companies With Biotechnology Programs

Company	Founded	Approximate employees 1979	Ph. D.s 1979	Research capacity Recombinant DNA	Hybridomas
Atlantic Antibodies	1973	50	2		X
Bethesda Research Laboratories	1976	130	30	X	X
Biogen	1978	30 (50[b])	(18[b])[(3)(5)]	X	X[b]
Centocor	1979	20[(1)] - 10[(4)]	?		X
Cetus	1972	250	50	X	X
Clonal Research	1979	6	1		X
Collaborative Research[c]	1961	85	15	X	X
(Collaborative Genetics)	(1979)	(4)	(3)	X	
Ens Bio Logicals	1979	15	10	X	X
Genentech	1976	90	30	X	
Genex	1977	30	12	X	
Hybritech	1978	33[(1)]	6		X
Molecular Genetics	1979	6[(4)]	2	X	X[b]
Monoclonal Antibodies	1979	6	3		X
New England Biolabs	1974	22[(22)] - 5[(4)]	?	X	

[a]F. Eberstadt & Co. estimates.
[b]Expected by December 1980.
[c]Collaborative Research is a major owner of Collaborative Genetics. The division between them is not yet distinct.

SOURCES: (1) *Science* 208, p. 692-693, 1980 (52 people to expand to 100 by 1981).
(2) *Science* 208, p. 692-693, 1980 (20 senior persons).
(3) *Science* 208, p. 692-693, 1980 (16 scientists, 30 employees).
(4) Dun & Bradstreet, Inc.
(5) *Chemical and Engineering News*, Mar. 19, 1980.
Office of Technology Assessment.

neering firm also intends to manufacture some products itself. It is likely that the products reserved for inhouse manufacture will be low-volume, high-priced compounds like interferon.

Genetic engineering by itself is a relatively small-scale laboratory operation. Consequently, genetic engineering firms will continue to offer services to companies that do not intend to develop this capacity in their own inhouse laboratories. Specifically, a genetic engineering company may contract with a firm to develop a biological production method for its products. At the same time, larger companies might establish inhouse staffs to develop biological methods for both old and new products. (Several larger companies already have more inhouse genetic engineering personnel than some of the independent genetic engineering companies.)

In addition, suppliers of genetic raw materials may decide to expand into the production of genetically engineered organisms. Suppliers of restriction endonuclease enzymes for example, which are used in constructing rDNA, have already entered the field. Diagnostic firms could develop new bioassays for which they themselves would guarantee a market. Finally, companies with byproducts or waste products are

beginning to examine the possibility of converting them into useful products. This approach (which is somewhat more developed in Europe) assumes that with the proper technology the waste materials can become a resource.

Some industries, including manufacturers of agitators (drives), centrifuges, evaporators, fermenters, dryers, storage tanks and process vessels, and control and instrumentation systems, might profit by producing equipment associated with fermentation.

Impacts on university research

From the beginning, genetic engineering firms established strong ties with universities. These were responsible for providing most of the scientific knowledge that formed the basis for applied genetics as well as the initial scientific workforce:

- Cetus Corp. established a pattern by recruiting a prestigious Board of Scientific Advisors who remain in academic positions.
- Genentech, Inc., cofounded by a professor at the University of California at San Fran-

cisco, initially depended largely on outside scientists.

- Biogen, S. A., was organized by professors at Harvard and MIT plus six European scientists, and placed R&D contracts with academic researchers.
- Collaborative Genetics has a Nobel prize winner from MIT as the chairman of its scientific advisory board.
- Hybritech, Inc., has as its scientific nucleus a University of California, San Diego, professor complemented by scientists at the Salk Institute.

In addition to these companies, others have also been establishing closer ties with the academic community.

Much of the research that will be useful to industry will continue to be carried out in university laboratories. At present, it is often difficult to decide whether a research project should be classified as "basic" (generally more interesting to an academic researcher) or "applied" (generally more interesting to industry). E.g., a change in the genetic code, which increases gene activity, would be just as exciting to a basic scientist as to an industrial one.

This dialog between the universities and industry—both through formal and informal arrangements—has fostered innovation. Although the number of patents applied for is not a direct reflection of the level of innovation, it is still one indication. By the end of 1980, several hundred patent applications were filed for genetically engineered micro-organisms, their products, and their processes.

University research has clearly affected industrial development, and has in turn been affected by industry. Although the benefits are easily recognized, some drawbacks have been suggested. The most serious is the concern that university scientists will be restrained in their academic pursuits and in their exchange of information and research material. To date, anecdotal information suggests that some scientists are being more circumspect about sharing information. Still, secrecy is not new to highly competitive areas of biomedical research. In addition, scientists in other academic disciplines

useful to industry—such as chemistry and physics—have managed to achieve a balance between secrecy and openness.

The social impacts of local industrial activity

Despite the extensive media coverage of rDNA and other forms of genetic engineering, there is little evidence that people who live near companies using such techniques are still greatly concerned about possible hazards. This may be partly owing to a lack of awareness that a particular company is doing genetic research and partly because companies thus far have adhered to the National Institutes of Health (NIH) Guidelines. Some companies have placed individuals on their institutional biosafety committees who are respected and trusted members of the local community. By involving the local citizens with no vested corporate interest, a mechanism for oversight has been provided. (For a more detailed discussion, see ch. 11.)

Impacts on manpower

Two types of impacts on workers can be expected:

- The creation of jobs that replace those held by others. E.g., a worker involved in chemical production might be replaced by one producing the same product biologically.
- The creation of new jobs.

Workers in three categories would be affected:

- those actually involved in the fermentation-production phase of the industry;
- those involved in the R&D phase of the industry, particularly professionals; and
- those in support industries.

Projections of manpower requirements are only as accurate as the projections of the level of industrial activity. In the past 5 years, about 750 new jobs have been created within the small genetic engineering firms (including monoclonal antibody producers). Of these, approximately one-third hold Ph. D. degrees.

Data obtained through an OTA survey of 284 firms indicate that the pharmaceutical industry employs the major share of personnel working in applied genetics programs. (See table 15.) The average number of Ph. D.s in each industry is given in table 16. A rough estimate of professional scientific manpower at this level includes: 6 in food, 45 in chemical, 120 in pharmaceutical, and 18 in specialty chemicals—a total of 189. If the number of research support personnel is approximately twice the number of Ph. D.s, the total rises to about 570. If $165,000 per year is required to support one Ph. D. in industry, the total value of such manpower is approximately $31 million.

Estimates of the number of companies engaged in applied genetics work in 1980 can be compared with the total number of firms with fermentation activities. A tabulation of firms on a worldwide basis in 1977 revealed 145 companies, of which 29 were American. (See table 17.) These companies produced antibiotics, enzymes, solvents, vitamins and growth factors,

Table 17.—Index to Fermentation Companies

1. Abbott Laboratories, North Chicago, Ill.
2. American Cyanamid, Wayne, N.J.
3. Anheuser-Busch, Inc., St. Louis, Mo.
4. Bristol-Myers Co., Syracuse, N.Y.
5. Clinton Corn Processing Co., Clinton, Iowa
6. CPC International, Inc., Argo, Ill.
7. Dairyland Laboratories, Inc., Waukesha, Wis.
8. Dawe's Laboratories, Inc., Chicago Heights, Ill.
9. Grain Processing Corp., Muscatine, Iowa
10. Hoffman-LaRoche, Inc., Nutley, N.J.
11. IMC Chemical Group, Inc., Terre Haute, Ind.
12. Eli Lilly & Co., Indianapolis, Ind.
13. Merck & Co., Inc., Rahway, N.J.
14. Miles Laboratories, Inc., Elkhart, Ind.
15. Parke, Davis & Co., Detroit, Mich.
16. S. B. Penick & Co., Lyndhurst, N.J.
17. Pfizer, Inc., New York, N.Y.
18. Premier Malt Products, Inc., Milwaukee, Wis.
19. Rachelle Laboratories, Inc., Long Beach, Calif.
20. Rohm & Haas, Philadelphia, Pa.
21. Schering Corp., Bloomfield, N.J.
22. G. D. Searle & Co., Skokie, Ill.
23. E. R. Squibb & Sons, Inc., Princeton, N.J.
24. Standard Brands, Inc., New York, N.Y.
25. Stauffer Chemical Co., Westport, Conn.
26. Universal Foods Corp., Milwaukee, Wis.
27. The Upjohn Co., Kalamazoo, Mich.
28. Wallerstein Laboratories, Inc., Morton Grove, Ill.
29. Wyeth Laboratories, Philadelphia, Pa.

SOURCE: Office of Technology Assessment.

Table 15.—Distribution of Applied Genetics Activity in Industry

Classification	Distribution of applied genetics activity by company class[a]	Percent of total
Food	(6/46)	13
Chemical	(9/52)	17
Pharmaceutical	(12/25)	48
Specialty chemical[b]	(6/68)	9

[a]Ignores small firms specializing in genetic research.
[b]Food ingredients, reagents, enzymes.

SOURCE: Office of Technology Assessment.

Table 16.—Manpower (low-(average)-high) Distribution of a Firm With Applied Genetics Activity

	Ph. D.	M.S.	Bachelors
Food	0-(1)-2	0-(1)-2	0-(2)-8
Chemical	3-(5)-7	0-(1)-2	2-(5)-7
Pharmaceutical	2-(10)-24	1-(4)-9	1-(8)-20
Specialty	1-(3)-8	1-(3)-4	2-(2)-4
Biotechnology			
Genetic engineering	3-(15)-32	2-(11)-20	5-(15)-25
Hybridoma	1-(3)-6	0-(20)-0	0-(20)-0
Other	0-(2)-4	2-(4)-6	8-(10)-13
Average	1-(6)-12	1-(4)-6	3-(8)-12

SOURCE: Office of Technology Assessment.

nucleosides, amino acids, and miscellaneous products. (See table 18.) The only chemical firm listed was the Stauffer Chemical Co. Ten firms are listed as having the ability to produce food and feed yeast. (See table 19.) Correcting for firms listed twice, at least 38 U.S. firms were engaged in significant fermentation activity for commercial products, excluding alcoholic beverages, in 1977. Not all have research expertise in fermentation or biotechnology, much less a regular genetics program: 10 to 20 were in the chemical industry; 25 to 40 in fermentation (enzyme, pharmaceutical, food, and specialized chemicals); and 10 to 15 in biotechnology (genetic engineering)—or about 45 to 75 firms in all.

If average manpower numbers are used, the total number of professionals involved in commercial applied genetics research is:

Ph. D.s: 300-450
Others: 600-900
900-1,350

The number of workers that will be involved in the production phase of biotechnology repre-

Table 18.—Fermentation Products and Producers

Product	Some producers[a]	Product	Some producers[a]
Amino acids		Capreomycin......................	
L-alanine.........................		Cephalosporins...................	4,12
L-arginine........................		Chromomycin A₃...................	
L-aspartic acid		Colistin...........................	
L-citrulline		Cycloheximide....................	27
L-glutamic acid	25	Cycloserine	11
L-glutamine		Dactinomycin.....................	13
L-glutathione		Daunorubicin.....................	
L-histidine		Destomycin	
L-homoserine....................		Enduracidin......................	
L-isoleucine.....................		Erythromycin.....................	17,27
L-leucine........................		Fortimicins.......................	
L-lysine.........................		Fumagillin	
L-methionine....................		Fungimycin	
L-ornithine......................		Fusidic acid	
L-phenylalanine.................		Gentamicins......................	21
L-proline........................		Gramicidin A	28
L-serine.........................		Gramicidin J (S)	
L-threonine......................		Griseofulvin	
L-tryptophan.....................		Hygromycin B	12
L-tyrosine		Josamycin	
L-valine		Kanamycins	4
		Kasugamycin.....................	
Miscellaneous products and processes		Kitasatamycin....................	
Acetoin..........................		Lasalocid	10
Acyloin	13	Lincomycin.......................	27
Anka-pigment (red)		Lividomycin	
Blue cheese flavor...............	7	Macarbomycin....................	
Desferrioxamine		Mepartricin.......................	
Dihydroxyacetone................	17,21,28	Midecamycin	
Dextran..........................		Mikamycins	
Diacetyl (from acetoin)		Mithramycin	17
Ergocornine......................		Mitomycin C......................	4
Ergocristine......................		Mocimycin	
Ergocryptine.....................		Monensin	
Ergometrine......................		Myxin	10
Ergotamine......................		Neomycins.......................	16,17,23,27
Bacillus thuringiensis insecticide......	1	Novobiocin.......................	27
Lysergic acid		Nystatin..........................	23
Paspalic acid		Oleandomycin....................	17
Picibanil		Oligomycin.......................	
Ribose...........................		Paromomycins....................	15
Scleroglucan		Penicillin G.......................	4,12,13,17,23,29
Sorbose (from sorbitol)...........	10,17	Penicillin V.......................	1,4,12,17,23,29
Starter cultures	7,13,14	Penicillins (semisynthetic)...........	4,13,17,23,29
Sterol oxidations	22,27	Pentamycin	
Steroid oxidations.................	21,23,27,29	Pimaricin	
Xanthan	13,17	Polymyxins.......................	17
		Polyoxins	
Antibiotics		Pristinamycins....................	
Adriamycin.......................		Quebemycin......................	
Amphomycin		Ribostamycin.....................	
Amphotericin B	23	Rifamycins	
Avoparcin........................	2	Sagamicin........................	
Azalomycin F.....................		Salinomycin......................	
Bacitracin........................	11,16,17	Siccanin	
Bambermycins....................		Siomycin.........................	
Bicyclomycin.....................		Sisomicin........................	21
Blasticidin S......................		Spectinomycin....................	27
Bleomycin		Streptomycins....................	13,17,29
Cactinomycin			
Candicidin B......................	16	***Tetracyclines***	
Candidin.........................		Clortetracycline..................	19

Table 18.—Fermentation Products and Producers (cont'd)

Product	Some producers[a]	Product	Some producers[a]
Demeclocycline	2	Lactase	28
Oxytetracycline	17,19	Lipase	20
Tetracycline	2,4,17,19,23,27	Microbial rennet	17,28
Tetranactin		Naringinase	28
Thiopeptin		Pectinase	20,28
Thiostrepton	23	Pentosanase	20,28
Tobramycin	12	Proteases	14,17,18,20,28
Trichomycin		Streptokinase-streptodornase	2
Tylosin	12	Uricase	
Tyrothricin	16,28	**Organic acids**	
Tyrocidine		Citric acid	14,17
Uromycin		Comenic acid	17
Validamycin		Erythorbic acid	
Vancomycin	12	Gluconic acid	4,17,18
Variotin		Itaconic acid	17
Viomycin		2-keto-D-gluconic acid	17
Virginiamycin		α-ketoglutaric acid	
Enzymes		Lactic acid	5
Amylases	5,19,20,24,28	Malic acid	
Amyloglucosidase	5,6,14,28	Urocanic acid	
Anticyanase		**Solvents**	
L-asparaginase		Ethanol	9
Catalase	8,14	2,3-butanediol	
Cellulase	6,20,28	**Vitamins and growth factors**	
Dextranase		Gibberellins	1,12,13
'Diagnostic enzymes'		Riboflavin	13
Esterase-lipase	28	Vitamin B₁₂	13
Glucanase	28	Zearalanol	11
Glucose dehydrogenase		**Nucleosides and nucleotides**	
Glucose isomerase	3,5,14,24	5-ribonucleotides and nucleosides	
Glucose oxidase	8,14	Orotic acid	
Glutamic decarboxylase	18	Ara-A-(9-β-D-arabino-furanosyl)	15
Hemi-cellulase	14,20,28	6-azauridine	
Hespiriginase			
Invertase	24,26,28		

[a]Blank means no U.S. producer in 1977; therefore, is produced by one or more foreign firms (from at least 120 different firms).

SOURCE: Compiled by Periman, *American Society for Microbiology News* 43:2, 1977, pp. 82-89.

sents a major impact of genetic engineering. To estimate this number these two calculations must be made:

- the value or volume of chemicals that might be produced by fermentation, and
- the number of production workers needed per unit volume of chemicals produced.

Any prediction of the potential volume of chemicals is necessarily filled with uncertainties. The approximate market value of organic chemicals produced in the United States is given in appendix I-B. Total U.S. sales in 1979 were calculated to be over $42 billion. On the basis of the assumptions made, $522 million worth of bulk organic chemicals could be commercially

produced by genetically engineered strains in 10 years and $7.1 billion in 20 years. Table I-B-10 in appendix I-B lists the potential markets for pharmaceuticals. Excluding methane production, the total potential market for products obtained from genetically engineered organisms is approximately $14.6 billion.

If the production of chemicals having this value is carried out by fermentation, it is possible to calculate how many workers will be needed. Data obtained from industrial sources reveal that 2 to 5 workers, including those in supervision, services, and production, are required for $1 million worth of product. Hence, 30,000 to 75,000 workers would be required for the estimated $14.6 billion market.

Table 19.—U.S. Fermentation Companies

Producers of Baker's yeast and food/feed yeast in the United States in 1977

Baker's yeast:
American Yeast Co., Baltimore, Md.
Anheuser-Busch, Inc., St. Louis, Mo.
Federal Yeast Co. (now Diamond Shamrock), Baltimore, Md.
Fleischmann Yeast Co., New York, N.Y.
Universal Foods Corp., Milwaukee, Wis.
Food/feed yeast:
Amber Laboratories, Juneau, Wis.
Amoco Foods Co., Chicago, Ill.
Boise-Cascade, Inc., Portland, Oreg.
Diamond Mills, Inc., Cedar Rapids, Iowa
Fleischmann Yeast Co., New York, N.Y.
Lakes States Yeast Co., Rhinelander, Wis.
Stauffer Chemical Co., Westport, Conn.

Enzyme producers, 1977
Clinton Corn Processing Co., Clinton, Iowa
Miles Laboratories, Inc., Elkhart, Ind.
Premier Malt Products, Inc., Milwaukee, Wis.

SOURCE: Compiled by Perlman, *American Society for Microbiology News* 43:2, 1977, pp. 82-89.

Since the chemicals considered above are currently being produced, any new jobs in biotechnology will displace the old ones in the chemical industry. Whether the change will result in a net loss or gain in the number of jobs is difficult to predict. However, a rough estimate indicates that approximately the same number of workers will be required per unit of output.

Estimates of the number of workers are divided into: 1) workers directly involved in the growth of the organisms; and 2) workers involved in the "recovery" phase, where the organisms are harvested and the chemical product is extracted, purified, and packaged. Based on industry data, the number of workers in the fermentation phase is approximately 30 percent of the total, and those in recovery approximately 50 percent. Hence, about 9,000 to 22,500 workers might be expected to hold jobs in the immediate fermentation area, and about 15,000 to 37,500 workers would be involved in handling the production medium (with or without the organisms).

Estimates of the number of totally new jobs that would be created are highly speculative; they should allow for estimates of increases in the quantity of chemicals currently being produced and the production of totally new compounds. According to estimates by Genex, the new and growth markets may reach $26 billion by the year 2000, which would add 52,000 to 130,000 jobs to the present number.

Chapter 6
The Food Processing Industry

Chapter 6

Figure

Tables

The Food Processing Industry

Introduction—the industry

The food processing industry comprises those manufacturers that transform or process agricultural products into edible products for market. It is distinguished from the production, or farming and breeding portions of the agricultural industry.

Genetics can be used in the food processing industry in two ways: to design micro-organisms that transform inedible biomass into food for human consumption or for animal feed; and to design organisms that aid in food processing, either by acting directly on the food itself or by providing materials that can be added to food.

Eight million to ten million people work in the meat, poultry, dairy, and baking industries; in canned, cured, and frozen food plants; and in moving food from the farm to the dinner table. In 1979, the payroll was over $3.2 billion for the meat and poultry industries, $2.6 billion for baking, and $1.9 billion for food processing.

Traditionally, micro-organisms have been used to stabilize, flavor, and modify various properties of food. More recently, efforts have been made to control microbial spoilage and to ensure that foods are free from micro-organisms that may be hazardous to public health. These are the two major ways in which microbiology has been useful.

Historically, most efforts have been devoted to improving the ability to control the harmful effects of micro-organisms. The industry recognized the extreme heat resistance of bacterial spores in the early 20th century and sponsored or conducted much of the early research on the mechanisms of bacterial spore heat resistance. Efforts to exploit the beneficial characteristics of micro-organisms, on the other hand, have been largely through trial-and-error. Strains that improve the quality or character of food generally have been found, rather than designed.

Single-cell protein

The interest in augmenting the world's supply of protein has focused attention on microbial sources of protein as food for both animals and humans.* Since a large portion of each bacterial or yeast cell consists of proteins (up to 72 percent for some protein-rich cells), large numbers have been grown to supply single-cell protein (SCP) for consumption. The protein can be consumed directly as part of the cell itself or can be extracted and processed into fibers or meat-like items. By now, advanced food processing technologies can combine this protein with meat flavoring and other substances to produce nutritious food that looks, feels, and tastes like meat.

*As an example of the potential significance of SCP, the Soviet Union, which is one of the largest producers, expects to produce enough fodder yeast from internally available raw materials to be self-sufficient in animal protein foodstuffs by 1990.

The idea of using SCP as animal feed or human food is not new; yeast has been used as food protein since the beginning of the century. However, in the past 15 years, there has been a dramatic increase in research on SCP and in the construction of large-scale plants for its production, especially for the production of yeast. (See table 20.) Interest in this material is reflected in the numerous national and international conferences on SCP, the increasing number of proceedings and reviews published, and the number of patents issued in recent years. (See table 21.)

The issues addressed have covered topics such as the economic and technological factors influencing SCP processes, nutrition and safety, and SCP applications to human or animal foods. Thus far, commercial use has been limited by

Table 20.—Estimated Annual Yeast Production, 1977
(dry tonnes)

	Baker's yeast	Dried yeast[a]
Europe.................	74,000[b]	160,000[b]
North America	73,000	53,000
The Orient..............	15,000	25,000
United Kingdom..........	15,500	(c)
South America	7,500	2,000
Africa.................	2,700	2,500
Totals	187,700	242,500

[a]Dried yeast includes food and fodder yeasts; data for petroleum-grown yeasts are not available.
[b]Production figures for U.S.S.R. not reported.
[c]None reported.

SOURCE: H. J. Peppler and D. Perlman (eds.), *Microbial Technology*, vol. 1 (London: Academic Press, 1979), p. 159.

Table 21.—Classification of Yeast-Related U.S. Patents (1970 to July 1977)

Category	Number issued
Yeast technology (apparatus, processing)	22
Growth on hydrocarbons...................	28
Growth on alcohols, acids, wastes...........	22
Production of chemicals...................	14
Use of baking and pasta products	24
Condiments and flavor enhancers	18
Reduced RNA.......................	11
Yeast modification of food products	13
Isolated protein	5
Texturized yeast protein	7
Lysates and ruptured cells	7
Animal feed supplements	12
Total...............................	183

SOURCE: H. J. Peppler, "Yeast," *Annual Report on Fermentation Processes*, D. Perlman (ed.), vol. 2 (London: Academic Press, 1978), pp. 191-200.

several factors. For each bacterial, yeast, or algal strain used, technological problems (from the choice of micro-organisms to the use of corresponding raw material) and logistical problems of construction and location of plants have arisen. But the primary limitation so far has been the cost of production compared with the costs of competing sources of protein. (The comparative price ranges in 1979 for selected microbial, plant, and animal protein products are shown in table 22.)

The costs of manufacturing SCP for animal feed in the United States are high, particularly relative to its major competing protein source, soybeans, which can be produced with little fertilizer and minimal processing. The easy availability of this legume severely limits microbial SCP production for animal feed or human food. In fact, according to the U.S. Department of

Table 22.—Comparison of Selling Price Ranges for Selected Microbial, Plant, and Animal Protein Products

Product, substrate, and quality	Crude protein content	Price range 1979 U.S. dollars/kg
Single-cell proteins		
Candida utilis, ethanol, food grade	52	1.32 - 1.35
Kluyveromyces fragilis, cheese whey, food grade............	54	1.32
Saccharomyces cerevisiae:		
Brewer's, debittered, food grade	52	1.00 - 1.20
Feed grade..................	52	0.39 - 0.50
Plant proteins		
Alfalfa (dehydrated)	17	0.12 - 0.13
Soybean meal, defatted	49	0.20 - 0.22
Soy protein concentrate.........	70 - 72	0.90 - 1.14
Soy protein isolate	90 - 92	1.96 - 2.20
Animal proteins		
Fishmeal (Peruvian)	65	0.41 - 0.45
Meat and bonemeal	50	0.24 - 0.25
Dry skim milk..................	37	0.88 - 1.00

SOURCE: Office of Technology Assessment.

Agriculture (USDA), total domestic and export supply for U.S. soybeans will grow 73 percent by 1985.

Soybeans are primarily consumed as animal feed. But while only 4 percent of their annual production are directly consumed by humans, the market is growing significantly. The introduction of improved textured soy protein in cereals, in meat substitutes and extenders, and in dairy substitutes has increased the use of soy products. Nevertheless, the market does not demand soy products in particular but protein supplements, vegetable oils, feed grain supplements, and meat extenders in general. Other protein and oil sources could replace soybeans if the economics were attractive enough. Fishmeal, dry beans, SCP, and cereals are all potential competitors. As long as a substitute can meet the nutritional, flavor, toxicity, and regulatory standards, competition will be primarily based on price.

The competition between soybeans and SCP illustrates one of the paradoxes of genetic engineering. While significant research is attempting the genetic improvement of soybeans, genetic techniques are also being explored to increase the production of SCP. Consequently, the same tool—genetic engineering—encourages competition between the two commodities.

Genetic engineering and SCP production

Despite the microbial screening studies that have been conducted and the wealth of basic genetic knowledge available about common yeast (a major source of SCP), genetic engineering has had little economic impact on SCP processes until recently. Today, a variety of substances are being considered as raw materials for conversion.

- *Petroleum-based hydrocarbons.*—Until recently, the wide availability and low cost of petrochemicals have made the *n*-alkane hydrocarbons (straight chain molecules of carbon and hydrogen), which are petrochemical byproducts, potential raw materials for SCP production. At British Petroleum, mutants of micro-organisms have been obtained having an increased protein content. Mutants have also been found with other increased nutritive values, e.g., vitamin content.
- *Methane or methanol.*—Relatively few genetic studies have been directed at investigating the genetic control of the microbial use of methane or methanol. However, one recent application of genetic engineering has been reported by the Imperial Chemical Industries (ICI) in the United Kingdom, where the genetic makeup of a bacterium (*Methylophilus methylotrophus*) has been altered so that the organism can grow more readily on methanol. The increase in growth provides increased protein and has made its production less expensive. The genetic alteration was accomplished by transferring a gene from *Escherichia coli* to *M. methylotrophus.*
- *Carbohydrates.*—Many carbohydrate substrates—from starch and cellulose to beets and papermill wastes—have been investigated. Forests are the most abundant source of carbohydrate in the form of cellulose. But before it can be used by microorganisms, it must be transformed into the carbohydrate, glucose, by chemical or enzymatic pretreatment. Many of the SCP processes that use cellulose employ organisms that produce the enzyme cellulase, which degrades cellulose to glucose.

Most of the significant genetic studies on the production of cellulase by micro-organisms are just beginning to appear in the literature. The most recent experiments have been successful in creating fungal mutants that produce excess amounts.

Commercial production

Of the estimated 2 million tons of SCP produced annually throughout the world, most comes from cane and beet molasses, with about 500,000 tons from hydrolyzed wood wastes, corn trash, and papermill wastes. (See table 23.)

Integrated systems can be designed to couple the production of a product or food with SCP production from wastes. E.g., the waste sawdust from the lumber industry could become a source of cellulose for micro-organisms. ICI's successful genetic engineering of a micro-organism to increase the usefulness of one raw material (methanol) should encourage similar attempts for other raw materials.

But while SCP can be obtained from a wide variety of micro-organisms and raw materials, the nutritional value and the safety of each micro-organism vary widely, as do the costs of competing protein sources in regional markets. Consequently, accurate predictions cannot be made about the likelihood that SCP will displace traditional protein products, overall. Displacements have and will continue to occur on a case-by-case basis.

Table 23.—Raw Materials Already Tested on a Laboratory or Small Plant Scale

Agave juices	Pulpmill wastes
Barley straw	Sawdust
Cassava	Sunflower seed husks
Citrus wastes	(treated)
Date carbohydrates	Wastes from chemical
Meatpacking wastes	production of maleic
Mesquite wood	anhydride
Peat (treated)	Waste polyethylene (treated)

SOURCE: Office of Technology Assessment.

Genetics in baking, brewing, and winemaking

The micro-organism of greatest significance in the baking, brewing, and winemaking industries is common yeast. Because of its importance, yeast was one of the first micro-organisms to be used in genetic research. Nevertheless, the surge in studies in yeast genetics has not been accompanied by an increase in its practical application, for three reasons:

- industries already have the desired efficient strains, mainly as a result of trial-and-error studies;
- new genetic strains are not easily bred; they are incompatible for mating and their genetic characteristics are poorly understood; and
- many of the important characteristics of industrial microbes are complex; several genes being responsible for each.

Changing technologies in the brewing industry and increased sophistication in the molecular genetics of yeast have made it possible for researchers to achieve novel goals in yeast breeding. One strain that has already been constructed can produce a low-carbohydrate beer suitable for diabetics. (See figure 26.)

The baking industry is also undergoing technological revolution, and yeasts with new properties are now needed for the faster fermentation of dough. New strains with improved biological activity, storage stability, and yield would allow improvements in the baking process.

In the past, most genetic applications have come in the formation of hybrid yeasts. The newer genetic approaches, which use cell fusion now open up the possibility of hybrids developed from strains of yeast that carry useful genes but cannot mate normally.

Classical genetic research has also been car-

Figure 26.—The Use of Hybridization To Obtain a Yeast Strain for the Production of Low-Carbohydrate Beer

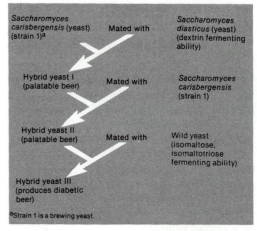

SOURCE: Office of Technology Assessment.

ried out with wine yeasts. Interestingly, within the past 10 years, scientists have isolated induced mutants of wine yeasts that have: 1) an increased alcohol tolerance and the capacity to completely ferment grape extracts of unusually high sugar content; 2) improved sedimentation properties, improving or facilitating separation of yeasts from the wine; and 3) improved performance in the production of certain types of wines. Hybridization studies of wine yeasts have been actively pursued only recently.

Progress in developing strains of yeast with novel properties is limited by the lack of enough suitable approved systems for using recombinant DNA (rDNA) technology. Eventual approval by the Recombinant DNA Advisory Committee is expected to boost applied research for the brewing, baking, and winemaking industries.

Microbial polysaccharides

The food processing industry uses polysaccharides (polymeric sugars) to alter or control the physical properties of foods. Many are incorporated into foods as thickeners, gelling agents, and agents to control ice crystal formation in frozen foods. They are used in instant foods, salad dressings, sauces, whips, toppings, processed cheeses, and dairy products. New uses are constantly appearing. The annual market in the United States is reported to be over 36,000 tons, not including starches and derivatives of cellulose.

Since many of the polymeric sugars now used in food processing are derived from plant sources, microbial polysaccharides have had limited use. To compete economically, a microbial polysaccharide must offer new properties, meet all safety requirements, and be readily available. Very few have reached the level of commercial applications; the only one in large-scale commercial production is xanthan gum.*

A wide variety of polysaccharides could theoretically be produced for use in food processing. Applied genetics may increase their production, modify those that are produced, eliminate the degradative enzymes that break them down, or change the microbes that produce them. However, as with other microbial processes, the application of genetics depends on an understanding of both the biochemical pathway for synthesis of a given polysaccharide and the systems that control microbial production. For many microbial polysaccharides, this information does not yet exist; furthermore, little is known about the enzymes that may be used to modify polysaccharides to more useful forms. Progress will only be able to occur when these information gaps are filled.

---

*The history of the development of xanthan gum indicates that the commercially significant organisms resulted from an extensive screening program for gum producers stored in the Northern Utilization Research and Development Division of USDA's large microbial culture collection. Xanthan gum produced by *Xanthomonas campestris* NRRL B-1459 was found to have characteristics that rendered it very promising as a commercial product. In 1960, the Kelco division of Merck & Co., Inc., carried out pilot plant feasibility studies, and substantial commercial production began in 1964. Although much of the work to date has been carried out with polysaccharides from one particular strain, there is increasing evidence to suggest that they could also be produced from other strains.

Enzymes

Enzymes are produced for industrial, medical, and laboratory use both by fermentation processes that employ bacteria, molds, and yeasts and by extraction from natural tissues. The present world market for industrial enzymes is estimated to be $150 million to $174 million; the technical (laboratory) market adds another $20 million to $40 million. Fewer than 50 microbial enzymes are of industrial importance today, but patents have been granted for more than a thousand. This reflects the increasing interest in developing new enzyme products; it also shows that it is easier to discover a new enzyme than to create a profitable application for it.*

Most industrial enzymes are used in the detergent industry and the food processing in-

---

*The enzyme literature is extensive and comprises well over 10,000 papers per year. Although less than 50 percent of these publications are concerned with microbial enzymes and most are found to have no industrial interest, a few thousand papers per year are of potential interest for the industrial development of enzymes. Less than 100 papers dealing with industrial processes appear every year, and few describe processes of great economic significance.

dustry, particularly for starch processing. Enzymes began to be used in quantity only 20 years ago. In the early 1960's, glucoamylase enzyme treatment began to replace traditional acid treatment in processing starch; around 1965, a stable protease (an enzyme) was introduced into detergent preparations to help break down certain stains; and in the 1970's, glucose isomerase was used to convert glucose to fructose, practically creating the high-fructose corn syrup industry.

Genetic engineering and enzymes in the food processing industry

Biotechnology applied to fermentation processes will make available larger quantities of existing enzymes as well as new ones. (See ch. 5.) The role of genetic engineering in opening commercial possibilities in the food processing industry is illustrated by the enzyme, pullulanase. This enzyme degrades pullulan, a polysaccharide, to the maltose or high-maltose syrups that give jams and jellies improved color and brilliance. They reduce off-color development produced by heat in candies and prevent sandiness in ice cream by inhibiting sugar crystallization. Maltose has several unique and favorable characteristics. It is the least water-absorbent of the maltose sugars and, although it is not as sweet as glucose, it has a more acceptable taste. It is also fermentable, nonviscous, and easily soluble. It does not readily crystallize and gives desirable browning reactions.

Pullulanase can also break down another carbohydrate, amylopectin, to produce high amylose starches. These starches are used in industry as quick-setting, structurally stable gels, as binders for strong transparent films, and as coatings. Their acetate derivatives are added to textile finishes, sizing, adhesives, and binders. In food, amylose starches thicken and give texture to gumdrop candies and sauces, reduce fat and grease in fried foods, and stabilize the protein, nutrients, colors, and flavors in reconstituted products like meat analogs.

In view of the current shortages of petroleum-derived plastics and the need for a biodegradable replacement, amylose's ability to form plastic-like wraps may provide its largest industrial market, although that market has not yet been developed.

If applications for the products made by pullulanase can be developed, genetic engineering can be used to insert this enzyme into industrially useful organisms and to increase its production. However, the food processing industry is permitted to use only enzymes that are obtained from sources approved for food use. Since the chief source of pullulanase is a pathogenic bacterium, *Klebsiella aerogenes,* no significant efforts have been made to apply genetics to improve its production or quality. Molecular genetics could ultimately transfer the pullulanase trait from *K. aerogenes* to a micro-organism approved for food use, if approved micro-organisms that manufacture pullulanase cannot be found.

Sweeteners, flavors, and fragrances

Biotechnology has already had a marked impact on the sweetener industry. The availability of the enzymes glucose isomerase, invertase, and amylase has made the production of high-fructose corn sweeteners (HFCS) profitable. Production of HFCS in the United States has increased from virtually nothing in 1970 to 10 percent of the entire production of caloric sweeteners in 1980 (11 lb per capita). The price advantage of HFCS is expected to cause its continued growth, particularly in the beverage industry. In fact, the Coca Cola Co. announced in 1980 that fructose will soon constitute as much as 50 percent of the sweetener used in its name brand beverage.

Biotechnology can be used to produce other sweeteners as well. While it is unlikely that sucrose will ever be made by micro-organisms (although improvements in sugarcane and sugar beet yields may result from agricultural genetic studies, see ch. 8.), the microbial production of low-caloric sweeteners is a distinct possibility. Three new experimental sweeteners—aspartame, monellin, and thaumatin—are candidates.

Aspartame is synthesized chemically from the amino acids, aspartic acid and phenylalanine, which can themselves be made by fermentation. The possibility of using microbes to couple the two amino acids is being investigated in at least one biotechnology research firm. Chemical production of aspartame is expensive and benefits from biotechnology are possible.

Monellin and thaumatin are natural substances—proteins obtained from West African plants. Both are intensely sweet—up to 100,000 times sweeter than table sugar—and the sensation of sweetness can last for hours. Their microbial production may be competitive with their extraction from plants. Since the physical and biological properties of thaumatin are known, it might also be produced through genetic engineering. Such an approach would not only increase the available supply, but would offer new molecules for investigating the physiology of taste.

Other flavors and fragrances show less promise at present. Although the chemistry of several flavors and aromas has been identified, too little research into their use has been conducted.*

*Recent work on the formation by micro-organisms of flavor and aroma chemicals known as lactones and terpenoids has been reported. Lactones occur as flavor-contributing components in many fermentation products, where they are formed by microbial reactions. Different pathways exist for their microbial formation. E.g., gamma-butyrolactone, which is formed during yeast fermentation, is found in sherry, wine, and beer. As early as 1930, an organism was isolated from orange leaves that had a peach-like odor and was thought to be *Sporobolomyces roseus*. The lactones, 4-decanolide and cis-6-dodecen-4-olide were found to be responsible.

Overview

The application of genetic engineering will affect the food processing industry in piecemeal fashion. Isolated successes can be expected for certain food additives, such as aspartame (not yet approved by the Food and Drug Administration (FDA) for sale in the United States) and fructose, and for improvements in SCP production. But an industrywide impact is not expected in the near future because of several conflicting forces:

- The basic genetic knowledge of characteristics that could improve food has not been adequately developed.
- The food processing industry is conservative in its research and development expenditures for improved processes, generally allocating less than half as much as more technologically sophisticated industries.
- Products made by new microbial sources must satisfy FDA safety regulations, which include undergoing tests to prove lack of harmful effects.* It may be possible to re-

duce the amount of required testing by transferring the desired gene into microorganisms that already meet FDA standards.

Nevertheless, the application of new genetic technologies will probably accelerate. Technologically sophisticated companies are being drawn into the business. Traditionally capital-intensive companies such as Union Carbide, ITT, General Electric, Corning Glass, and McDonnell-Douglas can be expected to introduce automation and more sophisticated engineering to food processing, modernizing the industry's technology. As has been noted by one industry observer:[1]

> You don't work on a better way to preserve fish. You try to change the system so that you no longer catch fish; you "manufacture" them and, if possible, do it right on top of your market so that you don't have to preserve them at all.

*E.g., all food additives and micro-organisms used in food processing must be approved as generally regarded as safe.

[1]M. L. Kastens, "The Coming Food Industry," *Chemtech*, April 1980, pp. 215-217.

You don't worry about processing bacon without nitrites, you engineer a synthetic bacon with designed-in shelf life.

You don't try to educate people to eat a "balanced diet;" you create a "whole" food with the proper balance of nutrients and supplements, and you make it taste like something people already like to eat.

Genetic engineering can be expected to aid in the creation of novel food preparations through effects on both the food itself and the additives used for texturizing, flavoring, and preserving.

The Use of Genetically Engineered Micro-Organisms in the Environment

Chapter 7

The Use of Genetically Engineered Micro-Organisms in the Environment

Although most genetically engineered micro-organisms are being designed for contained facilities like fermenters, some are being examined for their usefulness in the open environment for such purposes as mineral leaching and recovery, oil recovery, and pollution control.

All three applications are characterized by:

- the use of large volumes of micro-organisms;

- less control over the behavior and fate of the micro-organisms;
- a possibility of ecological disruption; and
- less basic research and development (R&D) —and a higher degree of speculation—than the industries previously discussed.

Mineral leaching and recovery

All micro-organisms interact with metals. Two interactions that are of potential economic and industrial interest are leaching metals from their ores, and concentrating metals from wastes or dilute mixtures. The first would allow the extraction of metals from large quantities of low-grade ores; the second would provide methods for recycling precious metals and controlling pollution caused by toxic metals.

Microbial leaching

In microbial or bacterial leaching, metals in ores are made soluble by bacterial action. Even before bacterial leaching systems became accepted industrial practice, it was known that dissolved metals could be recovered from mine and coal wastes. Active mining operations currently based on this process (such as those in Rio Tinto, Spain) date back to the 18th century. Presently, large-scale operations in the United States use bacterial leaching to recover copper from waste material. Estimates for the contribution of copper leaching to the total annual U.S. production range from 11.5 to 15 percent.

Leaching begins with the circulation of water through large quantities—often hundreds of tons—of ore. Bacteria, which are naturally associated with the rocks, then cause the metals to be leached by one of two general mechanisms: either the bacteria act directly on the ore to extract the metal or they produce substances, such as ferric iron and sulfuric acid, which then extract the metal. It appears that simply adding acid is not as efficient as using live bacteria. Although acid certainly plays a role in metal extraction, it is possible that direct bacterial attack on some ores is also involved. In fact, some of the bacteria that are known to be involved in mineral leaching have been shown to bind tenaciously to those minerals.

The application of the leaching process to uranium mining is of particular interest because of the possibility of in situ mining. Instead of using conventional techniques to haul uranium ore to the surface, microbial suspensions can extract the metal from its geological setting. Water is percolated through underground shafts where the bacteria dissolve the metals. The solution is then pumped to the surface where the metal is recovered. This approach, also called "underground solution mining," is already used in Canadian uranium mines, where it began almost by chance. In 1960, after only 2 years of operation, researchers at the Stanrock Uranium Mine found that the natural underground water contained large amounts of leached uranium. In 1962, over 13,000 kilograms (kg) of uranium oxide were obtained from the water. Thereafter, water was circulated through the mines as part of the mining operation. It has been suggested that extending this practice to most mines would have significant environmental benefits because of the minimal disruption of the land surface. Although the process is slower than the technology currently employed, the operating costs might be lower because of the simplicity of the system, since no grinding machinery is needed. Furthermore, deeper and lower grade deposits could be mined more readily.

Bacterial leaching can also extract sulfur-containing compounds, such as pyrite, from coal, producing coal with a lower sulfur content. Sulfur-containing coals from such areas as Ohio and the Appalachian Mountains are now less desirable than other coals because of the sulfur dioxide they release during burning. They often contain up to 6 percent sulfur, of which 70 percent can be in the form of pyrite. According to recent data, mixed populations of different bacteria, rather than a single species, are responsible for the most effective removal of sulfur—a finding that may lead to the genetic engineering of a single sulfur-removing bacterium in the future.

Applied genetics in strain improvement

The bacterium most studied for its leaching properties has been *Thiobacillus ferrooxidans* (which leaches copper), but others have also been identified in natural leaching systems. Although leaching ability is probably under genetic control in these organisms, practically nothing is known about the precise mechanisms. This is largely because little information exists in two critical areas: the chemistry of interaction between the bacteria and rock surfaces; and the genetic structure of the microorganisms. The finding that mixed populations of bacteria interact to increase leaching efficiency complicates the investigation.

Because of the lack of genetic and biochemical information about these bacteria, the application of genetic technologies to mineral leaching remains speculative. Progress in obtaining

more information is slow because less than a dozen laboratories in the Nation are actively performing research.

But even when the scientific knowledge is gathered, two obstacles to the use of genetically engineered micro-organisms will remain. The first is the need to develop engineered systems on a scale large enough to exploit their biological activities. A constant interchange must take place between microbial geneticists, geologists, chemists, and engineers. E.g., the geneticists must understand the needs identified by the geologists as well as the problems faced by the engineers, who must scale-up laboratory-scale processes. The complex nature of the problem can be approached most successfully by an interdisciplinary group that recognizes the needs and limitations of each discipline.

The second obstacle is environmental. Introducing large numbers of genetically engineered micro-organisms into the environment raises questions of possible ecological disruption, and liability if damage occurs to the environment or human health.

In summary, the present lack of sufficient scientific knowledge, scientists, and interdisciplinary teams, and the concerns for ecological safety present the major obstacles to the use of genetic engineering in microbial leaching.

Metal recovery

The use of micro-organisms to concentrate metals from dilute solutions such as individual waste streams has two goals: to recover metals as part of a recycling process; and to eliminate any metal that may be a pollutant. The process makes use of the ability of micro-organisms to bind metals to their surfaces and then concentrate them internally.

Studies at the Oak Ridge National Laboratory in Tennessee have shown that micro-organisms can be used to remove heavy metals from industrial effluents. Metals such as cobalt, nickel, silver, gold, uranium, and plutonium in concentrations of less than 1 part per million (ppm) can be recovered. The process is particularly useful for recovering metals from dilute solutions of

10 to 100 ppm, where nonbiological methods may be uneconomical. Organisms such as the common yeast *Saccharomyces cerevisiae* can accumulate uranium up to 20 percent of their total weight.

The economic competitiveness of biological methods has not yet been proven, but genetic improvements have been attempted only recently. The cost of producing the micro-organisms has been a major consideration. If it can be reduced, however, the approach might be useful.

As with other biological systems, genetic engineering may increase the efficiency of the extraction process. In the *Saccharomyces* system, differences in the ability to recover the metals have been demonstrated within populations of cells. Selection for cells with the genetic ability to accumulate large amounts of specific, desired metals would be an important step in designing a practical system.

Oil recovery

Since 1970, oil production in the United States has declined steadily. The supply can be increased by: accelerating explorations for new oilfields; by mining oil shale and coal and converting them to liquids; and by developing new methods for recovering oil from existing reservoirs.

In primary methods of oil recovery, natural expulsive forces (such as physical expansion) drive the oil out of the formation. In secondary methods of recovery, a fluid such as water or natural gas is injected into the reservoir to force the oil to the well. Approximately 50 percent of domestic crude in recent years has been obtained through secondary recovery.

Recently, new methods of oil recovery have been added to primary and secondary methods, which are called tertiary, improved, or enhanced oil recovery (EOR) techniques. They employ chemical and physical methods that increase the mobility of oil, making it easier for other forces to drive it out of the ground. The major target for EOR is the oil found in sandstone and limestone formations. It is here that applied genetics may play a major role, engineering micro-organisms to aid in recovery.

Oil susceptible to these processes is localized in reservoirs and pools at depths ranging from 100 ft to more than 17,000 ft. In these areas, the oil is adsorbed on grains of rock, almost always accompanied by water and natural gas. The physical association of the trapped oil and the surrounding geological formations varies significantly from site to site. The unknown characteristics of these variations are largely responsible for the economic risk in an attempted EOR.

Enhanced oil recovery

Of the original estimated volume of more than 450 billion barrels (bbl) of U.S. oil reserves, about 120 billion bbl have been recovered by primary and secondary techniques, and another 30 billion bbl are still accessible by these methods. The remaining 300 billion bbl however, are probably recoverable only by EOR methods. These figures include the oil remaining in known sandstone and limestone reservoirs and exclude tar sands and oil shale.

Four EOR processes are currently used. All are designed to dislodge the crude oil from its natural geological setting:

• In *thermal processes*, the oil reservoir is heated, which causes the viscosity of the oil to decrease, and with the aid of the pressure of the air introduced, supports the combustion that forces the petroleum to the producing well. Thermal processes will not be improved by genetic technologies.

• Various crude oils differ in their viscosity—ability to flow. Primary and secondary methods can easily remove those that flow as readily as water, but many of the reservoirs contain oil as viscous as road tar. *Miscible processes* use injected chemicals that blend with the crude oil to form mixtures that flow more readily. The chemicals used include alcohols, carbon dioxide, petroleum hydrocarbons such as propane and butane-propane mixtures, and petroleum gases. A fluid such as water is generally used to push a "slug" of these chemicals through the reservoir to mix with the crude oil and move it to the surface.

- Chemicals are also used in alkaline flooding, polymer flooding, and combined surfactant/polymer flooding.

In *alkaline flooding*, sodium hydroxide, sodium carbonate, or other alkaline materials are used to enhance the flow of oil. Neither natural nor genetically engineered micro-organisms are considered useful in this process.

Polymer flooding is a recent apparently successful method of recovery. It depends on the ability of certain chains of long molecules, known as polymers, to increase the viscosity of water. Instead of altering the characteristics of the crude oil, the aim is to make the injected water more capable of displacing it.

In the *combined surfactant/polymer flooding* technique, a detergent-like material (surfactant) is used to loosen the oil from its surrounding rock, while water that contains a polymer to increase its viscosity is used to drive the oil from the reservoir. (See figure 27.)

- *Other EOR methods* include many novel possibilities, such as the injection of live micro-organisms into a reservoir. These may produce any of the chemicals used in miscible and chemical processes, from surfactants and polymers to carbon dioxide. One target for EOR is the half million stripper wells (producing less than 10 barrels per day (bbl/d) in the United States.

MICROBIAL PRODUCTION OF CHEMICALS USED IN EOR

EOR methods that use chemicals tend to be expensive because of the cost of the chemicals. Nevertheless, potentially useful polymers were

Figure 27.—Chemical Flooding Process

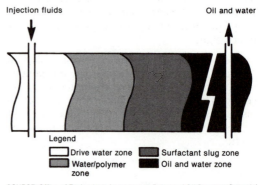

Legend

☐ Drive water zone   ■ Surfactant slug zone
▨ Water/polymer zone   ■ Oil and water zone

SOURCE: Office of Technology Assessment, *Enhanced Oil Recovery Potential in the United States* (Washington, D.C.: U.S. Government Printing Office, January 1978).

found in the early 1960's and have since been responsible for the recovery of more than 2 million bbl. Polymers such as polyacrylamide and xanthan gum can increase the viscosity of water in concentrations as low as one part in a thousand. Xanthan gum is readily made in large quantities by micro-organisms. Different strains of *Enterobacter aerogenes* produce a wide variety of other polymers. A useful biopolymer—one formed by a biological process—might be designed specifically to improve oil recovery.

Xanthan gum, produced by *Xanthomonas campestris* and currently marketed by the Kelco division of Merck & Co., Inc., is useful but far from ideal for oil recovery. While it has excellent viscous properties, it is also very expensive. Furthermore, unless it is exceptionally pure, it can plug reservoir pores, since the fluid often has to travel through hundreds of meters of fine pores. To avoid such plugging, the fluid must be filtered to remove bacterial debris before it is injected.

Nevertheless, micro-organisms can be selected or genetically engineered to overcome many obvious difficulties.* With improved properties, polysaccharides (polymeric sugars)

*A good organism, for example, might have the following desired properties: nonpathogenic to humans, plants, or animals; rapid growth on simple, cheap raw materials; ease of separation from its products; limited detrimental effect on reservoirs, e.g., plugging; easy disposal of cells, e.g., byproduct credits; ability to

obtained by microbial fermentation could compete with those obtained from alternative sources, especially seaweed. Controlled fermentation is not affected by marine pollution and weather, and production could be geared to market demand.

Biological processes have disadvantages primarily in the costs of appropriate raw materials and in the need for large quantities of solvent. Current efforts to find cheaper raw materials, such as sugar beet pulp and starch, show promise. The need for solvents to precipitate and concentrate the polymers before shipment from plant to field can be circumvented by producing them onsite.

Micro-organisms can also produce substances like butyl and propyl alcohols that can be used as cosurfactants in EOR. It has been calculated that if *n*-butanol were used to produce crude oil at a level of 5 percent of U.S. consumption, 2 billion to 4 billion lb per year—or four to eight times the current butanol production—would be required. Micro-organisms capable of producing such surfactants have been identified, and genetically superior strains were isolated several decades ago at the Northern Regional Research Laboratories in Illinois. Other chemicals, such as alcohols that increase the rate of formation and stability of chemical/crude oil mixtures and the agents that help prevent precipitation of the surfactants, have also been produced by microbial systems.

The uncertainties of the technical and economic parameters are compounded by the lack of sufficient field experiments. Laboratory tests cannot be equated with conditions in actual oil wells. Each oil field has its own set of characteristics—salinity, pH (acidity and alkalinity), temperature, porosity of the rock, and of the crude oil itself—and an injected chemical behaves differently in each setting. In most cases, not enough is known about a well's characteristics to predict the nature of the chemical/crude oil interaction and to forecast the efficiency of oil recovery.

---

use water available at site; growth under conditions that discourage the growth of unwanted micro-organisms; no major problems in culturing the bacterium; and genetic stability.

IN SITU USE OF MICRO-ORGANISMS

One alternative to growing micro-organisms in large fermenters then extracting their chemical products and injecting them into wells, is to inject the micro-organisms directly into the wells. They could then produce their chemicals in situ.

Unfortunately, the geophysical and geochemical conditions in a reservoir seldom favor the growth of micro-organisms. High temperature, the presence of sulfur and salt, low oxygen and water, extremes of pH, and significant engineering hurdles make it difficult to overcome these limitations. The micro-organisms must be fed and the microenvironment must be carefully adjusted to their needs at distances of hundreds to thousands of feet. The oil industry has already had discouraging experiences with micro-organisms in the past. In the late 1940's, for instance, the injection of sulfite-reducing micro-organisms, along with an inadvertently high-iron molasses as a carbon source, resulted in the formation of iron sulfide, which clogged the rock pores. One oil company developed a yeast to break down petroleum, but the size of the yeast cells (5 to 10 micrometers, μm) was enough to clog the 1-μm pores.

Nevertheless, information from geomicrobiology suggests that this approach is worth pursuing. Preliminary field tests have also been encouraging. The injection of 1 to 10 gal of *Bacillus* or *Clostridium* species, along with a water-suspended mixture of fermentable raw materials such as cattle feed molasses and mineral nutrients, has resulted in copious amounts of carbon dioxide, methane, and some nitrogen in reservoirs. The carbon dioxide made the crude less viscous, and the other gases helped to repressurize the reservoir. In addition, large amounts of organic acids formed additional carbon dioxide through reactions with carbonate minerals. The production of microbial surfactants further aided the process.

Although previous assessments have argued that reservoir pressure is a significant hindrance to the growth of micro-organisms, more recent studies indicate the contrary. The micro-organisms must, however, be selected for increased salt and pH tolerance.

EOR AND GENETIC ENGINEERING

The current research approach, funded by the Department of Energy (DOE) and, independently, by various oil companies, is a two-phase process. The first phase is to find a micro-organism that can function in an oil reservoir environment with as many of the necessary characteristics as possible. The second is to alter it genetically to enhance its overall capability.

The genetic alteration of micro-organisms to produce chemicals used in EOR has been more successful than the alteration of those that may be used in situ.* However, recombinant DNA (rDNA) technology has not been applied in either category. All efforts have employed artificially induced or naturally occurring mutations.

CONSTRAINTS TO APPLYING GENETIC ENGINEERING TECHNOLOGIES IN EOR

The genetic data base for micro-organisms that produce useful polysaccharides is weak. Few genetic studies have been done. Hence, theoretically plausible approaches such as transferring enzyme-coding plasmids (see ch. 2) for polysaccharide synthesis, cannot be seriously contemplated at present. Only the crudest methods of genetic selection for desirable properties have been used thus far. They remain the only avenue for improvement until more is learned about the micro-organism's genetic mechanisms.

The biochemical data base for the characteristics of both the micro-organisms and their products is also lacking. The wide potential for chemical reactions carried out by microbes remains to be explored. At the same time, a system must be devised to allow easy characterization, classification, and comparison of products derived from a variety of micro-organisms.

The physical data base for oil reservoirs is limited. The uniqueness of each reservoir suggests that no universal micro-organism or method of oil recovery will be found. Compounding

---

*Some of the goals have been to: improve polymer properties to enhance their commercial applicability; improve polymer production (a major mistake has been to reject a micro-organism in the initial screening because its level of production was too low); improve culture characteristics, e.g., resistance to phage, rapid growth, ability to use cheaper raw materials; and eliminate enzymes that naturally degrade the polymers.

this problem is the lack of sufficient physical, chemical, and biological information about the reservoirs, without which it is difficult to see how a rational genetic scheme can be constructed for strains. Clearly, the activities of micro-organisms under specified field conditions cannot be studied unless researchers know what the appropriate conditions are.

Three *institutional obstacles* exist. First, publication in this field is limited because most research is carried out in the commercial world and remains largely confidential. Second, neither the private nor the public sector has been enthusiastic about the potential role of micro-organisms in EOR. The biological approach has only recently been given consideration as a way to advance the state of the art of the technology, and most oil companies still have limited staffs in microbiology. To date, DOE's Division of Fossil Fuel Extraction has conducted the main Federal effort. Third, any effort to use micro-organisms must be multidisciplinary in nature. Geologists, microbiologists (including microbial physiologists and geneticists), chemists, and engineers must interact to evolve successful schemes of oil recovery. Thus far, such teams do not exist.

Environmental and legal concerns have also inhibited progress. Microbial EOR methods usually require significant quantities of fresh water and thus may compete with municipal and agricultural uses. Furthermore, the use of micro-organisms introduces concerns for safety. All strains of *Xanthomonas*, which produce xanthan gum polymer, are plant pathogens. Other micro-organisms with potential, such as *Sclerotium rolfii* and various species of *Aureobasidium* have been associated with lung disease and wound infections, respectively.

Immediate environmental and legal concerns, therefore, arise from the potential risks associated with the release of micro-organisms into the environment. When they naturally cause disease or environmental disruption, their use is clearly limited. And when they do not, genetic engineering raises the possibility that they might. Such concerns have reduced the private sector's enthusiasm for attempting genetic engineering. (See ch. 10 for a more detailed discussion of risk.)

GENETIC ENGINEERING OF MICRO-ORGANISMS FOR USE IN OTHER ASPECTS OF OIL RECOVERY AND TREATMENT

Two other aspects of microbial physiology deserve attention: the microbial production of oil muds or drill lubricants, and the treatment of oil once it has been recovered. Drilling muds are suspensions of clays and other materials that serve both to lubricate the drill and to counterbalance the upward pressure of oil. Microbially produced polysaccharides have been developed for this use. Exxon holds a patent on a formulation based on the production of xanthan gum, from *Xanthomonas campestris*, while the Pillsbury Co. has developed a polysaccharide (glucan) from various species of *Sclerotium*. At least two of the small genetic engineering firms have begun research programs to develop biologically produced polysaccharides with the desired lubricant qualities.

Interest in the postrecovery microbial treatment of oil after its extraction centers around the ability of micro-organisms to remove un-desirable contituents from the crude oil itself. As an indication of recent progress, three distinct microbial systems have been developed to help remove aromatic sulfur-containing material, a major impurity.

Overview of genetic engineering in mining and oil recovery

The underlying technical problem with the use of genetically engineered organisms in either mining or oil recovery is the magnitude of the effort. In both cases, large areas of land and large volumes of materials (chemicals, fluids, micro-organisms) must be used. The results of testing any new micro-organism in a laboratory cannot automatically be extrapolated to large-scale applications. The change in magnitude is further complicated by the lack of rigid controls. Unlike a large fermenter whose temperature, pH, and other characteristics can be carefully regulated, the natural environment cannot be controlled. Nevertheless, despite the formidable obstacles, the potential value of the products in these areas assures continuing efforts.

Pollution control

Life is a cycle of synthesis and degradation—synthesis of complex molecules from atoms and simple molecules and degradation by bacteria yeast, and fungi, back to simpler molecules and atoms when organisms die. The degradation of complex molecules is an essential part of life. Without it, ". . . we'd be knee-deep in dinosaurs."[1] A more quantitative statement is equally thought provoking. Livestock in the United States produce 1.7 billion tons of manure annually. Almost all of it is degraded by soil microorganisms.

For a long time people have exploited microbial life forms to degrade and detoxify human sewage. Now, on a smaller scale, science is beginning to use micro-organisms to deal with the pollution problems presented by industrial toxic wastes. Chemicals in their place can be useful and beneficial; out of place, they can be polluting.

Pollution problems can be divided into two categories: those that have been present for a long time in the biosphere—e.g., most hydrocarbons encountered in the petroleum industry and human and animal wastes—and those that owe their origin to human inventiveness—e.g., certain pesticides. Chemicals of both sorts, through mishap, poor planning, or lack of knowledge at the time of their application sometimes appear in places where they are potentially or actually hazardous to human health or the environment.

Pollution can be controlled by microbes in two ways: by enhancing the growth and activity

---

[1]R. B. Grubbs, "Bacterial Supplementation, What It Can and Cannot Do," oral presentation to the Ninth Engineering Foundation on Environmental Engineering in the Food Processing Industry, 1979 (Available from Flow Laboratories, Inc., Rockville, Md.)

of microbes already present at or near the site of the pollution problem, and by adding more (sometimes new) microbes to the pollution site. The first approach does not provide an opportunity for applying genetics, but an example will indicate how it functions.

Enhancing existing microbial degradation activity

Sun Oil successfully exploited indigenous microbes to clean up a 6,000 gal underground gasoline spill that threatened the water supply of a town in Pennsylvania.[2][3] First, engineers drilled wells to the top of the water table and used pumps to skim gasoline from the water surface. About half the gasoline was removed in this fashion, but company calculations showed that dissipating the remaining gasoline would require about 100 years. To speedup the process, it was decided to encourage the growth of indigenous bacteria that could degrade the gasoline.

Pollution-control microbes, like all organisms, require a number of different elements and compounds for growth. If the amount of any nutrient is limited, the microbe will not be able to metabolize the pollutant at the fastest rate. The cleanup depended on increasing the growth rate of the bacteria by supplying them with additional nutrients. In the case of the gasoline-degrading bacteria, the gasoline already supplied the hydrocarbon, but the water-gasoline environment was deficient in nitrogen, phosphate, and oxygen. Those three nutrients were pumped down to the water table, bacterial growth increased, and the gasoline was metabolized into innocuous chemicals by the bacteria. As a result, it was degraded in a single year.

Adding microbes to clean up pollution

Genetics may have important applications in approaches to pollution control that depend on

adding microbes to the pollution site. Three firms—Flow Laboratories, Polybac Corp., and Sybron/Biochemicals Corp.—sell microbes for such use. Two companies select bacteria for enhanced degradation activity and two mutate bacteria to the same end, but none of the three firms currently uses genetic engineering techniques.

Some "formulations" (mixtures) of bacteria are designed to degrade particular pollutants, such as one that was used to digest the 800,000 gal of oily water that lay in the bilges of the Queen Mary. After a 6-week treatment with the formulation, the water from the bilges was judged safe for disposal into the Long Beach, Calif., harbor. It was discharged without causing an oil slick or harming marine life.[4] Flow Laboratories markets its services to companies with industrial pollution problems. It investigates the problem, develops a formulation to degrade the pollutants, and sells it.

In addition to industrial pollution problems, Flow markets its products and services for use in sewerage systems, which collect and hold human wastes to facilitate degradation and detoxification. Sludge bacteria in sewerage plants degrade the waste, but they are not present in the lines that carry wastes to the treatment plant. As a result, greases and oils from fat discarded through garbage disposals and from cosmetic oils and creams coat the inside of sewerage lines and reduce their carrying capacity.[5]

Cities have resisted using added microbes in sewerage systems. Standard textbooks simply state that the ideal bacteria will establish themselves in a well-planned and well-managed system. The idea that "better" bacteria can be added to improve the plant operation is not readily accepted.

The value of adding bacteria to large sewerage sytems has not been adequately tested. Because of the size of municipal systems (which already contain tons of sludge bacteria), some have argued that adding a few additional

[2]R. L. Raymond, V. W. Jamison, J. O. Hudson, "Beneficial Stimulation of Bacterial Activity in Groundwaters Containing Petroleum Products," AIChE symposium series 73:390-404, 1976.

[3]V. W. Jamison, R. L. Raymond, J. O. Hudson, "Biodegradation of High-Octane Gasoline," Proceedings of the Third International Biodegradation Symposium, J. M. Sharpley and A. M. Kaplan (eds.) (City???? : Applied Science Publishers, 1976).

[4]Anon., Environmental Science and Technology 13:1180, 1979.

[5]R. E. Kirkup and L. R. Nelson, "City Fights Grease and Odor Problems in Sewer Systems," Public Works Magazine, October 1977.

pounds of bacteria is unlikely to have any effect. Thus far, the Environmental Protection Agency (EPA) has not recommended adding bacteria to municipal systems; however, EPA suggests that they might be useful in smaller installations and for specific problems in large systems.

Dry formulations are available for use in cleaning drains and pipes in smaller installations, such as restaurants and other food processing facilities. In restaurants, the bacteria are added to the drain at the end of the workday. Bacteria have been selected for their inability to produce hydrogen sulfide, which means that the degrading process does not produce the unpleasant odors frequently encountered in the digestion of oils and fats.[6]

As of November 1979, the pollution control industry had few plans for the genetic manipulation of bacteria, except for the selection of naturally occurring better performers. Consumer resistance to "mutants" is a factor that discourages the move to microbial genetics. Probably even more important is the high cost of establishing and maintaining microbial genetics laboratories. It has been estimated that the cost of carrying a single Ph. D. microbial geneticist is over $100,000 annually.[7] This expense is quite high relative to the $2 million to $4 million sales of all biological pollution control companies in 1978.[8]

Resistance to the use of genetically manipulated bacteria is not universal. Many industrial wastes are oxidized to nontoxic chemicals by biological treatment in aerated lagoons. The process depends on the presence of microbes in the lagoons; over time, those that grow best on the wastes come to dominate the microbial populations. Three companies now sell bacteria that they claim outperform the indigenous strains found in the lagoons. E.g., the Polybac

Corp. has sold its products to all seven Exxon biological waste treatment plants to treat chemical wastes. One of its formulations has been used to degrade toxic dioxins from an herbicide spill. One month's treatment with the bacterial formulation reduced the orthochlorophenol concentration from 600 to 25 ppm in a 20,000-gal lagoon.[9]

Sybron/Biochemical, a division of Sybron Corp., sells cultures of bacteria that are intended to aid in the biological oxidation of industrial wastewater; this company also lists 20 different cultures for application to specific wastes. Patent number 4,199,444 was granted on April 22, 1980, for a process involving the use of a mutant bacterial culture to decolor waste water produced in Kraft paper processing.[10] Other patents are pending on a mixture of two strains that degrade grease and a strain that degrades "nonbiodegradable" detergents.[11]

There is disagreement about the value of adding microbes to decontaminate soils or waters. One point of view argues that serious spills frequently sterilize soils, and that adding microbes is necessary for any biodegradation. The other contends that encouraging indigenous microbes is more likely to succeed because they are acclimated to the spill environment. Added bacteria have a difficult time competing with the already-present microbial flora. In the case of marine spills, bacteria, yeast, and fungi already present in the water participate in degradation, no one has been able to demonstrate the usefulness of added microbes.

Commercial applications—market size and prospects

The estimated market size of pollution-control biological products in 1978 was $2 million to $4 million, divided among some 20 companies,

[6]Anon., "Clean That Sewage System With Bugs!" *Environmental Science and Technology* 13:1198-1199, 1979.

[7]Anon., "Biotechnology DNA Research Expenditures in U.S. May Reach $500 Million in 1980, With About $150-200 Million for Commercial Products," Hill told. *Drug Research Reports,* "The Blue Sheet," May 28, 1980, p. 22.

[8]Anon., *Business Week,* July 5, 1976, p. 280; *Chemical Week* 121:47, 1977; and *Food Engineering* 49:138, 1977, cited in T. Gassner, "Microorganisms for Waste Treatment," *Microbial Technology,* 2 ed., vol. II, (London: Academic Press, 1979), pp. 211-222.

[9]See footnote 6.

[10]L. Davis, J. E. Blair, and C. W. Randall, "Communication: Development of Color Removal Potential in Organisms Treating Pulp and Paper Wastewater," *J. Water Pollution Control Fed.,* February 1978, pp. 382-385.

[11]P. Spraher and N. Tekeocgak, "Foam Control and Degradation of Nonionic Detergent," *Industrial Wastes,* January/February 1980; L. David, J. E. Blair, and C. Randall, "Mixed Bacterial Cultures Leak 'Non-Biodegradable' Detergent," *Industrial Wastes,* May/June 1979.

and the potential market was estimated to be as much as $200 million.[12] These estimates can be compared to Polybac's own sales records. In 1976, its first year, its sales totaled $0.5 million and in 1977, $1.0 million. It expects to reach $5 million in 1981.

To date genetically engineered strains have not been applied to pollution problems. At least one prominent genetic engineering company has decided not to enter the pollution control field, concluding that it was improbable that added microbes could compete with indigenous organisms. More specifically, the possiblity of liability problems make the approach even less attractive. Pollution control requires that "new" life forms be released into the environment, which is already seen as precariously balanced. Such new forms might cause health, economic, or environmental problems. The problems of liability that might arise from such applications are enough to deter entrepreneurs from contemplating work in the field at this time.

An additional reason for the reluctance of some companies to engage in this activity is that the opportunities for making money are limited. Selling microbes, rather than their products, may well be a one-shot opportunity. The microbes, once purchased, might be propagated by the buyer. Nevertheless, at least two small companies have announced that they are pursuing efforts to use genetic engineering.

The low-key efforts in this field might accelerate quickly if a significant breakthrough occurred. To date, no "new" organism has appeared that will degrade previously intractable chemicals. The effect of such a development might be enormous.

Genetic research in pollution control

The Oil and Hazardous Materials Spills Branch of EPA currently supports research aimed at isolating organisms to degrade three specific chemical compounds. The work is being carried out on contract; as of November 1979, no field trials of the organisms had been under-

taken. Two of the toxic chemicals, pentachorophenol and hexachlorocyclopentadiene, are relatively long-lived compounds and present long-term problems. A fungus and a bacterium that can degrade the first compound have been isolated,[13] and Sybron/Biochemical already sells a culture specifically for pentachlorophenol degradation. The third toxic compound is methyl parathion. Its inclusion is more difficult to understand, since it is degraded within a few days after its application as a pesticide.

Efforts have been made to isolate bacteria that can degrade (2,4-dichlorophenoxy) acetic acid (2,4-D) and (2,4,5-trichlorophenoxy) acetic acid (2,4,5-T), the components of Agent Orange.[14] Strains of the bacterium *Alcaligenes paradoxus* rapidly degrade 2,4-D, and the genetic information for the degradation activity has been located on a plasmid. The investigator who found that strain, while optimistic about the opportunities for isolating and transferring other resistance genes, has been unable to find a bacterium that degrades 2,4,5-T or its very toxic contaminant, 2,3,7,8-tetrachlorodibenzo-para-dioxin (TCDD or dioxin).

By far the best known research in this area is that of Dr. Ananda M. Chakrabarty who engineered two strains of *Pseudomonas*, each of which has the ability to degrade the four classes of chemicals found in oil spills. Chakrabarty began with four different strains of *Pseudomonas*. None of them presented a threat to human health, and each could degrade one of the four classes of chemicals. His research showed that the genes controlling the degrading activities were located on plasmids. Taking advantage of the relative ease of moving such genes among bacteria, he produced two recombinant bacteria.

Chakrabarty has presented evidence that his bacterium degrades complex petroleum mixtures such as crude oil or "Bunker C" oil, and he

[13]N. K. Thuma, P. E. O'Neill, S. G. Brownlee, and R. S. Valentine, "Laboratory Feasibility and Pilot Plant Studies: Novel Biodegradation Processes for the Ultimate Disposal of Spilled Hazardous Materials," National Environment Research Center, U.S. Environmental Protection Agency, Cincinnati, Ohio, 1978.
[14]J. M. Pemberton, "Pesticide Degrading Plasmids: A Biological Answer to Environmental Pollution by Phenoxyherbicides," *Ambio* 8:202-205, 1979.

has proposed a method for using it to clean up oil spills. The bacteria are to be grown in the laboratory, mixed with straw, and dried. The bacteria-coated straw can be stored until needed, then dropped from a ship or aircraft onto oil spills. The straw absorbs the oil and the bacteria degrades it.[15] To completely cleanup a spill will probably require mechanical efforts in addition to the biological attack. It was the production of one of Chakrabarty's strains that led to the Supreme Court decision on "the patenting of life." (See ch. 12 for further details.)

The essential difference between the well-publicized Chakrabarty approach and a less well-known one is that all the desired activities in Chakrabarty's approach are combined in a single organism; while in the other method, bacteria bearing single activities are mixed together to yield a desired "formulation." In yet another approach, Sybron/Biochemical uses mutation and selection to produce specialized degradation activities. It also sells mixed cultures for some applications.

The single-organism, multiple-enzyme system has the advantage that every bacterium can attack a number of compounds. The mixed formulations allow the preferential proliferation of bacteria that feed on the most abundant chemical; then, as that chemical is exhausted, other bacteria, which flourish on the next most abundant chemical, become dominant. The preferential survival of only one or a few strains in a mixed formulation might result in no bacteria being available to degrade some compounds. The multienzyme bacteria, on the other hand, can degrade one chemical after another, or alternatively, more than one at the same time.

Federal research support for engineering microbes to detoxify hazardous substances

EPA currently limits its support to research aimed at selecting indigenous microbes, an area

that has already attracted some commercial research support. Commercial firms are looking for large-scale markets, such as sewerage systems, or commonly occurring smaller markets, such as gasoline spills and common industrial wastes.

Whatever potential exists in identifying, growing, and using naturally occurring microbes for pollution control pales beside the opportunities offered by engineering new ones. Unfortunately, the potential risks increase as well. EPA has taken a preliminary step toward assessing the risks by soliciting studies to determine what environmental risks may exist from accidentally or deliberately released engineered microbes.

Summary

While some unreported efforts may be underway, genetics has apparently been little applied to pollution abatement. Nevertheless, the production of "new" life forms that offer a significant improvement in pollution control is a possibility. The constraints are questions of liability in the event of health, economic, or environmental damage; the contention that added organisms are not likely to be a significant improvement; and the assumption that selling microbes rather than products or processes is not likely to be profitable.

The factors that have discouraged developments in this area would probably become less deterring if convincing evidence were found that microbes could remove or degrade an intractable pollutant. In the meantime, the research necessary to produce marked improvements has been inhibited. Overcoming this inhibition may require a governmental commitment to support the research, to buy the microbes, and to provide for protection against liability suits. Such a governmental role would be in keeping with its commitment to protecting health and the environment from the toxic effects of pollutants.

[15]Patent Specification 1 436 573, May 19, 1976, Patent Office, London, England.

Issue and Options—Biotechnology

ISSUE: How can the Federal Government promote advances in biotechnology and genetic engineering?

The United States is a leader in applying genetic engineering and biotechnology to industry. One reason is the long-standing commitment by the Federal Government to the funding of basic biological research; several decades of support for some of the most esoteric basic research has unexpectedly provided the foundation for a highly useful technology. A second is the availability of venture capital, which has allowed the formation of small innovative companies that can build on the basic research.

The argument for Government promotion of biotechnology and genetic engineering is that Federal help is needed in those high priority areas *not* being developed by industry.

The argument against such assistance is that industry will develop everything of commercial value without Federal help.

A look at what industry is now attempting indicates that sufficient investment capital is available to pursue specific manufacturing objectives, such as for interferon and ethanol, but that some high-risk areas that might be of interest to society, such as pollution control, may need promotion by the Government. Other areas, such as continued basic biological research, might not be profitable soon enough to attract industry's investment. Specialized education and training are areas in which the Government has already played a major role, although industry has both supported university training and conducted its own inhouse training.

OPTIONS:

A. *Congress could allocate funds specifically for genetic engineering and biotechnology R&D in the budget of appropriate agencies, such as the National Science Foundation (NSF), the U.S. Department of Agriculture (USDA), the Department of Health and Human Services (DHHS), the Department of Energy (DOE), the* Department of Commerce (DOC), and the Department of Defense (DOD).

Congress has a long history of recognizing areas of R&D that need priority treatment in the allocation of funds. Biotechnology has not been one of these. Even though agencies like NSF receive congressional funding, its Alternative Biological Sources of Materials program is one of the few applied programs that is not congressionally mandated. As a result, the fiscal year 1980 budget saw a *reduction* in the allocation of funds, from $4.1 million in 1979 to $2.9 million. A congressionally mandated program, analogous to the successful NSF Earthquake Hazard Mitigation program, could be written into law. Other programs, such as the competitive grants program at USDA (or the Office of Basic Biological Research at DOE), are also modestly funded.

Increasing the amount of money in an agency's biotechnology program could bring criticism from other programs within each agency if their levels of funding are not increased commensurately. The Competitive Grants Program at USDA has similar problems; those who are most critical of it argue that it should not take funds from traditional programs. Nevertheless, Congress could promote two types of programs: those with long-range payoffs (basic research), and those which industry is not willing to undertake but that might be in the national interest.

B. *Congress could establish a separate Institute of Biotechnology as a funding agency.*

The merits of a separate institution lie in the possibility of coordinating a wide range of efforts, all related to biotechnology. Among present organizations, biotechnology and applied genetics cut across several institutes and divisions within them. Medically oriented research falls primarily under the domain of the National Institutes of Health (NIH). EPA is concerned with the prevention of pollution; while NSF's effort in biotechnology has been restricted to modest support scattered through several divisions.

The creation of an organization such as the National Technology Foundation (H.R. 6910) would represent the kind of commitment to engineering, in general, that currently does not exist.

Competition for funds within other agencies would be avoided, since funding would now occur at the level of congressional appropriations. A separate institute, carrying the stamp of Government recognition, would make it clear to the public that this is a major new area with great potential. This might foster greater academic and commercial interest in biotechnology and genetic engineering.

On the other hand, biotechnology and genetic engineering cover such a broad range of disciplines that a single agency would overlap the mandates of existing agencies. Furthermore, the creation of yet another agency carries with it all the disadvantages of increased bureaucracy and competition for funds at the agency level.

C. *Congress could establish research centers in universities to foster interdisciplinary approaches to biotechnology. In addition, a program of training grants could be offered to train scientists in biological engineering.*

The successful use of biological techniques in industry depends on a multidisciplinary approach involving biochemists, geneticists, microbiologists, process engineers, and chemists. Little is now being done, publicly or privately, to develop expertise in this interdisciplinary area.

In 1979, President Carter proposed the creation of generic technology centers (useful to a broad range of industries) as one way to stimulate innovation. The centers would conduct the kind of research that an individual company might not consider cost effective, but that might ultimately benefit several companies. Each center would be jointly funded by Government and industry, with Government providing the seed money and industry carrying most of the costs within 5 years. If the centers were established at universities, startup costs could be minimized.

Several congressional bills contain provisions for centers similar to these. For example, on

October 21, 1980, President Carter signed into law a bill (S. 1250) that would establish Centers for Industrial Technology to foster research links between industry and universities. They would be affiliated with a university or nonprofit institution.

One or more of these centers could be specifically designated to specialize in biotechnology. In addition, training grants could be used to support the education of biotechnologists at the centers or elsewhere. Currently, there is no nationwide training program to train students in this discipline. Education programs, especially for the postgraduate and graduate training of engineers, could further the idea of using biological techniques to solve engineering problems.

D. *Congress could use tax incentives to stimulate biotechnology.*

The tax laws could be used to stimulate biotechnology in several ways. First, they could expand the supply of capital for small high-risk firms, which are generally considered more innovative than established firms, because of their willingness to undertake the risks of innovation. Much of the pioneering work in the industrial application of genetic techniques has been done by such firms. By nature, they are speculative, high-risk investments. Second, the tax law could provide special subsidies to new high-technology firms, which cannot use the standard investment incentives, such as the investment tax credit, because they usually have no taxable profits for the first several years against which to apply the tax credit. Third, tax incentives could be provided for both established and new firms to make the investment of money for R&D more attractive.

There are a number of ways to expand the supply of venture capital. One is to decrease the tax rate on capital gains or the period an asset must be held for it to be considered a capital gain rather than ordinary income. This change could be limited to stocks in high-technology firms in order to focus its impact and minimize revenue loss. Other options involving the stock of high-technology companies are: a tax credit to the investor who purchases the stock; deferment of capital gains taxes on the sales of these stocks if the proceeds are reinvested into similarly qualifying stock; and more liberal capital loss provisions.

In addition to focusing on the supply of capital, tax policy could attempt to directly increase the profitability of potential growth companies. Since most are not profitable for several years, they cannot take full advantage of the investment tax credit—or even the provision for carrying net operating losses back 3 years and forward 7 years to offset otherwise taxable profits. Two proposals may remedy this situation. First, the investment tax credit could be refundable to the extent it exceeded any tax liability of the firm. A preliminary estimate of the revenue loss for this proposal was $1 billion for 1979. Second, new companies could be permitted to carry net operating losses forward for 10 years. This change would give new firms the same number of years over which to deduct losses as established firms.

The final type of tax incentive is directed at increasing R&D expenditures. Two major proposals would permit companies to take tax credits on a certain percentage of their R&D expenses, and on contributions to universities for research.

The R&D credit has been advocated for several reasons. First, it would increase the after-tax return on R&D investments, making them more attractive. Second, it would reduce the degree of risk on such investments; with a 10-percent credit, the real after-tax expense of a $1 million investment is $900,000. Finally, it would give firms maximum flexibility in selecting projects for investment.

Questions have been raised about the cost effectiveness of the credit. For calendar year 1980, the Treasury Department estimated the cost of a 10-percent R&D credit to be $1.9 billion. Since R&D costs average only 10 to 20 percent of the total cost of bringing a new product or process to the market, the net reduction in the cost of commercializing an invention would be 1 to 2 percent. Moreover, the commercial stage of innovation is thought to be riskier and costlier than the technical stage. Another prob-

lem is that the credit may be a windfall for firms that would be investing in R&D anyway. Finally, the credit would subsidize R&D devoted to minor product changes or incremental improvements in addition to R&D directed to more fundamental breakthroughs.

One of the provisions of a pending congressional bill (H.R. 5829) provides for a credit of 25 percent for incremental research expenditures above those for a base period. By limiting the credit to incremental expenditures, the bill would create a more cost-effective credit, if passed.

The final type of tax credit would be for corporate contributions to university research. The Treasury Department estimated that a 25 percent credit for research in all fields would cost $40 million in 1980. This credit would be targeted to more fundamental research and not to the subsidy of short-term, incremental projects that are usually a significant part of corporate R&D budgets.

E. *Congress could improve the conditions under which U.S. companies can collaborate with academic scientists and make use of the technology developed in universities in whole or in part at the taxpayer's expense.*

Developments in genetic engineering have kindled interest in this option. Nevertheless, the Government's role in fostering university-academic interaction is far from accepted. Such a role may limit the flexibility of a cooperative effort. At the very least, disincentives such as patent restrictions could be removed.

The controversy has been summed up as follows:[1]

> At the next level of involvement, the Government could identify potential partners, and facilitate negotiations. A more active role would involve the Government's providing startup funds. Finally, the Government could be a third partner, sharing costs with industry and the university. In this case, too large a Government role could lead to Federal intervention in activities that should be the responsibility of business and industry.

---

[1]Dennis Prager, G. S. Omenn, *Science* 207: 379-384, 1980.

Certainly the Government can facilitate communication; in the health field, NIH, for instance, is an effective stimulus for contacts among scientists.

The possible advantages and disadvantages of university-industry interaction is illustrated by a recent case involving a plan by Harvard University to collaborate with a genetic engineering company. The plan had called for the establishment of a corporation to commercialize the results of research being done in the laboratory of a Harvard molecular biologist, who would have been a principal in the firm. The University would not have been involved in financing or managing the firm, which would also have been housed separately from the campus. However, Harvard would have derived substantial income if the company proved successful through a gift of 10 to 15 percent of the equity and a royalty on sales. After much debate among the Harvard faculty and educators nationwide, the administration decided not to implement the plan because of concerns about possible adverse impacts on academic values.

Proponents of such arrangements argue that the universities should reap some return from the commercialization of research conducted by their staff. In addition, many universities are pressed for money, and joint ventures or research funding arrangements with industry provide an attractive source of funds for research programs, especially when Federal support may decline. In return, industry would gain access to the kind of fundamental research that is the foundation for innovation and appears to be especially crucial in the field of genetic engineering, where the gap between basic research and product development is smaller than for other fields.

Opponents of these arrangements, especially ones involving significant interaction as in the Harvard plan, fear that the profit-seeking goals of industry may be incompatible with academic values. The following possible adverse impacts, among others, have been articulated: 1) increase in secrecy, to the detriment of the free exchange of ideas so important in academia; 2) discrimination by the university in its hiring and promotion policies in favor of those doing the

revenue-producing research; and 3) distortion in the direction of research and in the training of graduate students.

F. Congress could mandate support for specific research tasks, such as pollution control using microbes.

Investment in creating microbes to degrade pollutants is slow because the potential market is thought to be small and because of the severe liability problems that might arise from intentional release of commercially supplied microbes.

But microbes may be useful in degrading intractable waste and pollutants. Genetic determinants for desired degradation activities may be present in naturally occurring organisms, or scientists may have to combine genes from different sources into a single organism. Current research, however, is limited to isolating organisms from natural sources or from mutated cultures. More elaborate efforts, involving recombinant DNA (rDNA) techniques or other forms of microbial genetic exchange, will require additional effort.

A decision by the Federal Government to support research and to reduce liability concerns is probably needed before the potential of microbial control of pollution can be realized. Federal activity might depend on the results of an evaluation of the technical feasibility of microbial pollution control, which could be made by either an interagency task force or a special commission. If the evaluation is negative, Congress might elect to do nothing to encourage the technology. If the evaluation is positive, Congress might select from the following suboptions:

1. Initiate no research support nor any Federal relief from or limit on potential liability claims. This option would not foreclose private commercial efforts, but it would limit them because of restricted research funds and large liability questions. If sufficiently large markets were anticipated or found, the limitations would be overcome.
2. Initiate research support programs. Research might be directed at problems posed by particular pollutants (contract re-

search). Federal support of biological research is managed by several agencies, and this course would create few, if any, major administrative problems.

3. Guarantee markets for particular products. In addition to patent protection, which would be of little value in the case of an organism purposefully disseminated into the environment, the Government could offer to buy desirable microbes. This public sector market might provide enough incentive to research to make Federal funding unnecessary, or the market incentive and research support might be used jointly.

4. Fix a limit on liability and set up liability insurance, funded partly or wholly by the Government. This option would reduce the financial risk for entrepreneurs who venture to clean up pollutants with microbes. Such an insurance scheme would require that a Federal agency (EPA, for instance) be satisfied that little risk was attendent in the use of the microbe.

5. Arrange a scheme to test micro-organisms for known and anticipated risks before they are released. The Federal Government might have to bear these costs as part of a research program.

6. Leave most efforts to industry and allow each Government agency to develop programs in the fields of genetic engineering and biotechnology as it sees fit.

This option, currently the status quo, seems to be favored by some industry officials. If it is worth doing, they argue, industry will do it. To a large extent, the availability of venture capital in the United States has allowed many companies to pursue projects that are deemed practical and economically important. The production of interferon, insulin, ethanol, ethylene glycol, and fructose are cited as examples of successful applications that were motivated by industry.

Generic research, or research that is fundamentally useful to a broad range of companies, will probably not be undertaken by any one company. When the payoff does not come soon enough, the Government has traditionally taken the responsibility for funding the work. E.g., NIH supported 717 basic research projects involving rDNA in fiscal year 1980 at a cost of $91.5 million. Similarly, high-risk research with high capital costs would be likely targets for Government support.

Leaving all R&D in industry's hands would still produce major commercial successes, but would not ensure the development of generic knowledge or the undertaking of high-risk projects.

Part II
Agriculture

Chapter 8
The Application of Genetics to Plants

Chapter 8

The Application of Genetics to Plants

Perspective on plant breeding

As primitive people moved from hunting and gathering to farming, they learned to identify broad genetic traits, selecting and sowing seeds from plants that grew faster, produced larger fruit, or were more resistant to pests and diseases. Often, a single trait that appeared in one plant as a result of a mutation (see Tech. Note 1, p. 162.) was selected and bred to increase the trait's frequency in the total crop population.

Mendel's laws of trait segregation enabled breeders to predict the outcomes of hybridization and refinements in breeding methods. (See app. II-A.) Consequently, they achieved breeding objectives faster and with more precision, significantly increasing production. During the past 80 years classical applied genetics has been responsible for:

- increased yields;
- overcoming natural breeding barriers;
- increased genetic diversity for specific uses;
- expanded geographical limits where crops can be grown; and
- improved plant quality.

Since the beginning of the 20th century, plant breeders have helped increase the productivity (see Tech. Note 2, p. 162.) of many important crops for food, feed, fiber, and pharmaceuticals by successfully developing cultivars (cultivated varieties) to fit specific environments and production practices. Some breeding objectives have met the needs of the local farmer, while other genetic improvements have been applied worldwide. The commercial development of hybrid corn in the 1920's and 1930's and of "green revolution" wheats in the 1950's and 1960's are but two examples of how plant breeding has affected the supply of food available to the world market. (See Tech. Note 3, p. 162.) A comparison of average yields per acre in

1930 and 1975 in table 24 gives a measure of the contribution of genetics.[1]

It is impossible to determine exactly to what degree applied genetics has directly contributed to increases in yield, because there have been simultaneous improvements in farm management, pest control, and cropping techniques using herbicides, irrigation, and fertilizers. Various estimates, however, indicate that applied genetics has accounted for as much as 50 percent of harvest increases in this century. The yield superiority of new varieties has been a major impetus to their adoption by farmers. Historically, the primary breeding objective has been to maintain and improve crop yields. Other

---

[1]G. F. Sprague, D. E. Alexander, and J. W. Dudley, "Plant Breeding and Genetic Engineering: A Perspective," *BioScience* 30(1):17, 1980.

Table 24.—Average Yield per Acre of Major Crops in 1930 and 1975

	Average yield per acre			Percent increase
	1930	1975	Unit	
Wheat	14.2	30.6	Bushels	115
Rye	12.4	22.0	Bushels	77
Rice.	46.5	101.0	Bushels	117
Corn	20.5	86.2	Bushels	320
Oats	32.0	48.1	Bushels	50
Barley	23.8	44.0	Bushels	85
Grain sorghum.	10.7	49.0	Bushels	358
Cotton	157.1	453.0	Pounds	188
Sugar beets	11.9	19.3	Tons	62
Sugarcane	15.5	37.4	Tons	141
Tobacco	775.9	2,011.0	Pounds	159
Peanuts	649.9	2,565.0	Pounds	295
Soybeans	13.4	28.4	Bushels	112
Snap beans	27.9	37.0	Cwt	33
Potatoes.	61.0	251.0	Cwt	129
Onions	159.0	306.0	Cwt	92
Tomatoes:				
Fresh market	61.0	166.0	Cwt	172
Processing.	4.3	22.1	Tons	413
Hops.	1,202.0	1,742.0	Pounds	45

SOURCE: U.S. Department of Agriculture, *Plant Genetic Resources: Conservation and Use* (Washington, D.C.: USDA, 1979).

breeding objectives are specific responses to the needs of local growers, to consumer demands, and to the requirements of the food processing firms and marketing systems.

Developing new varieties does the farmer little good unless they can be integrated profitably into the farming system either by increasing yields and the quality of crops or by keeping costs down. The three major goals of crop breeding are often interrelated. They are:

- to maintain or increase yields by selecting varieties for:
  —pest (disease) resistance;
  —drought resistance;
  —increased response to fertilizers; and
  —tolerance to adverse soil conditions.
- to increase the value of the yield by selecting varieties with such traits as:
  —increased oil content;
  —improved storage qualities;
  —improved milling and baking qualities; and
  —increased nutritional value, such as higher levels of proteins.
- to reduce production costs by selecting varieties that:
  —can be mechanically harvested, reducing labor requirements;
  —require fewer chemical protectants or fertilizers; and
  —can be used with minimum tillage systems, conserving fuel or labor by reducing the number of cultivation operations.

The plant breeder's approach to commercialization of new varieties

The commercialization of new varieties strongly depends on the genetic variability that can be selected and evaluated. A typical plant breeding system consists of six basic steps:

1. Selecting the crop to be bred.
2. Identifying the breeding goal.
3. Choosing the methodological approach needed to reach that goal.
4. Exchanging genetic material by breeding.
5. Evaluating the resulting strain under field conditions, and correcting any deficiencies in meeting the breeding goal.

6. Producing the seed for distribution to the farmer.

The responsibilities for the different breeding phases are distributed but interactive. In the United States, responsibility for crop improvement through plant breeding is shared by the Federal and State governments, commercial firms, and foundations.[2] Although some specific genes have been identified for breeding programs, most improvements are due to gradual selection for favorable combinations of genes in superior lines. The ability to select promising lines is often more of an art (involving years of experience and intuition) than a science.

The plant breeder's approach is determined for the most part by the particular biological characteristics of the crop being bred—e.g., the breeder may choose to use a system of inbreeding or outbreeding, or the two in combination, as an approach to controlling and manipulating genetic variability. The choice is influenced by whether a particular plant in question naturally fertilizes itself or is fertilized by a neighboring plant. To a lesser degree, the breeding objectives influence the choice of methods and the sequence of breeding procedures.

Repeated cycles of self-fertilization reduce the heterozygosity in a plant, so that after numerous generations, the breeder has homozygous, pure lines that breed true. (See Tech. Note 4, p. 162.) Cross-fertilization, on the other hand, results in a new mixture of genes or increased genetic variability. Using these two approaches in combination produces a hybrid—several lines are inbred for homozygosity and then crossed to produce a parental line of enhanced genetic potential. More vigorous hybrids can be selected for further testing. The effects of hybrid vigor vary and include earlier germination, increased growth rate or size, and greater crop uniformity.

A second method for exchanging or adding genes is achieved through altering the number of chromosomes, or ploidy (see Tech. Note 5, p. 162.), of the plant. Since chromosomes are

---

[2]National Academy of Sciences, *Conservation of Germplasm Resources: An Imperative*, Washington, D.C., 1978.

generally inherited in sets, plants whose ploidy is increased usually gain full sets of new chromosomes. Over one-third of domesticated species are polyploids.[3] Generally, crop improvement due to increased ploidy corresponds to an overall enlargement in plant size; leaves can be broader and thicker with larger flowers, fruits, or seeds. A well-known example is the cultivated strawberry, which has four times more chromosomes than the wild type, and is much fleshier.

Another technique, called *backcrossing*, can improve a commercially superior variety by lifting one or more desirable traits from an inferior one. Generally, this is accomplished by making a series of crosses from the inferior to the superior plant while selecting for the desired traits in each successive generation. Self-fertilizing the last backcrossed generation results in some progeny that are homozygous for the genes being transferred and that are identical with the superior variety in all other respects. Single gene resistance to plant pests and disease-causing agents has been successfully transferred through backcrossing.

Major constraints on crop improvement

Two of the many constraints on crop breeding are related to genetics.

Many important traits are determined by several genes.

The genetic bases for improvements in yield and other characteristics are not completely defined, mainly because most biological traits, such as plant height, are caused by the interaction of numerous genes. Although many—perhaps thousands—of genes contribute to quantitative traits, much variation can be explained by a few genes that have major impact on the observable appearance (phenotype)[4]—e.g., the height of some genetic dwarfs in wheat can be doubled by a single gene. Many other genes contribute to the general health of the

plant (such as resistance to pests and diseases), although some of their contributions are small and difficult to assess. Favorable combinations of genes result in plants well-adapted to particular growing conditions and agronomic practices. With thousands of genes in a single plant contributing to overall fitness, the possible combinations are almost infinite.

Most poor combinations of genes are eliminated by selection of the best progeny; initially favorable combinations are preserved and improved. Literally millions of plants may be examined each year to find particularly favorable genotypes for development into new breeding stocks. Increasingly sophisticated field testing procedures, as well as advanced statistical analyses, are now used to evaluate the success of breeding efforts. Overall yield is still the most important criterion for success, although considerable care is taken to test stress tolerance, pest and disease resistances, mechanical harvestability, and consumer acceptability. Breeding programs with specialized goals often use rapid and accurate chemical procedures to screen lines and progeny for improvements.

Because the vigor of the plant depends on the interaction of many genes, it has been difficult to identify individual genes of physiological significance in whole plants. As a result, many important genes have not been mapped in major crop species. There is little doubt that breeders would select traits like photosynthetic efficiency (the ability to convert light to such organic compounds as carbohydrates) or mineral uptake if the genes could be identified and manipulated in the same ways that resistance is selected for pathogens.

It is uncertain how much genetic variation for improvement exists.

Although the world's germplasm resources have not been completely exploited, it has become more difficult for breeders to improve many of the highly developed varieties now in use—e.g., height reduction in wheat has made enormous contributions to its productivity, but further improvement on this basis seems to be limited.[5] A parallel condition in the potato crop

[3]W. J. C. Lawrence, *Plant Breeding* (London: Edward Arnold Ltd., 1968).

[4]J. N. Thompson, Jr., "Analysis of Gene Number and Development in Polygenic Systems," *Stadler Genetics Symposium* 9:63.

[5]N. F. Jensen, "Limits to Growth in World Food Production," *Science* 201:317, 1978.

Photo credit: U.S. Department of Agriculture

Bundles of wheat showing variance in height

was recognized by the National Research Council's Committee on Genetic Vulnerability of Major Crops:[6]

> If we bear in mind the fairly recent origin of modern potato varieties and that they are, for the most part, derived from the survivors of the late blight epidemics of the 1840's in Europe and North America, it seems likely that the genetic

base was already somewhat narrow by the time modern potato breeding got under way. The five-fold increase in yield resulting from selection during the last 100 years of potato improvement has produced a group of varieties that are genetically similar and unlikely to respond to further selection for yield. In the long run response to selection for other characteristics is also likely to be limited.

As these examples indicate, the level of genetic homogeneity of some crops may make selection for higher yields in general more difficult. Nevertheless, while the genetic basis for overall crop improvement is poorly understood, refinements in plant breeding techniques may increase the potential for greater efficiency in the transfer of genetic information for more precise selection methods, and as a new source of genetic variation.

Besides these two constraints, other pressures and limitations may also affect crop productivity; some are biological (see Tech. Note 6, p. 162.), requiring technological breakthroughs, while others are related to environmental, social, and political factors. (See Tech. Note 7, p. 162.)—e.g., it has been argued that the agricultural rate of growth is declining: In 1976, the U.S. Department of Agriculture (USDA) estimated that the *total-factor productivity* of U.S. agriculture increased by 2 percent per year from 1939 to 1960, but by only 0.9 percent from the period of 1960 to 1970.[7]

---

[6]National Academy of Sciences, *Genetic Vulnerability of Major Crops*, Washington, D.C., 1972.

[7]U.S. Department of Agriculture, Economics, Statistics, and Cooperation Services, *Agricultural Productivity: Expanding the Limits*, Agriculture Information Bulletin 431, Washington, D.C., 1979.

Genetic technologies as breeding tools

The new technologies may provide potentially useful tools, but they must be used in combination with classical plant breeding techniques to be effective. The technologies developed for classical plant breeding and those of the new genetics are not mutually exclusive, they are both tools for effectively manipulating genetic information through methods that have been adapted from genetic recombination observed in nature. Plant breeders have many techniques for artificially controlling pollina-

tion—some are capable of overcoming natural barriers such as incompatibility. Yet even though one new technology—protoplast fusion —allows breeders to overcome incompatibility, the new plant must still be selected, regenerated from single-cell culture, and evaluated under field conditions to ensure that the genetic change is stable and the attributes of the new variety meet commercial requirements. Evaluation is still the most expensive and time-consuming step.

New genetic technologies for plant breeding

The recent breakthroughs in genetic engineering permit the plant breeder to bypass the various natural breeding barriers that have limited control of the transfer of genetic information. While the new technologies do not necessarily offer the plant breeder the radical changes that recombinant DNA (rDNA) technology provides the microbiologist, they will, in theory, speed up and perfect the process of genetic refinement.

The new technologies fall into two categories: those involving genetic transformations through cell fusion, and those involving the insertion or modification of genetic information through the cloning (exactly copying) of DNA and DNA vectors (transfer DNA). Most genetic transformations require that enzymes digest the plant's impermeable cell wall, a process that leaves behind a cell without a wall, or a protoplast. Protoplasts can fuse with each other, as well as with other components of cells. In theory, their ability to do this permits a wider exchange of genetic information.

The approach exploiting the new technologies is usually a three-phase program.

Phase I. Isolated cells from a plant are established in tissue culture and kept alive.
Phase II. Genetic changes are engineered in those cells to alter the genetic makeup of the plant; and desired traits are selected at this stage, if possible.
Phase III. The regeneration of the altered single cells is initiated so that they grow into entire plants.

This approach contains similarities to the genetic manipulation of micro-organisms. However, there is one major conceptual difference. In micro-organisms, the changes made on the cellular level are the goals of the manipulation. With crops, changes made on the cellular level are meaningless unless they can be reproduced in the entire plant. Therefore, unless single cells in culture can be grown into mature plants that have the new, desired characteristics—a procedure which, at this time, has had limited success—the benefits of genetic engineering will

not be widespread. If the barriers can be overcome, the new technologies will offer a new way to control and direct the genetic characteristics of plants.

PHASE I: TISSUE CULTURE TO CLONE PLANTS

Tissue culture involves growing cells from a plant in a culture or medium that will support them and keep them viable. It can be started at three different levels of biological organization: with plant organs (functional units such as leaves or roots);* with tissues (functioning aggregates of one type of cell, such as epidermal cells (outermost layer) in a leaf; and with single cells. Tissue cultures by themselves offer specific benefits to plant breeders; just as fermentation is crucial to microbial genetic technologies, tissue culture is basic to the application of the other new genetic technologies for plants.

The idea of growing cells from higher plants or animals and then regenerating entire plants from these laboratory-grown cells is not new. However, a better scientific understanding now exists of what is needed to keep the plant parts alive.

In tissue culture, isolated single plant cells are typically induced to undergo repeated cell divisions in a broth or gel, the resulting amorphous cell clump is known as a callus. If culture conditions are readjusted when the callus appears, its cells can undergo further proliferation. As the resulting cells differentiate (become specialized), they can grow into the well-organized tissues and organs of a complete normal plant. The callus can be further subcultured, allowing mass propagation of a desired plant.

At this time, it is not uncommon to produce as many as a thousand plants from each gram of starting cells; 1 g of starting carrot callus routinely produces 500 plants. The ultimate goal of tissue culturing is to have these plantlets placed in regular soil so that they can grow and develop into fully functional mature plants. The complete cycle (from plant to cell to plant) permits production of plants on a far more massive scale, and in a far shorter period, than is possible by conventional means. (See table 25 for a

---

*Also referred to as organ culture.

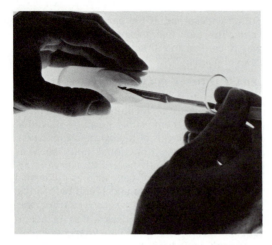

First stage in plant tissue culturing: inoculation of plant tissue

Photo credits: Flow Laboratories

Transfer of plantlets grown on agar to soil

Shows the gradual development of the plant tissue on an agar medium

list of some plants propagated through tissue culture.)

Each of the four stages of the complete cycle—establishment in culture, organogenesis, plantlet amplification, and reestablishment in soil—requires precise biological environments that have to be determined on a species-by-spe-cies basis. However, several commercial uses of tissue culture already exist. (See table 26.)

Storage of Germplasm.—Tissue culture can be used in the long-term storage of specialized germplasm, which involves freezing cells and types of shoots. The culture provides stable genetic material, reduces storage space, and decreases maintenance costs.

Carrot tissues have been frozen in liquid nitrogen, thawed 2 years later, and regenerated into normal plants. This technique has also proved successful with morning glories, sycamores, potatoes, and carnations. Generally, the technique is most useful for plant material that is vegetatively propagated, although if it can be generally applied it could become important for other agriculturally important crops.

Production of Pharmaceuticals and Other Chemicals From Plant Cells.—Because plant cells in culture are similar to microorganisms in fermentation systems, they can be engineered to work as "factories" to produce

Table 25.—Some Plants Propagated Through Tissue Culture for Production or Breeding

Agriculture and horticulture
Vegetable crops
  Asparagus
  Beets
  Brussels sprouts
  Cauliflower
  Eggplant
  Onion
  Spinach
  Sweet potato
  Tomato
Fruit and nut trees
  Almond
  Apple
  Banana
  Coffee
  Date
  Grapefruit
  Lemon
  Olive
  Orange
  Peach
Fruit and berries
  Blackberry
  Grape
  Pineapple
  Strawberry
Foliage
  Silver vase
  Begonia
  Cryptanthus
  Dieffenbachia
  Dracaena
  Fiddleleaf

Pointsettia
Weeping fig
Rubber plant
Flowers
  African violet
  Anthruium
  Chrysanthemum
  Gerbera daisy
  Gloxinia
  Petunia
  Rose
  Orchid
Ferns
  Australian tree fern
  Boston fern
  Maidenhair fern
  Rabbitsfoot fern
  Staghorn fern
  Sword fern
Bulbs
  Lily
  Daylily
  Easter lily
  Hyacinth

Pharmaceutical
Atropa
Ginseng
Pyrethium

Silviculture (forestry)
Douglas fir
Pine
Quaking aspen
Redwood
Rubber tree

SOURCE: Office of Technology Assessment.

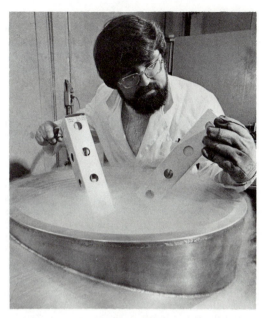

Photo credit: U.S. Department of Agriculture

Seed samples being withdrawn from a tank of liquid nitrogen where they had been stored at − 190° C for 6 months. In addition to testing these seeds for retained germination potential, some will be grown into fully mature plants to determine if any genetic changes occurred during storage

Table 26.—Representative List of Tissue Culture Programs of Commercial Significance in the United States

Industry	Application	Economic benefits
Asparagus industry..........	Rapid multiplication of seed stock	Improved productivity, earliness, and spear quality
Chemical and pharmaceutical .	Biosynthesis of chemicals	Reduced production costs
	Propagation of medicinal plants	High volumes of plants for planting
Citrus industry..............	Virus elimination	Improved quality, high productivity
Coffee industry	Disease resistance breeding	Disease resistance
Land reclamation............	Mass propagation	Availability of select clones of wild species for revegetation
Ornamental horticulture......	Mass propagation	Reduced costs of certain species
		Virus elimination of certain species
		Introduction of new selections
		Increased volumes of difficult selections
Pineapple industry	Mass propagation	Improved quality in higher volumes
Strawberry industry..........	Mass propagation	Rapid introduction of new strains

SOURCE: Office of Technology Assessment.

plant products or byproducts. In recent years, economic benefits have been achieved from the production of plant constituents through cell culture. Among those currently produced commercially are camptothecin (an alkaloid with antitumor and antileukemic activity), proteinase inhibitors (such as heparin), and antiviral substances. Flavorings, oils, other medicinals, and insecticides will also probably be extracted from the cells.

The vinca alkaloids—vincristine and vinblastine, for instance—are major chemotherapeutic agents in the treatment of leukemias and lymphomas. They are derived from the leaves of the Madagascar periwinkle (*Catharanthus roseus*). Over 2,000 kilograms (kg) of leaves are required for the production of every gram of vinca alkaloid at a cost of about $250/g. Plant cells have recently been isolated from the periwinkle, immobilized, and placed in culture. This culture of cells not only continues to synthesize alkaloids at high rates, but even secretes the material directly into the culture medium instead of accumulating it within the cell, thus removing the need for extensive extraction procedures.

Similarly, cells from the Cowage velvetbean are currently being cultured in Japan as a source of L-Dopa, an important drug in the treatment of Parkinson's disease. Cells from the opium poppy synthesize both the plant's normal alkaloids in culture and, apparently, some alkaloids that have not as yet found in extracts from the whole plant.

Another pharmaceutical, diosgenein, is the major raw material for the production of corticosteroids and sex steroids like the estrogens and progestins used in birth control pill. The large tuberous roots of its plant source, *Dioscorea*, are still collected for this purpose in the jungles of Central America, but its cells have been cultured in the laboratory.

Other plant products, from flavorings and oils to insecticides, industrial organic chemicals, and sweeteners, are also beginning to be derived from plants in cell-cultures. Glycyrrhiza, the nonnutritive sweetener of licorice, has been produced in cultures of *Glycyrrhiza glabra*, and

anthraquinones, which are used as dye bases, accumulate in copious amounts over several weeks in cultures of the mulberry, *Morinda citrifolia*.

PHASE II: ENGINEERING CHANGES TO ALTER GENETIC MAKEUP; SELECTING DESIRED TRAITS

The second phase of the cycle involves the genetic manipulation of cells in tissue culture, followed by the selection of desired traits. Tissue culturing, in combination with the new genetic tools, could allow the insertion of new genetic information directly into plant cells. Several approaches to exchanging genetic information through new engineering technologies exist:

- culturing plant sex cells and embryos;
- protoplast fusion; and
- transfer by DNA clones and foreign vectors.

These are then followed by:

- screening for desired traits.

Culturing Plant Cells and Embryos.— Culturing the plant's sex cells—the egg from the ovary and the pollen from the anther (pollen-secreting organ)—can increase the efficiency of creating pure plant lines for breeding. Since sex cells contain only a single set of unpaired chromosomes per cell, plantlets derived from them also contain only a single set. Thus, any genetic change will become apparent in the regenerated plant, because a second paired gene cannot mask its effect. Large numbers of haploid plants (cells contain half the normal number of chromosomes) have been produced for more than 20 species. Simple treatment with the chemical, colchicine, can usually induce them to duplicate their genomes (haploid set of chromosomes)—resulting in fully normal, diploid plants. The only major crop that has been bred by this technique is the asparagus.[8]

If the remaining technical barriers can be overcome, the technique can be used to enhance the selection of elite trees and to create hybrids of important crops. Although still

---

[8]J. G. Torrey, "Cytodifferentiation in Cultured Cells and Tissues," *HortScience* 12(2):138, 1977.

primarily experimental, successful plant sex-cell cultures have been achieved for a variety of important cultivars, including rice, tobacco, wheat, barley, oats, sorghum, and tomato. However, because the technique can lead to bizarre unstable chromosomal arrangements, it has had few applications.

Embryo cultures have been used to germinate, in vitro, those embryos that might not otherwise survive because of basic incompatibilities, especially when plants from different genera are crossed. Embryos may function as starting material in tissue culture systems requiring juvenile material. They are being used to speed up germination in such species as oil palms, which take up to 2 years to germinate under natural conditions.

Protoplast Fusion.—In protoplast fusion, either two entire protoplasts are brought together, or a single protoplast is joined to cell components—or organelles—from a second protoplast. When the components are mixed under the right conditions, they fuse to form a single hybrid cell. The hybrids can be induced to proliferate and to regenerate cell walls. The functional plant cell that results may often be cultured further and regenerated into an entire plant—one that contains a combination of genetic material from both starting plant cell progenitors. When protoplasts are induced to fuse, they can, in theory, exchange genetic information without the restriction of natural breeding barriers. At present, protoplast fusion still has many limitations, mainly due to the instability of chromosome pairing.

Organelles are small, specialized components within the cell, such as chloroplasts and mitochondria. Some organelles, called plastids, carry their own autonomously replicating genes, as a result, they may hold promise for gene transfer and for carrying new genetic information into protoplasts in cultures, or possibly for influencing the functions of genes in the cell nucleus. (See Tech. Note 8, p. 163.)

The feasibility of protoplast fusion has been borne out in recent work with tobacco—a plant that seems particularly amenable to manipulation in culture. An albino mutant of *Nicotiana*

tabacum was fused with a variety of a sexually incompatible *Nicotiana* species. The resultant hybrids were easily recognized by their intermediate light green color. They have now been regenerated into adult plants, and are currently being used as a promising source of hornworm resistance in tobacco plants.

Transfer by DNA Clones and Foreign Vectors.—Recombinant DNA technology makes possible the selection and production of more copies (amplification) of specific DNA segments. Several basic approaches exist. In the "shotgun" approach, the whole plant genome is cut by one or more of the commercially available restriction enzymes. The DNA to be transferred is then attached to a plasmid or phage, which carries genetic information into the plant cell.—E.g., a gene coding for a protein (zein) that is a major component of corn seeds has been spliced into plasmids and cloned in micro-organisms. It is hoped that the zein-gene sequence can be modified through this approach to increase the nutritional quality of corn protein before it is reintroduced into the corn plant.

Foreign vectors are nonplant materials (viruses and bacterial plasmids) that can be used to transfer DNA into higher plant cells. Transformation through foreign vectors might improve plant varieties or, by amplifying the desired DNA sequence, make it easier to recover a cell product from culture. In addition, methods have been discovered that eliminate the foreign DNA from the transformed mixture, leaving only the desired gene in the transformed plant. The most promising vector so far seems to be the tumor-inducing (Ti) plasmid carried by *Agrobacterium tumefaciens*. This bacterium causes tumorous growths around the root crowns of plants. It infects one major group of plants—the dicots (such as peas and beans), so-called because their germinating seeds initially sprout double leaves. Its virulence is due to the Ti plasmid, which, when it is transferred to plant cells, induces tumors. Once inside the cell, a smaller segment of the Ti plasmid, called T-DNA, is actually incorporated into the recipient plant cell's chromosomes. It is carried in this form, replicating right along with the rest of the chromosomal DNA as plant cell proliferation proceeds. Researchers have been wondering whether new genetic material for plant improvement can be inserted into the T-DNA region and carried into plant cell chromosomes in functional form.

Adding foreign genetic material to the T-DNA region has proved successful in several experiments. Furthermore, it has been found that one type of plant tumor cell that contains mutagenized T-DNA can be regenerated into a complete plant. This new discovery supports the use of the *Agrobacterium* system as a model for the introduction of foreign genes into the single cells of higher plants.

Many unanswered questions remain before *Agrobacterium* becomes a useful vector for plant breeding. Considerable controversy exists about exactly where the Ti plasmid integrates into the host plant chromosomes; some insertions might disrupt plant genes required for growth. In addition, these transformations may not be genetically stable in recipient plants; there is evidence that the progeny of Ti-plasmid-containing plants do not retain copies of the Ti sequence. Finally, *Agrobacterium* does not readily infect monocots (a second group of plants), which limits its use for major grain crops.

Another promising vector is the cauliflower mosaic virus (CaMV). Since none of the known plant DNA viruses has ever been found in plant nuclear DNA, CaMV may be used as a vector for introducing genetic information into plant cytoplasm. Although studies of the structural organization, transcription, and translation of the CaMV are being undertaken, information available today suggests that the system needs further evaluation before it can be considered an alternative to the *Agrobacterium* system.

Although work remains to be done on Ti-plasmid and CaMV genetic mechanisms, these systems have enormous potential. Most immediately, they offer ways of examining basic mechanisms of differentiation and genetic regulation and of delineating the organization of the genome within the higher plant cell. If this can be accomplished, the systems may provide a way of incorporating complex genetic traits into whole plants in stable and lasting form.

Screening for Desired Traits.—The benefits of any genetic alteration will be realized only if they are combined with an adequate system of selection to recover the desired traits. In some cases, selection pressures can be useful in recovery.[9] The toxin from plant pathogens, for example, can help to identify disease resistance in plants by killing those that are not resistant. So far, this method has been limited to identifying toxins excreted by bacteria or fungi and their analog; after sugarcane calluses were exposed to toxins of leaf blight, the resistant lines that survived were then used to develop new commercial varieties. In theory, however, it is possible to select for many important traits. Tissue culture breeding for resistance to salts, herbicides, high or low temperatures, drought, and new varieties that are more responsive to fertilizers is currently under study.

Five basic problems must be overcome before any selected trait can be considered beneficial (see figure 28):

- the trait itself must be identified;
- a selection scheme must be found to identify cells with altered properties;
- the properties must prove to be due to genetic changes;
- cells with altered properties must confer similar properties on the whole plant; and
- the alteration must not adversely affect such commercially important characteristics as yield.

While initial screens involving cells are easier to carry out than screening tests involving entire plants, tolerance at the cellular level must be confirmed by inoculations of mature plants with the actual pathogen under field conditions.

PHASE III: REGENERATING WHOLE PLANTS FROM CELLS IN TISSUE CULTURE

New methods are being developed to:

- increase the speed with which crops are multiplied through mass propagation, and
- create and maintain disease-free plants.

Mass Propagation.—The greatest single use of tissue culture systems to date has been for mass propagation, to establish selected

[9]J. F. Shepard, D. Bidney, and E. Shahin, "Potato Protoplasts in Crop Improvement," *Science* 208:17, 1980.

Figure 28.—The Process of Plant Regeneration From Single Cells in Culture

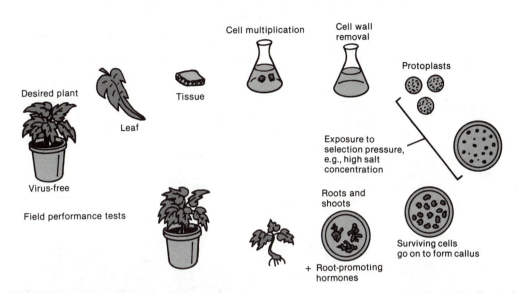

The process of plant propagation from single cells in culture can produce plants with selected characteristics. These selections must be tested in the field to evaluate their performance.

SOURCE: Office of Technology Assessment.

Photo credit: Plant Resources Institute

Multiplying shoots of jojoba plant in tissue culture on a petri dish. These plants may potentially be selected for higher oil content

culture because of the increased speed with sources of improved seed or cutting material. (See table 26.) In some cases, producing plants by other means is simply not economically competitive. A classic example is the Boston fern, which, while it is easy to propagate from runner tips, is commercially propagated through tissue which it multiplies and the reduced costs of stock plant maintenance. A tissue culture stock of only 2 square feet (ft^2) can produce 20,000 plants per month.[10]

Currently, mass production of such cultivars as strawberries (see Tech. Note 9, p. 163.), asparagus, oil palms, and pineapples is being carried out through plant tissue cultures.[11] Very recently, alfalfa was propagated in the same way, giving rise to over 200,000 plants, several thousand of which are currently being tested in field trials. Also, 1,300 oil palms, selected for high yield and disease resistance, are being tested in Malaysia.[12] Other crops not produced by this method but for which cell culture is an important source of breeding variation include

---

[10]D. P. Holdgate, "Propagation of Ornamentals by Tissue Culture," in *Plant Cell, Tissue, and Organ Culture*, J. Feinert and Y. P. S. Bajaj (eds.) (New York: Springer-Verlag, 1977).

[11]T. Murashige, "Current Status of Plant Cell and Organ Cultures," *HortScience* 12(2):127, 1977.

[12]"The Second Green Revolution," special report, *Business Week*, Aug. 25, 1980.

beets, brussels sprout, cauliflower, tomatoes, citrus fruits, and bananas. Various horticultural plants—such as chrysanthemums, carnations, African violets, foliage plants, and ferns—are also being produced by in vitro techniques.

Accelerating propagation and selection in culture is especially compelling for economically important forest species for which traditional breeding approaches take a century or more. Trees that reach maturity within 5 years require approximately 50 years to achieve a useful homozygous strain for further breeding. Species such as the sequoia, which do not flower until they are 15 to 20 years old, require between 1 and 2 centuries before traits are stabilized and preliminary field trials are eval-

uated. Thus, tissue culture production of trees has become an area of considerable interest. Already, 2,500 tissue-cultured redwoods have been grown under field conditions for comparison with regular, sexually produced seedlings. (See app. II-B.) Loblolly pine and Douglas fir are also being cultured; the number of trees that can be grown from cells in 100 liters (l) of media in 3 months are enough to reforest roughly 120,000 acres of land at a 12 x 12 ft spacing.[13] To date, 3,000 tissue-cultured Douglas firs have actually been planted in natural soil conditions. (See figure 29.)

[13]D. J. Durzan, "Progress and Promise in Forest Genetics," in Proceedings, 50th Anniversary Symposium Paper, *Science and Technology . . . The Cutting Edge* (Appleton, Wis: Institute of Paper Chemistry, 1980).

Photo credits: Weyerhaeuser Co.

A plantlet of loblolly pine grown in Weyerhaeuser Co.'s tissue culture laboratory. The next step in this procedure is to transfer the plantlet from its sterile and humid environment to the soil

A young Douglas fir tree propagated 4 years ago from a small piece of seedling leaf tissue. Three years ago this was at the test-tube stage seen in the loblolly pine photograph

Figure 29.—A Model for Genetic Engineering of Forest Trees

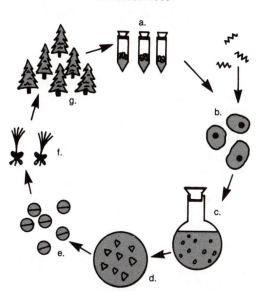

a. Selection of genetic material from germplasm bank

b. Insertion of selected genes into protoplasts

c. Regeneration of cells from protoplasts and multiplication of cell clones

d. Mass production of embryos from cells

e. Encapsulation to form 'seeds'

f. Field germination of 'seeds'

g. Forests of new trees

SOURCE: Office of Technology Assessment.

Creation and Maintenance of Disease-Free Plants.—Cultivars maintained through standard asexual propagation over long periods often pick up viruses or other harmful pathogens, which while they might not necessarily kill the plants, may cause less healthy growth. A plant's true economic potential may be reached only if these pathogens are removed—a task which culturing of a plant's meristem (growing point) and subsequent heat therapy can perform. Not all plants produced through these methods are virus-free, so screening cells for viruses must be done to ensure a pathogen-free plant. In horticultural species, the advantages of

virus-free stock often appear as larger flowers, more vigorous growth, and improved foliage quality.

Today, virus-free fruit plants are maintained and distributed from both private and public repositories. Work of commercial importance has been done with such plants as strawberries, sweet potatoes, citrus, freesias, irises, rhubarbs, gooseberries, lilies, hops, gladiolus, geraniums, and chrysanthemums.[14] Over 134 virus-free potato cultures have also been developed by tissue culture.[15]

Constraints on the new genetic technologies

Although genetic information has been transferred by vectors and protoplast fusion, no DNA transformations of commercial value have yet been performed. The constraints on the successful application of molecular genetic technologies are both technical and institutional.

TECHNICAL CONSTRAINTS

Molecular engineering has been impeded by a lack of understanding about which genes would be useful for plant breeding purposes, as well as by insufficient knowledge about cytogenetics. In addition, the available tools—vectors and mutants—and methods for transforming plant cells using purified DNA are still limited.

Cells carrying traits important to crop productivity must be identified after they have been genetically altered. Even if selection for an identified trait is successful, it must be demonstrated that cells with altered properties confer similar properties on tissues, organs, and, ultimately on the whole plant, and that the genetic change does not adversely affect yield or other desired characteristics. Finally, only limited success has been achieved in regenerating whole plants from individual cells. While the list of plant species that can be regenerated from tissue culture has increased over the last 5 years, it includes mostly vegetables, fruit and

---

[14]M. Misawa, K. Sakato, M. Tanaka, M. Havashi, and H. Samejima, "Production of Physiologically Active Substances by Plant Cell Suspension Cultures," H. E. Street (ed.), *Tissue Culture and Plant Science* (New York: Academic Press, 1974).

[15]Murashige, op. cit.

nut trees, flowers, and foliage crops. Some of the most important crops—like wheat, oats, and barley—have yet to be regenerated. In addition, cells that form calluses in culture cannot always be coaxed into forming embryos, which must precede the formation of leaves, shoots, and roots. Technical breakthroughs have come on a species-by-species basis; key technical discoveries are not often applicable to all plants. And even when the new technologies succeed in transferring genetic information, the changes can be unstable.

The hope that protoplast fusion would open extensive avenues for gene transfer between distantly related plant species has diminished with the observation of this instability. However, if whole chromosomes or chromosome fragments could be transferred in plants where sexual hybridization is presently impossible, the possibilities would be enormous.

INSTITUTIONAL CONSTRAINTS

Institutional constraints on molecular genetics include those in funding, in regulation, in manpower, and in industry.

Federal funding for plant molecular genetics in agriculture has come from the National Science Foundation (NSF) and from USDA. Research support in USDA is channeled primarily through the flexible Competitive Grants Program (fiscal year 1980 budget of $15 million) for the support of new research directions in plant biology. The panel on genetic mechanisms (annual budget less than $4 million) is of particular significance for developing new genetic technologies. The panel's charter specifically seeks proposals on novel genetic technologies. The remaining three panels concerned with plants—nitrogen, photosynthesis, and stress—also support projects to define the molecular basis of fundamental plant properties. The success of the USDA Competitive Grants Program is hard to assess after just 2 years of operation; however, its budget over the past 2 years has severely limited expansion of the program into new areas of research.

Some private institutions[16] argue that the

---

[16]V. Walbot, *Past, Present and Future Trends in Crop Breeding*, Vol. II, Working Papers, *Impact of applied Genetics*, NTIS, 1981.

Competitive Grants Program is shifting support from ongoing USDA programs to new genetics research programs that are not aimed at the important problems facing agriculture today. There is no opposition to supporting the molecular approaches as long as they do not come at the expense of traditional breeding programs, and as long as both molecular biologists and classical geneticists working with major crop plants are assured of enough support to foster research groups of sufficient size.

At present, funds from nine programs at the NSF—primarily in the Directorate for Biological, Behavioral, and Social Sciences—support plant research. The total support for the plant sciences may be as high as $25 million, of which only about $1 million is designated specifically for molecular genetics.

The regulation of the release of genetically altered plants into the environment has not had much effect to date. As of November 1980, only one application [which requested exception from the NIH Guidelines (see ch. 11) to release rDNA-treated corn into the environment] has been filed with the Office of Recombinant DNA Activities (ORDA). Whether regulation will produce major obstacles is difficult to predict at present. It is also unclear whether restrictions will be placed on other genetic activities, such as protoplast fusion. Currently, at least one other nation (New Zealand) includes such restrictions in its guidelines. It is not clear how much the uncertainty of possible ecological disruption and the attendent liability concerns from intentional release of genetically engineered plants has prevented the industrial sector from moving toward commercial application of the new technology.

Only a few universities have expertise in both plant and molecular biology. In addition, only a few scientists work with modern molecular techniques related to whole plant problems. As a result, a business firm could easily develop a capability exceeding that at any individual U.S. university. However, building industrial laboratories and hiring from the universities could easily deplete the expertise at the university level. With the recent investment activity in bioengineering firms, this trend has already begun; in the long-run it could have serious consequences for the quality of university research.

Despite these constraints, progress in overcoming the difficulties is continuing. At the prestigious 1980 Gordon Conference, where scientists meet to exchange ideas and recent findings, plant molecular biology was added to the list of meetings for the first time. In addition, four other recent meetings have concentrated on plant molecular biology.[17] Up to 50 percent of the participants at these meetings came from nonplant-oriented disciplines searching for future research topics. This influx of investigators from other fields can be expected to enrich the variety of approaches used to solve the problems of the plant breeder.

[17]*Genome Organization and Expression in Plants*, NATO symposium held in Edinburgh, Scotland, July 1979; *Genetic Engineering of Symbiotic Nitrogen Fixation and Conservation of Fixed Nitrogen*, June 29-July 2, 1980, Tahoe City, Calif; "Molecular Biologists Look at Green Plants," *Sixth Annual Symposium*, Sept. 29-Oct. 2, 1980, Heidelberg, West Germany; and *Fourth International Symposium on Nitrogen Fixation*, Dec. 1-5, 1980, Canberra, Australia.

Finally, as a general rule, tradeoffs arise in the use of the new technologies that may interfere with their application. It is impossible to get something for nothing from nature—e.g., in nitrogen fixation the symbiotic relationship between plant and micro-organism requires energy from the plant; screening for plants that can produce and transfer the end products of photosynthesis to the nodules in the root more efficiently may reduce inorganic nitrogen requirements but may also reduce the overall yield. This was the case for the high lysine varieties of corn. (See Tech. Note 10, p. 163.) Farmers in the United States tended to avoid them because improving the protein quality reduced the yield, an unacceptable tradeoff at the market price. Thus, unless the genetic innovation fits the requirements of the total agricultural industry, potentials for crop improvement may not be realized.

Impacts of generating new varieties

Progress in the manipulation of gene expression in eukaryotic (nucleus-containing) cells, which include the cells of higher plants, has been enormous. Most of the new methodologies have been derived from fruit flies and mammalian tissue culture lines; but many should be directly applicable to studies with plant genes. There has been great progress in isolating specific RNA from plants, in cloning plant DNA, and in understanding more about the organization of plant genomes. Techniques are available for manipulating organs, tissues, cells, or protoplasts in culture; for selecting markers; for regenerating plants; and for testing the genetic basis of novel traits. So far however, these techniques are routine only in a few species. Perfecting procedures for regenerating single cells into whole plants is a prerequisite for the success of many of the novel genetic technologies. In addition, work is progressing on viruses, the Ti plasmid of *Agrobacterium*, and engineered cloning vehicles for introducing DNA into plants in a directed fashion. There

have been few demonstrations in which the inheritance of a new trait was maintained over several sexual generations in the whole plant.

Because new varieties have to be tested under different environmental conditions once the problems of plant regeneration are overcome, it is difficult to assess the specific impacts of the new technologies.—E.g., it is impossible to determine at this time whether technical and biological barriers will ever be overcome for regenerating wheat from protoplasts. Nevertheless, the impact of genetics on the structure of American agriculture can be discussed with some degree of confidence.

Genetic engineering can affect not only what crops can be grown, but where and how those crops are cultivated. Although it is a variable in production, it usually acts in conjunction with other biological and mechanical innovations, whose deployment is governed by social, economic, and political factors.

Examples of new genetic approaches

The ways in which the new genetic approaches could aid modern agriculture are described in the following two examples:

SELECTION OF PLANTS FOR METABOLIC EFFICIENCY

Because terrestrial plants are immobile, they live and die according to the dictates of the soil and weather conditions in which they are planted; any environmental stress can greatly reduce their yield. The major soil stresses faced by plants include insufficient soil nutrients and water or toxic excesses of minerals and salts. The total land area with these conditions approaches 4 billion hectares (ha), or about 30 percent of the land area of the Earth.

Traditionally, through the use of fertilizers, lime, drainage, or freshwater irrigation, environments have been manipulated to suit the plant. Modern genetic technologies might make it easier to modify the plant to suit the environment.

Many micro-organisms and some higher plants can tolerate salt levels equal to or greater than those of sea water. While salt tolerance has been achieved in some varieties of plants, the classical breeding process is arduous and limited. If the genes can be identified, the possibility of actually transferring those for salt tolerance into plants makes the adaptation of plants to high salt, semiarid regions with high mineral toxicities or deficiencies a more feasible prospect. In the future, selecting among tissue cultures for metabolic efficiency could become important. Tissue culture systems could be used to select cell lines for resistance to salts and for responsiveness to low-nutrient levels or less fertilizer. However, too little is known about the biochemistry and physiology of plants to allow a more directed approach at this time. Chances for success would be increased with a better understanding of plant cell biology.

Such techniques could be applied to agricultural programs in less developed countries, where, commonly, supplies of fertilizers and lime are scarce, the potential for irrigation is small, and adequate support for technological innovation is limited. In addition, the United States itself contains marginal land that could be exploited for forest products and biomass. The semiarid lands of the Southwest, impoverished land in the Lake States, and reclaimed mining lands could become cost-effective areas for production.

NITROGEN FIXATION

It has been known since the early 1800's that biological fixation of nitrogen is important to soil fertility. In fixation, micro-organisms, such as the bacterium *Rhizobium*, transform atmospheric nitrogen into a form that plants can use. In some cases—e.g., with legumes this process occurs through a symbiotic relationship between the micro-organism and the plant in specialized nodules on the plant roots. Unfortunately, the major cereal crops such as wheat, corn, rice, and forage grasses do not have the capacity to fix atmospheric nitrogen, thus are largely dependent on chemically produced nitrogen fertilizers. Because of these crops, it has been estimated that the world demand for nitrogen fertilizers will grow from 51.4 million metric tonnes (1979 estimate) to 144 million to 180 million tonnes by the year 2000.[18] Therefore, geneticists are looking into the possibility that the genes for nitrogen fixation present in certain bacteria (called "nif genes") can be transferred to the major crops.

Laboratory investigation has focused on the molecular biology of nitrogen fixation in the free living bacterium, *Klebsiella pneumoniae*. A cluster of 15 nif genes has been successfully cloned onto bacterial plasmids using rDNA technology. These clones are being used to study the molecular regulation of nif gene expression and the physical organization of the nif genes on the *Klebsiella* chromosome. In addition they have aided the search for nitrogen fixation genes in other bacteria.

It is thought that a self-sufficient package of nitrogen-fixing genes evolved during the course of plant adaptation, and that this unit has been transferred in a functional form to a variety of different bacterial species, including *Klebsiella* and *Rhizobium*. If the right DNA vector can be found, the nif genes might be transferred from bacteria to plants. The chloroplasts, the cauliflower mosaic virus, and the *Agrobacterium* Ti-plasmid are being investigated as possible vectors.

The way that *Agrobacteria*, in particular, infect cells is similar to the way *Rhizobia* infect plants and form nitrogen-fixing nodules. In both cases, the physical attachment between bacterium and plant tissue is necessary for successful infection. In the case of *Agrobacteria*, tumors form when a segment of the Ti-plasmid is inserted into the nuclear genome of the plant cell. Scientists do not yet know exactly how a segment of the rhizobial genome is transferred into the root tissue to induce the formation of nodules; nevertheless, it is hoped that *Agrobacteria* will act as vectors for the introduction and expression of foreign genes into plant cells, just as *Rhizobia* do naturally.

Other researchers have been investigating the requirements for getting nif genes to express themselves in plants. Nif genes from *Klebsiella* have already been transferred into common yeast, an organism that can be grown in environments without oxygen. Unfortunately, the presence of oxygen destroys a major enzyme for nitrogen fixation and severely limits the potential applications in higher plants. Nevertheless, it is hoped that nif gene expression in yeast will be applicable to higher plants.

An approach that does not involve genetic engineering, uses improved *Rhizobia* strains that are symbiotic with soybeans. Through selection, *Rhizobia* mutants are being found that out-perform the original wild strains. Further

---

[18]F. Ausubel, "Biological Nitrogen Fixation," *Supporting Papers: World Food and Nutrition Study* (Washington, D.C.: National Academy of Sciences, 1977).

testing is needed to determine whether the improvement can be maintained in field trials, where the improved strains must compete against wild-type *Rhizobia* already present in the soil.

Another way to improve nitrogen fixation is to select plants that have more efficient symbiotic relationships with nitrogen-fixing organisms. Since the biological process requires a large amount of energy from the plant, it may be possible to select for plants that are more efficient in producing, and then to transfer the end products of photosynthesis to the nodules in the roots. Also existing nitrogen-fixing bacterial strains that can interact with crop plants which do not ordinarily fix nitrogen could be searched for or developed.

Reducing the amount of chemically fixed nitrogen fertilizer—and the cost of the natural gas previously used in the chemical process—would be the largest benefit of successfully fixing nitrogen in crops. Environmental benefits, from the smaller amount of fertilizer runoff into water systems, would accrue as well. But is it difficult to predict when these will become reality. Experts in the field disagree: some feel the breakthrough is imminent; others feel that it might take several decades to achieve.

The refinements in breeding methods provided by the new technologies may allow major crops to be bred more and more for specialized uses—as feed for specific animals, perhaps, or to conform to special processing requirements. In addition, since the populations in less developed countries suffer more often from major nutritional deficiencies than those in industrialized countries, a specific export market of cereal grains for human consumption, like wheat with higher protein levels, may be developed.

But genetic methods are only the tools and catalysts for the changes in how society produces its food; financial pressures and Federal regulation will continue to direct their course. E.g., the automation of tissue culture systems will decrease the labor needed to direct plant propagation and drastically reduce the cost per plantlet to a level competitive with seed prices for many crops. While such breakthroughs may increase the commercial applications of many technologies, the effects of a displaced labor force and cheaper and more efficient plants are hard to predict.

Although it is difficult to make economic projections, there are several areas where genetic technologies will clearly have an impact if the predicted breakthroughs occur:

- Batch culture of plant cells in automated systems will be enhanced by the ability to engineer and select strains that produce larger quantities of plant substances, such as pharmaceutical drugs.
- The technologies will allow development of elite tree lines that will greatly increase yield, both through breeding programs similar to those used for agricultural crops and by overcoming breeding barriers and lengthy breeding cycles. Refined methods of selection and hybridization will increase the potential of short-rotation forestry, which can provide cellulosic substrates for such products as ethanol or methanol.
- The biological efficiency of many economically important crops will increase. Advances will depend on the ability of the techniques to select for whole plant characteristics, such as photosynthetic soil and nutrient efficiency.[19]
- Besides narrowing breeding goals, the techniques will increase the potential for faster improvement of underexploited plants with promising economic value.

For such advances to occur, genetic factors must be selected from superior germplasm, the genetic contributions must be integrated into improved cultural practices, and the improved varieties must be efficiently propagated for distribution.

---

[19]For the soybean and tomato crops, the research area for improved biological efficiency received the *highest* allotment of funds in fiscal year 1978. Total funding was $12.9 million for soybeans and $2.1 million for tomatoes. The second largest category to be funded was control of diseases and nematodes of soybeans at $5.1 million and for tomato at $1.6 million.

Genetic variability, crop vulnerability, and storage of germplasm

Successful plant breeding is based on the amount of genetic diversity available for the insertion of new genes into plants. Hence, it is essential to have an adequate scientific understanding of how much genetic erosion has taken place and how much germplasm is needed. Neither of these questions can be satisfactorily answered today.

The amount of genetic erosion that has taken place

Most genetic diversity is being lost because of the displacement of vegetation in areas outside the United States. The demand for increased agricultural production is a principal pressure causing deforestation of tropical latitudes (see Tech. Note 11, p. 163), zones that contain extensive genetic diversity for both plants and animals.

It has been estimated that several hundred plant species become extinct every year and that thousands of indigenous crop varieties (wild types) have already been lost. However, it is difficult to measure this loss, not only because resources are on foreign soil but because erosion must be examined on a species-by-species basis. In theory, an adequate evaluation would require knowledge of both the quantity of diversity within a species and the breadth of that diversity; this process has in practice, just begun. What is known is that the lost material cannot be replaced.

The amount of germplasm needed

Germplasm is needed as a resource for improving characteristics of plants and as a means for guaranteeing supplies of known plant derivatives and potential new ones. Even if plant breeders adequately understand the amount of germplasm presently needed, it is dif-

ficult to predict future needs. Because pests and pathogens are constantly mutating, there is always the possibility that some resistance will be broken down. Even though genetic diversity can reduce the severity of economic loss, an epidemic might require the introduction of a new resistant variety. In addition, other pressures will determine which crops will be grown for food, fiber, fuel, and pharmaceuticals, and how they will be cultivated; genetic diversity will be fundamental to these innovations.

Even if genetic needs can be adequately identified, there is disagreement about how much germplasm to collect. In the past, its collection has been guided by differences in morphology (form and structure), which have not often been directly correlated to breeding objectives. Furthermore, the extent to which the new genetic technologies will affect genetic variability, vulnerability, or the storage of germplasm, has not been determined. (See app. II-A.)

In addition to its uses in plant improvement, germplasm can provide both old and new products. Recent interest in growing guayule as a source of hydrocarbons (for rubber, energy materials, etc.) has focused attention on plants that may possibly be underutilized. It has been found that past collections of guayule germplasm have not been adequately maintained, making current genetic improvements more difficult. In addition, half of the world's medicinal compounds are obtained from plants; maintaining as many varieties as possible would ensure the availability of compounds known to be useful, as well as new, and as yet undiscovered compounds—e.g., the quinine drugs used in the treatment of malaria were originally obtained from the *Cinchona* plant. A USDA collection of superior germplasm established in 1940 in Guatemala was not maintained. As a consequence, difficulties arose during the Vietnam War when the new antimalarial drugs became less effective on resistant strains of the parasite and natural quinines were once again used.

An important distinction exists between preserving genetic resources in situ and preserving germplasm stored in repositories. Although genetic loss can occur at each location, evolution will continue only in natural ecosystems. With better storage techniques, seed loss and genetic "drift" can be kept to a minimum. Nevertheless, species extinction in situ will continue.

The National Germplasm System

USDA has been responsible for collecting and cataloging seed (mostly from agriculturally important plants) since 1898. Yet it is important to realize that other Federal agencies also have responsibilities for gene resource management. (See table 27.) Over the past century, over 440,000 plant introductions from more than 150 expeditions to centers of crop diversity have been cataloged.

The expeditions were needed because the United States is gene poor. The economically im-

Table 27.—Gene Resource Responsibilities of Federal Agencies

Agency	Type of ecosystems under Federal ownership/control	Responsibilities
U.S. Department of Agriculture Animal & Plant Health Inspection Service	—	Controls insect and disease problems of commercially valuable animals and plants.
Forest Service	Forestlands and rangelands (U.S. National Forest)	Manages forestland and rangeland living resources for production.
Science & Education Administration	—	Develops animal breeds, crop varieties, and microbial strains. Manages a system for conserving crop gene resources.
Soil Conservation Service	—	Develops plant varieties suitable for reducing soil erosion and other problems.
Department of Commerce National Oceanic & Atmospheric Administration	Oceans—between 3 and 200 miles off the U.S. coasts	Manages marine fisheries.
Department of Energy	—	Develops new energy sources from biomass.
Department of Health & Human Services National Institutes of Health	—	Utilizes animals, plants, and micro-organisms in medical research.
Department of the Interior Bureau of Land Management	Forestlands, rangelands, and deserts	Manages forest, range, and desert living resources for production.
Fish & Wildlife Service	Broad range of habitats, including oceans up to 3 miles off U.S. coasts	Manages game animals, including fish, birds, and mammals.
National Park Service	Forestlands, rangelands, and deserts (U.S. National Parks)	Conserves forestland, rangeland, and desert-living resources.
Department of State	—	Concerned with international relations regarding gene resources.
Agency for International Development	—	Assists in the development of industries in other countries including their agriculture, forestry, and fisheries.
Environmental Protection Agency	—	Regulates and monitors pollution.
National Science Foundation	—	Provides funding for genetic stock collections and for research related to gene resource conservation.

SOURCE: David Kafton, National Association for Gene Resource Conservation.

portant food plants indigenous to the continental United States are limited to the sunflower, cranberry, blueberry, strawberry, and pecan. The centers of genetic diversity, found mostly in tropical latitudes around the world, are believed to be the areas where progenitors of major crop plants originated. Today, they contain genetic diversity that can be used for plant improvement.

It is difficult to estimate the financial return from the germplasm that has been collected, but its impact on the breeding system has been substantial. A wild melon collected in India, for instance, was the source of resistance to powdery mildew and prevented the destruction of California melons. A seemingly useless wheat strain from Turkey—thin-stalked, highly susceptible to red rust, and with poor milling properties—was the source of genetic resistance to stripe rust when it became a problem in the Pacific Northwest. Similarly, a Peruvian species contributed "ripe rot" resistance to American pepper plants, while a Korean cucumber strain provided high-yield production of hybrid cucumber seed for U.S. farmers. And a gene for resistance to Northern corn blight transferred to Corn Belt hybrids has resulted in an estimated savings of 30 to 50 bushels (bu) per acre, with a seasonal value in excess of $200 million.[20] (See table 28.)

The effort to store and evaluate this collected germplasm was promoted by the Agricultural Marketing Act of 1946, which authorized regional and interregional plant introduction stations (National Seed Storage Centers) run cooperatively by both Federal and State Governments. The federally controlled National Seed Storage Laboratory in Fort Collins, Colo., was established in 1958 to provide permanent storage for seed. In the 1970's, it was recognized that the system should include clonal material for vegetatively propagated crops, which cannot be stored as seed. Although their storage requires more space than comparable seed stor-

[20]U.S. Department of Agriculture, Agricultural Research Service, *Introduction, Classification, Maintenance, Evaluation, and Documentation of Plant Germplasm*, (ARS) National Research Program No. 20160 (Washington, D.C., U.S. Government Printing Office, 1976).

Table 28.—Estimated Economic Rates of Return From Germplasm Accessions

1. A plant introduction of wheat from Turkey was found to have resistance to all known races of common and dwarf bunts, resistance to stripe rust and flag smut, plus field resistance to powdery and snow mold. It has contributed to many commercial varieties, with estimated annual benefits of $50 million.
2. The highly successful variety of short-strawed wheat, 'Gaines' has in its lineage three plant introductions that contributed to the genes for the short stature and for resistance to several diseases. During the 3 years, 1964-66, about 60 percent of the wheat grown in the State of Washington was with the variety 'Gaines'. Increased production with this variety averaged slightly over 13 million bu or $17.5 million per year in the 3-year period.
3. Two soybean introductions from Nanking and China were used for large-scale production, because they are well-adapted to a wide range of soil conditions. All major soybean varieties now grown in t e Southern United States contain genes from one or both of these introductions. Farm gate value of soybean crop in the South exceeded $2 billion in 1974.
4. Two varieties of white, seedless grapes resulted from crosses of two plant introductions. These varieties ripen 2 weeks ahead of 'Thompson Seedless'. Benefits to the California grape industry estimated to be more than $5 million annually.

SOURCE: U.S. Department of Agriculture, Agricultural Research Service, *Introduction, Classification, Maintenance, Evaluation, and Documentation of Plant Germplasm*, (ARS) National Research Program No. 20160 (Washington, D.C., U.S. Government Printing Office, 1976).

age, 12 new repositories for fruit and nut crops as well as for other important crops, from hops to mint, were proposed by the National Germplasm Committee as additions to the National Germplasm System (see Tech. Note 12, p. 163). (The development of tissue culture storage methods may reduce storage costs for these proposed repositories.)

The National Germplasm System is a vital link in ensuring that germplasm now existing will still be available in the future. However, the present system was challenged after the Southern corn blight epidemic of 1970. Many scientists questioned whether it was large enough and broad enough in its present form to provide the genetic resources that might be needed.

The devastating effects of the corn blight of 1970 actually led to the coining of the term crop vulnerability. During the epidemic, as much as 15 percent of the entire yield was lost. Some fields lost their whole crop, and entire sections of some Southern States lost 50 percent of their corn. Epidemics like this one are, of course, not new. In the 19th century, the phylloxera disease of grapes almost destroyed the wine industry of France, coffee rust disrupted the economy of Ceylon, and the potato famine triggered extensive local starvation in Ireland and mass emigration to North America. In 1916, the red rust destroyed 2 million bu of wheat in the United States and an additional million in Canada. Further epidemics of wheat rust occurred in 1935 and 1953. The corn blight epidemic in the United States stimulated a study that led to the publication of a report on the "Genetic Vulnerability of Major Crops".[21] It contained two central findings: that vulnerability stems from genetic uniformity, and that some American crops are, on this basis, highly vulnerable. (See table 29.)

However, genetic variability is only a hedge against vulnerability. It does not guarantee that an epidemic will be avoided. In addition, pathogens from abroad can become serious problems when they are introduced into new environments. As clearly stated in the study, a triangular relationship exists between host, pathogen, and environment, and the coincidence of their interaction dictates the severity of disease.

---

[21]National Academy of Sciences, *Genetic Vulnerability of Major Crops*, Washington, D. C., 1972.

The basis for genetic uniformity

Crop uniformity results most often from societal decisions on how to produce food. The structure of agriculture is extremely sensitive to changes in the market. Some of the basic factors influencing uniformity are:

* the consumer's demand for high-quality produce;
* the food processing industry's demand for harvest uniformity;
* the farmer's demand for the "best" variety that offers high yields and meets the needs of a mechanized farm system; and
* the increased world demand for food, which is related to both economic and population growth.

New varieties of crops are bred all the time, but several can dominate agricultural production—e.g., Norman Borlaug and his colleagues in Mexico pioneered the "green revolution" by developing high-yielding varieties (HYV) of wheat that required less daylight to mature and possessed stiffer straw and shorter stems. Since the new varieties (see Tech. Note 13, p. 163) gave excellent yields in response to applications of fertilizer, pesticides, and irrigation, the innovation was subsequently introduced into countries like India and Pakistan. When a single

Table 29.—Acreage and Farm Value of Major U.S. Crops and Extent to Which Small Numbers of Varieties Dominate Crop Average (1969 figures)

Crop	Acreage (millions)	Value (millions of dollars)	Total varieties	Major varieties	Acreage (percent)
Bean, dry	1.4	143	25	2	60
Bean, snap	0.3	99	70	3	76
Cotton	11.2	1,200	50	3	53
Corn[a]	66.3	5,200	197[b]	6[c]	71
Millet	2.0	?	—	3	100
Peanut	1.4	312	15	9	95
Peas	0.4	80	50	2	96
Potato	1.4	616	82	4	72
Rice	1.8	449	14	4	65
Sorghum	16.8	795	?	?	?
Soybean	42.4	2,500	62	6	56
Sugar beet	1.4	367	16	2	42
Sweet potato	0.13	63	48	1	69
Wheat	44.3	1,800	269	9	50

[a]Corn includes seeds, forage, and silage.
[b]Released public inbreds only.
[c]There were six major public lines used in breeding the major varieties of corn, so the actual number of varieties is higher.

SOURCE: National Academy of Sciences, *Genetic Vulnerability of Major Crops*, Washington, D.C., 1972.

variety dominates the planting of a crop, there is some loss of genetic variability, the resulting uniformity causes crop vulnerability—and the displacement of indigenous varieties—a real problem.

The rate of adoption of HYVs levels off below 100 percent in most countries, mainly because of the numerous factors affecting supply and demand:[22]

- supply factors:
  - —the present HYVs are not suitable for all soil and climatic conditions;
  - —they require seeds and inputs (such as fertilizers, water, and pesticides) that are either unavailable or not fully utilized by every farmer; and
  - —in some regions, a strong demand still exists for the longer straw of traditional varieties.
- demand factors:
  - —consumers may not prefer the HYVs over traditional food varieties;
  - —Government price policies may not encourage the production of HYVs.

For these and other reasons, countries already using a great deal of HYVs will continue to adopt them more slowly.

Six factors affecting adequate management of genetic resources

1. Estimating the potential value of genetic resources is difficult.

Of the world's estimated 300,000 species of higher plants, only about 1 percent have been screened for their use in meeting the diverse demands for food, animal feed, fiber, and pharmaceuticals.[23] Genetic resources not yet collected or evaluated are valuable until proven otherwise, and the efforts to conserve, collect, and evaluate plant resources should reflect this assumption. This point of view was strongly re-flected in a 1978 recommendation by the National Plant Genetic Resources Board. It's recommendation was that four major areas of genetic storage—collection, maintenance, evaluation, and distribution—be viewed as a "continuum that sets up a gene flow from source to end use".[24]

2. The management of genetic resources is complex and costly.

The question of how much germplasm to collect is difficult and strongly influenced by cost. Thus far, only a fraction of the available diversity has been collected. A better scientific understanding of the genetic makeup and previous breeding history of major crops will help determine just how much germplasm should be collected. Efforts to give priorities for collection[25] have been hindered by the scientific gaps in knowledge about what is presently stored worldwide. And while attempts have been made to estimate the economic return from introduction of specific plants (see table 28), the degree to which agricultural production and stability are dependent on genetic variability has not been adequately analyzed.

Evaluation of genetic characteristics must be conducted at different ecological sites by multi-disciplinary teams. The data obtained will only be useful if adequately assessed and made available to the breeding community (see Tech. Note 14, p. 163).

Germplasm must be adequately maintained to assure viability, "working stocks" must be made available to the breeding community. The primary objective of storing germplasm is to make the genetic information available to breeders and researchers.

3. How much plant diversity can be lost without disrupting the ecological balances of natural and agricultural systems is not known.

[22]D. G. Dalrymple, *Development and Spread of High-Yielding Varieties of Wheat and Rice in the Less Developed Nations*, 6th ed. (Washington D.C.: U.S. Department of Agriculture, Office of International Cooperation and Development in cooperation with U.S. Agency for International Development, 1978).

[23]N. Myers, "Conserving Our Global Stock," *Environment* 21(9):25, 1979.

[24]Report to the Secretary of Agriculture, by the Assistant Secretary for Conservation, Research, and Education based on the deliberations and recommendations, National Plant Genetic Resources Board, July 1978.

[25]Secretariat, International Board for Plant Genetic Resources, *Annual Report 1978*, Rome; Consultative Group on International Agricultural Research, 1979.

The arguments parallel those previously discussed in Congress for protection of endangered species (see Tech. Note 15, p. 163). The last decade has shown that modes of production and development can severely affect the ecological balance of complex ecosystems. What is not known is how much species disruption can take place before the quality of life is also affected.

4. The extent to which the new genetic technologies will affect genetic variability, germplasm storage methodologies, and crop vulnerability has not been determined.

The new genetic technologies could either increase or decrease crop vulnerability. In theory, they could be useful in developing early warning systems for vulnerability by screening for inherent weaknesses in major crop resistance. However, the relationship between the genetic characteristics of plant varieties and their pests and pathogens is not understood (see Tech. Note 16, p. 164).

The new technologies may also enhance the prospects of using variability, creating new sources of genetic diversity and storing genetic material by:

* increasing variability during cell regeneration,
* incorporating new combinations of genetic information during cell fusion,
* changing the ploidy level of plants, and
* introducing foreign (nonplant) material and distantly related plant material by means of rDNA.

With the potential benefits, however, come risks. Because genetic changes during the development of new varieties are often cumulative, and because superior varieties are often used extensively, the new technologies could increase both the degree of genetic uniformity and the rate at which improved varieties displace indigenous crop types. Furthermore, it has not been determined how overcoming natural breeding barriers by cell fusion or rDNA will affect a crop's susceptibility to pests and diseases.

5. Because pests and pathogens are constantly mutating, plant resistance can be broken down, requiring the introduction of new varieties.

Historically, success and failure in breeding programs are linked to pests and pathogens overcoming resistance. Hence, plant breeders try to keep one step ahead of mutations or changes in pest and pathogen populations; a plant variety usually lasts only 5 to 15 years on the market. There is some evidence that pathogens are becoming more virulent and aggressive—which could increase the rate of infection, enhancing the potential for an epidemic (see Tech. Note 17, p. 164).

6. Other economic and social pressures affect the use of genetic resources.

The Plant Variety Protection Act has been criticized for being a primary cause of planting uniform varieties, loss of germplasm, and conglomerate acquisition of seed companies. In its opponents' view, such ownership rights provide a strong incentive for seed companies to encourage farmers to buy "superior" varieties that can be protected, instead of indigenous varieties that cannot. They also make plant breeding so lucrative that the ownership of seed companies, is being concentrated in multinational corporations—e.g., opponents claim that 79 percent of the U.S. patents on beans have been issued to four companies and that almost 50 once-independent seed companies have been acquired by The Upjohn Co., ITT, and others.[26] One concern raised about such ownership is that some of these companies also make fertilizer and pesticides and have no incentive to breed for pest resistance or nitrogen-fixation. For the above reasons, one public interest group has concluded:[27]

> [t]hanks to the patent laws, the bulk of the world's food supply is now owned and developed by a handful of corporations which alone, without any public input, determine which strains are used and how.

Numerous arguments have been advanced against the above position. Planting of a single variety, for instance, is claimed to be a function of the normal desires of farmers to purchase the best available seed, especially in the com-

[26]P. R. Mooney, *Seed of the Earth* (London: International Coalition for Development Action, 1979).
[27]Brief for Peoples' Business Commission as Amicus Curiae, *Diamond v. Chakrabarty*, 100 S. Ct. 2204 (1980), p. 9.

petitive environment in which they operate. Moreover, hybrid varieties (such as corn), are not covered by the plant protection laws; yet they comprise about 90 percent of the seed trade.

As for the loss of varieties by vegetation displacement, statutory protection has been too recent to counter a phenomenon that has occurred over a 30- to 40-year period, and available evidence indicates that some crops are actually becoming more diverse. Since most major food crops are sexually produced, they have only been subject to protection since 1970 when the Plant Variety Protection Act was passed; the first certificates under that Act were not even issued until 1972. Moreover, at least in the case of wheat, as many new varieties were developed in the 7 years after the passage of the Plant Variety Protection Act as in the previous 17.[28]

It is clear that large corporations have been acquiring seed companies. However, the con-

nection between this trend and the plant variety protection laws is disputed. One explanation is that the takeovers are part of the general takeover movement that has involved all parts of the economy during the past decade. Since the passage of the 1970 Act, the number of seed companies, especially soybean, wheat, and cereal grains, has increased.[29] While there were six companies working with soybean breeding prior to 1970, there are 25 at this time.[30]

Thus, to date, although no conclusive connection has been demonstrated between the two plant protection laws and the loss of genetic diversity, the use of uniform varieties, or the claims of increasing concentration in the plant breeding industry; the question is still controversial and these complex problems are still unresolved.

---

[28]H. Rept. No. 96-1115, 96th Cong., 2d sess., p. 5 (June 20, 1980).

[29]Hearings on H.R. 2844, supra *note 35* (Statement of Harold Loden, Executive Director of the American Seed Trade Association).

[30]Brief for Pharmaceutical Manufacturers' Association as Amicus Curiae, *Diamond* v. *Chakrabarty*, 100 S. Ct. 2204 (1980), p. 26.

Summary

The science and structure of agriculture are not static. The technical and industrial revolutions and the population explosion have all contributed to agricultural trends that influence the impacts of the new technologies. Several factors affect U.S. agriculture in particular:

• To some degree, the United States depends on germplasm from sources abroad, which are, for the most part, located in less developed countries; furthermore, the amount of germplasm from these areas that should be collected has not been determined.

• Genetic diversity in areas abroad is being lost. The pressures of urbanization, industrial development, and the demands for more efficient, more intensive agricultural production are forcing the disappearance of biological natural resources in which the supply of germplasm is maintained.

• This lost genetic diversity is irreplaceable.

• The world's major food crops are becoming more vulnerable as a result of genetic uniformity.

The solutions—examining the risks and evaluating the tradeoffs—are not limited to securing and storing varieties of seed in manmade repositories; genetic evolution—one of the keys to genetic diversity and a continuous supply of new germplasm—cannot take place on storage shelves. Until specific gaps in man's understanding of plant genetics are filled, and until the breeding community is able to identify, collect, and evaluate sources of genetic diversity, it is essential that natural resources providing germplasm be preserved.

Issues and Options—Plants

ISSUE: Should an assessment be conducted to determine how much plant germplasm needs to be maintained?

An understanding of how much germplasm should be protected and maintained would make the management of genetic resources simpler. But no complete answers exist; nobody knows how much diversity is being lost by vegetation displacement in areas mostly outside the United States.

OPTIONS:

A. *Congress could commission a study on how much genetic variability is needed or desirable to meet present and future needs.*

A comprehensive evaluation of the National Germplasm System's needs in collecting, evaluating, maintaining, and distributing genetic resources for plant breeding and research could serve as a baseline for further assessment. This evaluation would require extensive cooperation among the Federal, State, and private components linked to the National Germplasm System.

B. *Congress could commission a study on the need for international cooperation to manage and preserve genetic resources both in natural ecosystems and in repositories.*

This investigation could include an evaluation of the rate at which genetic diversity is being lost from natural and agricultural systems, and an estimate of the effects this loss will have. Until such information is at hand, Congress could:

- Instruct the Department of State to have its delegations to the United Nations Educational, Scientific, and Cultural Organization (UNESCO) and United Nations Environmental Program (UNEP) encourage efforts to establish biosphere reserves and other protected natural areas in less developed countries, especially those within the tropical latitudes. These reserves would serve as a source for continued natural mutation and variation.

- Instruct the Agency for International Development (AID) to place high priority on, and accelerate its activities in, assisting less developed countries to establish biosphere reserves and other protected natural areas, providing for their protection, and support associate research and training.

- Instruct the International Bank for Reconstruction and Development (World Bank) to give high priority to providing loans to those less developed countries that wish to establish biosphere reserves and other protected natural areas as well as to promote activities related to biosphere reserve preservation, and the research and management of these areas and resources.

- Make a one-time special contribution to UNESCO to accelerate the establishment of biosphere reserves.

Such measures for in situ preservation and management are necessary for long-term maintenance of genetic diversity. Future needs are difficult to predict; and the resources, once lost are irreplaceable.

C. *Congress could commission a study on how to develop an early warning system to recognize potential vulnerability of crops.*

A followup study to the 1972 National Academy of Science's report on major crop vulnerability could be commissioned. Where high genetic uniformity still exists, proposals could be suggested to overcome it. In addition, the avenues by which private seed companies could be encouraged to increase the levels of genetic diversity could be investigated. The study could also consider to what extent the crossing of natural breeding barriers as a consequence of the new genetic technologies will increase the risks of crop vulnerability.

ISSUE: What are the most appropriate approaches for overcoming the various technical constraints that limit the success of molecular genetics for plant improvement?

Although genetic information has been transferred by vectors and protoplast fusion, DNA transformations of commercial value have not yet been performed. Molecular engineering has been impeded by the lack of vectors that can transfer novel genetic material into plants, by insufficient knowledge about which genes would be useful for breeding purposes, and by a lack of understanding of the incompatibility of chromosomes from diverse sources. Another impediment has been the lack of researchers from a variety of disciplines.

OPTIONS:

A. *Increase the level of funding for plant molecular genetics through:*
   1. *the National Science Foundation (NSF),and*
   2. *the Competitive Grants Program of the U.S. Department of Agriculture (USDA).*

B. *Establish research units devoted to plant molecular genetics under the auspices of the National Institutes of Health (NIH), with empha-*

sis on potential pharmaceuticals derived from plants.

C. *Establish an institute for plant molecular genetics under the Science and Education Administration at USDA that would include multidisciplinary teams to consider both basic research questions and direct applications of the technology to commercial needs and practices.*

The discoveries of molecular plant genetics will be used in conjunction with traditional breeding programs. Therefore, each of the three options would require additional appropriations for agricultural research. Existing funding structures could be used for all three, but institutional reorganization would be required for options B and C. The main argument for increasing USDA funding is that it is the lead agency for agricultural research, for increasing NSF and NIH funding, that they currently have the greatest expertise in molecular techniques. Option C emphasizes the importance of the *interdisciplinary* needs of this research.

Technical notes

1. A recent example of such a mutation was the opaque-2 gene in corn, which was responsible for increasing the corn's content of the amino acid lysine.
2. There is disagreement about what is meant by productivity and how it is measured. Statistical field data can be expressed in various ways—e.g., output per man-hour, crop yield per unit area, or output per unit of total inputs used in production. A productivity measurement is a relationship among physical units of production. It differs from measurements of efficiency, which relate to economic and social values.
3. Nevertheless, some parts of the world continue to lack adequate supplied of food. A recent study by the Presidential Commission on World Hunger[31] estimates that "at least one out of every eight men, women, and children on earth suffers malnutrition severe enough to shorten life, stunt physical growth, and dull mental ability."
4. In theory, pure lines produce only identical *gametes*, which makes them true breeders. Successive cross-breeding will result in a mixture of gametes with varying combinations of genes at a given locus on homologous chromosomes.

5. Normally, chromosomes are inherited in sets. The more frequent diploid state consists of two sets in each plant. Because chromosome pairs are homologous (have the same linear gene sequence), cells must maintain a degree of genetic integrity between chromosome pairs during cell division. Therefore, increases in ploidy involve entire sets of chromosomes—diploid (2-set) is manipulated to triploid (3-set) or even to tetraploid (4-set).
6. The estimated theoretical limit to efficiency of photosynthesis during the growth cycle is 8.7 percent. However, the record U.S. State average (116 bu/acre, Illinois, 1975) for corn, having a high photosynthetic rate in comparison to other major crops, approaches only 1 percent efficiency.[32] Since a major limiting step in plant productivity lies in this efficiency for the photosynthetic process, there is potential for plant breeding strategies to improve the efficiency of photosynthesis of many other important crops. This would have a tremendous impact on agricultural productivity.
7. It is difficult to separate social values from the economic structures affecting the productivity of American agriculture. Social pressures and decisions are complex

[31]Report of the Presidential Commission on World Hunger, *Overcoming World Hunger: The Challenge Ahead*, Washington, D.C., March 1980.

[32]Office of Technology Assessment, U.S. Congress, *Energy for Biological Processes, Volume II: Technical Analysis* (Washington, D.C.: U.S. Government Printing Office, July 1980).

and integrated—e.g., conflicts between developing maximum productivity and environmental concerns are typified by the removal of effective pesticides from the market. Applications of existing or new technologies may be screened by the public for acceptable environmental impact. Conflict also exists between higher productivity and higher nutritive content in food, since selection for one often hurts the other.

8. A critical photosynthetic enzyme (ribulose biphosphate carboylase) is formed from information supplied by different genes located independently in the chloroplast (a plastid) and the nucleus of the cell. It is composed of two separate protein chains that must link together within the chloroplast. The larger of these chains is coded for by a gene in the chloroplast—and it is this gene that has been recently isolated and cloned. The smaller subunit, however, derives from the plant nucleus itself. This cooperation between the nucleus and the chloroplast to produce the functional expression of a gene is an interesting phenomenon. Because it exists, the genetics of the cell could be manipulated so that cytoplasmically introduced genes can influence nuclear gene functions. Perhaps most importantly at this stage, plastid genes are prime candidates to clarify the basic molecular genetic mechanisms in higher plants.

9. The advantages to using mass propagation techniques for strawberry plants are that those produced from tissue culture are virus-free, and a plantlet produced in tissue culture can produce more shoots or runners for transplanting.

The disadvantages are that during the first year the fruit tends to be smaller and, therefore, less commercially acceptable; the plants from tissue culture may have trouble adapting to soil conditions, which can affect their vigor, especially during the first growing season; and the price per plantlet ready for planting from tissue culture systems may be more expensive than commercial prices for rooted shoots or runners bought in bulk.

10. Wheat protein is deficient in several amino acids, including lysine. Considerable attention has been devoted in the past 5 to 10 years to improving the nutritional properties of wheat. Thousands of lines have been screened for high protein, with good success, and high lysine genes with poor success. Some high protein varieties have been developed, but adoption by the farmer has been mediocre at best, partly because of reduced yield levels. There are some exceptions;—e.g., the Variety "Plainsman V" has maintained *both* high protein and yield levels, which indicates tha there is no consistent relationship between low protein and high yields in some varieties.

11. Some 42 percent of the total land area in the tropics, consisting of 1.9 billion hectares, contains significant forest cover. It is difficult to measure precisely the amount of permanent forest cover that is being lost; however, it has been estimated that 40 percent of "closed" forest (having a continuous closed canopy) has already been lost, with 1 to 2 percent cleared annually.

If the highest predicted rate of loss continues, half of the remaining closed forest area will be lost by the year 2000.[33] The significance of this loss is expressed by Norman Myers in his report, *Conversion of Tropical Moist Forests*, prepared for the Committee on Research Priorities in Tropical Biology of the National Academy of Science's National Research Council: "Extrapolation of figures from well-known groups of organisms suggest that there are usually twice as many species in the tropics as temperate regions. If two-thirds of the tropical species occur in TMF (tropical moist forests), a reasonable extrapolation from known relationships, then the species of the TMF should amount to some 40 to 50 percent of the planet's stock of species—or somewhere between 2 million and 5 million species altogether. In other words, nearly half of all species on Earth are apparently contained in a biome that comprises only 6 percent of the globe's land surface. Probably no more than 300,000 of these species—no more than 15 percent and possibly much less—have ever been given a Latin name, and most are totally unknown."[34]

12. In 1975, the Committee estimated that $4 million would be necessary for capital costs of each repository, with recurring annual expenses of $1.4 million for salaries and operations. USDA has allocated $1.16 million for its share of the construction costs for the first facility to be constructed at the Oregon State University in Corvallis.

13. High yielding varieties (HYVs) can be defined as potentially high-yielding, usually semidwarf (shorter than conventional), types that have been developed in national research programs worldwide. Wheat varieties were developed by the International Maize and Wheat Improvement Center and rice varieties by International Rice Research Institute. Many improved varieties of major crops of conventional height are not currently considered HYV types, but they have often been incorporated into HYV breeding. HYVs, because of biological and management factors, rarely reach their full harvest potential.

14. Although the National Germplasm System sucessfully handles some 500,000 units to meet annual germplasm requests, many accessions—like the 35,000 to 40,000 wheat accessions stored at the Plant Genetics and Germplasm Institute at Beltsville, Md.—have yet to be examined. Furthermore, the varieties released for sale by the seed companies are not presently evaluated for their comparative genetic differences.

15. For comparison, the National Germplasm System functions on less than $10 million annually, whereas the Endangered Species Program had a fiscal year 1980 budget of over $23 million. The funds allocated to the En-

[33]Report to the President by a U.S. Interagency Task Force on Tropical Forests, *The World's Tropical Forests: A Policy, Strategy, and Program for the United States*, State Department publication No. 9117, Washington, D.C., May, 1980.

[34]N. Myers, *Conversion of Tropical Moist Forests*, report for the Committee on Research Priorities in Tropical Biology of the National Research Council, National Academy of Sciences, Washington, D.C., 1980.

dangered Species Program are used for such activities as listing endangered species, purchasing habitats for protection, and law enforcement.

16. The uses of pest-resistant wheat and corn cultivars on a large scale for both diseases and insects are classic success stories of host- plant resistance. However, recent trends in the Great Plains Wheat Belt are disturbing. The acreage of Hessian fly-resistant wheats in Kansas and Nebraska has decreased from about 66 percent in 1973 to about 42 percent in 1977. Hessian fly infestations have increased where susceptible cultivars have been planted. In South Dakota in 1978, in an area not normally heavily infested, an estimated 1.25 million acres of spring wheat were infested resulting in losses of $25 million to $50 million. An even greater decrease in resistant wheat acreage is expected in the next 2 to 5 years as a result of releases of cultivars that have improved agronomic traits and disease resistance but that are susceptible to the Hessian fly. Insect resistance has not been a significant component of commercial breeding programs.[35]

17. Expressed in genetic terms, cases exist "where the introduction of novel sources of major gene resistance into commercial cultivars of crop plants has resulted in an increase in their frequency of corresponding virulence genes in the pathogen".[36] This has been reported in Australia with wheat stem rust, barley powdery mildew, tomato leaf mold, and lettuce downy mildew. Evidence suggests that there is considerable gene flow in the various pathogen populations—e.g., asexual transfer can quickly alter the frequency of virulence genes. Furthermore, pressures brought about in the evolutionary process have developed such a high degree of complexity in both resistance and virulence mechanisms, that breeding approaches, especially those only using single gene resistance, can be easily overcome.

[35]Office of Technology Assessment, U.S. Congress, *Pest Management Strategies in Crop Protection* (vol. I, Washington, D.C.: U.S. Government Printing Office, October 1979), p. 73.

[36]R. C. Shattock, B. D. Janssen, R. Whitbread, and D. S. Shaw, "An Interpretation of the Frequencies of Host Specific Phenotypes of *Phytophthora infestans* in North Wales," *Ann. Appli. Biol.* 86:249, 1977.

Advances in Reproductive Biology and Their Effects on Animal Improvement

Chapter 9

Advances in Reproductive Biology and Their Effects on Animal Improvement

Background

During the past 30 years, new technologies have led to a fundamental shift in the way the United States produces meat and livestock. One set of these technologies—the subject of this section—uses knowledge of the reproductive process in farm animals to increase production. The impacts of existing breeding technologies have been great, and much progress is still possible through their continued use. The development of new technologies is inevitable as well.

In a market economy like that of the United States, the factor that most influences the adoption of technology is economics. New technologies in reproductive physiology will be used widely only if they increase the efficiency of breeding programs—i.e., only if they provide greater control over breeding than present methods do, and only if the economic advantages of the increased control can be recovered.*

But economic factors are not the only ones that influence technological change—e.g., poultry and livestock production have influenced and have been influenced by:

- Government regulation such as meat grading standards;
- increased awareness of health effects, such

*As discussed in app. III-B, very early adopters of a technology often do so for other than economic reasons.

as from the use of antibiotics in livestock feed;
- environmental concerns, such as the problems of waste removal, especially near factory farms;
- the growth of knowledge, in—e.g., the reproductive processes of farm animals and the accuracy of evaluating the genetic merit of breeding animals; and
- complementary technologies such as refrigerated storage and transportation.

New technologies, from breeding to food delivery systems, have reshaped the traditional American farm into a modern production system that is increasingly specialized, capitalized, and integrated with off-farm services. Applied genetics in animal production has been one of the forces behind these changes. The technologies that have sprung from it include not only the new, esoteric techniques for cellular manipulation discussed in other parts of this report, but also more well-known technologies, like artificial insemination.*

*Technologies selected for discussion in this part of the report involve direct manipulation of sex cells. More speculative technologies for manipulations at the subcellular level are assessed here as well. No effort was made to cover all technologies with potential for improving the genetic qualities of livestock—e.g., management techniques like estrus detection and pregnancy diagnosis were omitted, as were various other methods for improving reproduction efficiency.

The scientific era in livestock production

Producing purebred beef livestock has been the dominant breeding objective throughout most of the 20th century. The open range of the American West and Southwest—the "romantic" era in beef cattle production—lasted until about 1890. (See figure 30.) Then the range was

fenced-in and the longhorn was replaced with new breeds by the turn of the century—the beginning of the "purebred" era.

Pedigree records and visual comparison of conformation to breed type were the basic tools

Figure 30.—Eras in U.S. Beef Production

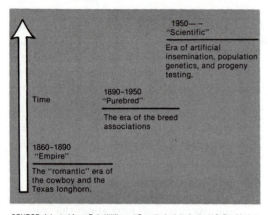

SOURCE: Adapted from R. L. Willham, "Genetic Activity in the U.S. Beef Industry," Journal paper No. J-7923 of the Iowa Agriculture and Home Economics Experiment Station, Ames, Iowa, Project No. 2000, n.d. See also Yao-chi Lu and Leroy Quance, *Agriculture Productivity: Expanding the Limits*, USDA, ESCS, Agriculture Information Bulletin 431.

of breeding programs. They were reinforced by an institutional system of breed associations, and yearly competitions at county fairs and stock shows, and by import regulations and prohibitions against artificial insemination (AI) that restricted innovation. In rearing animals for sale to the slaughterhouse, early breeders and farmers more often than not were satisfied with producing a calf or pig that survived, weaned early, and grew rapidly. Because of the high rate of newborn deaths, the production of an "average" animal was a considerable achievement in its own right; the intricacies of sophisticated breeding methods were beyond the capacity of small operations and were difficult to carry out on large spreads. Producing a prize-winning purebred was left to the farmer with the time, money, or luck to breed animals that met the strict standards of the breed associations and the trained eyes of the judges at stock shows.

During the first half of the 20th century, breeding objectives became more complex; farmers and breeders began to look at qualities other than mere external physical attributes. Breeding for multiple-purposes led directly to the beginning of the "scientific" era in breeding.

The increased use of AI for dairy cattle,

which took place about 30 years ago—the beginning of the scientific era—was an uncertain start for applied genetics in animal breeding. While practitioners and purchasers of AI were quick to grasp its promise of immediate benefits, and while using AI was cheaper than owning a bull, its expected genetic effects were not realized immediately. Dairymen had assumed that semen from bulls selected from the best herds and chosen on the basis of ancestral performance would result in rapid genetic improvement. They were wrong; progress was much less than projected. Because milk production is a sex-limited trait, records on female relatives were needed for the evaluation of sires. Unfortunately, the records on relatives were usually limited to comparisons within one herd, were confounded by management and other environmental factors, and were weakened by small sample sizes. The major factor responsible for the difference between top- and mediocre-performing herds turned out to be management, not genetics; separating the effects of genetics from the effects of generally improved husbandry was extremely difficult.

Controlled breeding

The objective of any breeding program is to increase production. The scientific era has provided the breeder with a variety of new technologies that help in manipulating and controlling the reproductive processes of the animals to increase genetic gain. The breeder's basic tool is selection, or deciding which animals to mate—e.g., in beef cattle, a breeder can now select for a wide variety of performance or economic traits. (See table 30.) However, simply "breeding better beef cattle" is not a workable objective from a manager's point of view. Tender meat, lean steaks and roasts, high fertility, or heavy weight at weaning are all specific, measurable objectives of breeding.[1][2] Other goals, such as those pertaining to temperament, disease resistance, food efficiency, and carcass quality,

---

[1] T. C. Cartwright, "Selection Criteria for Beef Cattle for the Future," *Journal of Animal Science* 30:706, 1970.

[2] Larry V. Cundiff and Keith E. Gregory, *Beef Cattle Breeding*, USDA, Agriculture Information Bulletin No. 286, revised November 1977.

Table 30.—Heritability Estimates of Some Economically Important Traits

Trait	Heritability
Calving interval (fertility)	10%
Birth weight	40
Weaning weight	30
Cow maternal ability	40
Feedlot gain	45
Pasture gain	30
Efficiency of gain	40
Final feedlot weight	60
Conformation score:	
Weaning	25
Slaughter	40
Carcass traits:	
Carcass grade	40
Ribeye area	70
Tenderness	60
Fat thickness	45
Retail product (percent)	30
Retail product (pounds)	65
Susceptibility to cancer eye	30

SOURCE: Larry V. Cundiff and Keith E. Gregory, *Beef Cattle Breeding*, USDA, Agriculture Information Bulletin No. 286, revised November 1977, p. 9.

may also have economic value,[3] but they are much harder to measure.

The extent to which important economic or performance traits are genetically determined and heritable varies from trait to trait and from animal to animal. (See table 30.) Heritability is defined as the percentage of the difference among animals in performance traits passed from parent to offspring*—e.g., bulls and heifers with superior weight at weaning might average 5 pounds (lb) more than their herdmates. Because weaning weight has an average heritability estimate of 30 percent, the offspring of these top performing animals can be expected to average 1.5 lb heavier at weaning than their contemporaries (0.30 × 5 = 1.5). This improvement can normally be expected to be permanent and cumulative as it is passed on to the next generation. The improvement accumulates like compound interest in a savings account; gains made in each generation are compounded on the gains of previous generations.

[3]Michael I. Lerner and H. P. Donald, *Modern Developments in Animal Breeding* (New York: Academic Press, 1966).

*Heritability and genetic association are important in decisions about individual matings. Most breeding programs are concerned with spreading genetic gain rapidly throughout a population (herd, flock); thus two other refinements for selection enter the picture—generation interval, and selection differential.

Like land, equipment, and cash, breeding stock represents capital available to the commercial farmer. Because all inputs must be used efficiently, modern herd or flock managers cannot afford to leave reproduction to chance mating in the pen or on the range. These pressures for efficient production have been described as follows:[4]

> Where dairymen are judged by the number of cows milked in an hour, there is no place for the slow milking cow or the man who will patiently milk her out. There is no place for the time-consuming hurdle flock of sheep, for the small flock of chickens maintained under extensive conditions, or for the sow that must be watched while she farrows. By degrees all classes of stock are being subjected to selection which favors animals that need a minimum of individual attention.

The scientific basis for modern breeding has developed slowly over the last century. Applied genetics—one part of today's programs—has helped modernize livestock and poultry breeding by elaborating on the variation of continuously distributed traits in a population; carrying over what was known about rapidly reproducing laboratory species, like fruit flies or mice, to the much slower reproduction of large farm animals; and developing the statistical techniques for predicting breeding values or merit and analyzing breeding programs.[5]

Two examples show the power of breeding tools and the increased efficiency and productivity of today's breeders' stocks.

- Over the past 30 years, the average milk yield of cows in the United States has more than doubled. At the same time, the number of dairy cows in the United States has been reduced by more than 50 percent. (See figure 31.) Of this increase in output and efficiency, more than one-fourth can be attributed to permanent genetic change for at least one breed (Holsteins) participating in the Dairy Herd Improvement Program. (See figure 32.)
- Poultry production in the United States has become the most intensive industry among

[4]Ibid., p. 20.
[5]Ibid., p. 126.

Figure 31.—Milk Yield/Cow and Cow Population, United States, 1875-1975

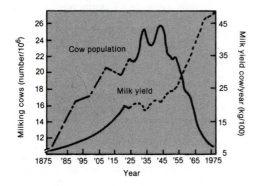

SOURCE: J. T. Reid, "Progress in Dairy Cattle Production," *Agricultural and Food Chemistry: Past, Present, and Future*, R. Teranishi (ed.) (Westport, Conn.: Avi Press, 1978).

Figure 32.—Milk Production per Cow (Holsteins) in 1958-78 (New York and New England)

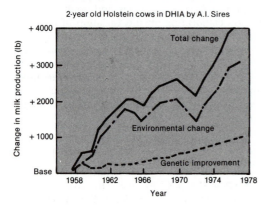

SOURCE: R. H. Foote, Department of Animal Science, Cornell University, Ithaca, N.Y. from unpublished data of R. W. Everett, Cornell University.

those for farm species. For turkeys, the use of AI in breeding for breast meat has been so successful that commercial turkeys can no longer breed naturally. The big-breasted male, even when inclined to do so, finds it physically impossible to mount the female. As a result, a full 100 percent of the commercial turkey flock in the United States is replaced each year using AI. In

other species of poultry as well, production processes have become equally efficient. As A. W. Nordskog has noted:

Compared with the breeding of other economically important animals, poultry breeding has been the first to leave the farm . . . to become part of a sophisticated breeding industry. On a commercial level, chickens have been the first to be commercially exploited by the application of inbreeding-hybridization techniques, as earlier used in corn, as well as by methods of selective improvement using the principles of quantitative genetics. Thus, the poultry industry, compared to other animal industries, seems to have been the quickest to apply modern methods of genetic improvement, including the employment of formally trained geneticists to handle breeding technology plus the use of computers and other modern business methods.[6]

Scientific production

Farm resources include land, labor, capital, and, increasingly, new knowledge. Today, those who innovate recapture the costs of innovating by maintaining output while lowering costs or by increasing output while holding costs down. Some results of the drive toward efficiency have included increasing specialization, intensified use of capital and land relative to labor, and integration of production phases.

Poultry and livestock operations have slowly become specialized over the past 50 years. The farmer who used to do his own breeding, raising, feeding, and slaughtering is disappearing. Now, the beef cattle industry in the United States consists of: the purebred breeder who provides breeding stock, the commercial producer, the feeder, the packer, and the retailer. Similar specialization has occurred for most other species—e.g., less than 15 primary breeders maintain the breeding stock that produces the 3.7 billion chickens consumed each year in the United States. The emergence of other specialized services—such as AI providers, manage-

[6]A. W. Nordskog, "Success and Failure of Quantitative Genetic Theory in Poultry" in *Proceedings of the International Conference on Quantitative Genetics*, Edward Pollacket, et al. (ed.) (Amers, Iowa: Iowa State University Press, 1977), pp. 47-51.

ment consultants, equipment manufacturers—has accelerated the trend toward specialization, and has given the commercial operator more time to concentrate on his specific contribution to the chain of production.

Intensification is the increasing use of some inputs to production in comparison to others. Increasing the use of land and capital relative to labor describes the development of U.S. agriculture, including livestock raising, in this century. The "factory" farm typifies this trend. Herds and flocks are bred, born, and raised in enclosed areas, never seeing a barnyard or the open range. The best examples of land- and capital-intensive systems are those of poultry (layers, broilers, and turkeys), confined hog production, drylot dairy farming, and some veal production.

The greater use of land has been encouraged by several factors, including improved corn production for confined hog feeding, programs of preventive medicine curtailing the spread of diseases in close spaces, and environmental control (light, temperature, water, humidity) to increase output under closely controlled conditions. However, extensive ranching for beef and sheep is still common in the United States; the difficulties associated with detecting estrus ("heat") in these species and their relatively slow rates of reproduction have made it uneconomical to invest in them the capital necessary for intensive farming. Furthermore, beef and sheep on extensive systems forage on marginal land that might otherwise have no use. Beeflot feeding, or the fattening of cattle before slaughter at a centralized location, is the only aspect of the beef industry that is land-intensive; in 1977, approximately one-fourth of U.S. beef cattle were "fed."[7]

Linking phases of production to eliminate waste or inefficiencies in the system has progressed with great speed. For some species, such linkages now extend from breeding to the supermarket (and, in the case of fast food chains, to the dinner table). Integration includes the linking of supply industries (feeds, medicines, breeding stock) with production and then with marketing services (slaughtering, dressing, packaging). Entire industries and the Government in combination have produced a complex chain of operations that makes use of Government inspectors, the pharmaceutical industry, equipment manufacturers, the transportation industry, and the processed feed industry in addition to the traditional commercial farmer.

Because of this complex linkage, meat grades, cuts, and packaging have become fairly standard in the American supermarket. Shoppers have come to expect these standards; consumers wanting special services have learned to pay more for them. Thus, the American farm has changed radically over the past 30 years. This change has been described as follows:[8]

> As farming enterprises grow larger, their management have to equip themselves with information and resort to technologists to help them reach decisions and plan for more distant goals. Industrial developments of this kind widen the range of farming activities, since the old style farmer, sensitive to local markets and operating on hunches, remains as a contrast to those for whom farming is rapidly becoming more of a programme than a way of life.

Resistance to change

New technologies in U.S. agriculture and new ways of producing food and fiber have been both a cause and an effect of the movement from farms to cities in the 20th century. Commercial farmers, operating on thin or nonexistent profits and under extreme competition, have had strong reason to innovate. They have been forced by the availability of new technologies either to do so or to watch their potential earnings go to the neighboring farmer. Various policies that have been adopted to soften the impacts of the "technological treadmill," have somewhat slowed the exodus from the farms. They may have been adopted for social reasons, but they have also become increasingly costly to society. The taxpayer pays for them; the consumer pays as well for every failure to innovate on the farms.

---

[7]Lyle P. Schertz, et al., *Another Revolution in U.S. Farming?* USDA, ESCS, Agricultural Economic Report No. 441, December 1979.

---

[8]Cundiff, et al., op. cit., p. 9.

Besides a lack of capital or a lack of interest in innovating, some farmers have resisted applied genetics because efficiency is not their most important priority. This attitude has been described as follows:[9]

> It is easy to see why breeders are unreceptive to the science of genetics. The business of breeding pedigree stock for sale is not just a matter of heredity, perhaps not even predominantly so. The devoted grooming, feeding and fitting, the propaganda about pedigrees and wins at fairs and shows, the dramatics of the auction ring, the trivialities of breed characters, and the good company of fellow breeders, constitute a vocation, not a genetic enterprise.

Farmers are traditionally an independent group. Many believe that they may not directly recapture the benefits of participating in a breeding program based on genetics; having no records on one's animals is often preferable to discovering proof that one's herd is performing poorly. On the other hand, one impact of AI has been to demonstrate to farmers the value of adopting new technologies. Furthermore, the economic reward of production records has increased, since AI organizations purchase only dairy sires with extensive records on relatives.

Some future trends

Applied genetics in poultry and livestock breeding comprise a group of powerful technologies that have already strongly influenced prices and profits. Nevertheless, the effect of genetics is only just beginning to be felt; much improvement remains to be made in all species. It has been observed that modern genetics:[10]

> . . . provides a verifiable starting point for the development of the complex breeding operation that many populations now require . . . (which) are as far removed from simple selection as the motor car is from the bicycle.

Of these technologies, some are already in regular use, some are in the process of being applied, and others must await further research and development before they become generally available.

Societal pressures are one of the many factors that influence the introduction of these technologies. Several developments around the world will have a clear impact on innovation in general and on genetics in particular:

- An expanding population, with its growing demand for food products of all kinds.
- The growth in income for parts of the population, which may increase the demand for sources of meat protein.
- Increasing competition for the consumer's dollar among various sources of protein, which could reduce demand for meat.
- Increasing competition for prime agricultural land among agricultural, urban, and industrial interests. Less-than-prime land may also be brought back into production as demand rises, and the same pressures may cause land prices to rise high enough to encourage greater, or intensified, use of land in livestock production.
- Increasing demand for U.S. food and fiber products from abroad, leading to opportunities for increased profits for successful producers.

Changes like these will strongly affect the way American farmers produce food and fiber products. The economics of efficiency and a growing world population will continue to place pressure on the agricultural sector to innovate. In animals and animal products, efficiencies will be found in all steps of production. Efforts will be made to increase the number of live births and to reduce neonatal calf fertility, presently one of the costliest steps—in terms of animals lost—throughout the world. Estimates of the potential monetary benefits of the application of knowledge obtained from prior research in reproductive physiology range as high as $1 billion per year. Another area for great economies in production is genetic gain. Much genetic progress remains to be made in all species.

Certain technologies promise to increase the ability of farmers to capitalize on the genetic improvement of economically important traits. Suppliers of genetic material (semen, embryos) will focus increased attention on the value of their products for sale both in the United States and abroad.

[9]Ibid., p. 170.
[10]E. P. Cunningham, "Current Developments in the Genetics of Livestock Improvement," in *15th International Conference on Animal Blood Groups and Biochemistry, Genetics* 7:191, 1976.

The development and application of certain key technologies will affect related technologies—e.g., the availability of reliable estrus detection and estrus synchronization methods should increase the use of AI and embryo transfer in beef and dairy cattle, thereby spreading genetic advantage. Further progress in the freezing of embryos should facilitate the genetic evaluation of cows and heifers.

Other trends that may influence technological change include the shifting availability of research funds, changing consumer tastes, and growth of regulations (for instance, stricter controls on environmental quality or hormonal treatments). The expansion of an animal rights movement may influence the degree to which confinement housing, and therefore controlled breeding, is acceptable. And increased energy costs may either encourage development of the technologies (through efforts for greater efficiency) or discourage them (through greater use of forage and extensive systems).

Technologies

Sexual reproduction is a game of chance. Because sperm and ova each contain only a random half of the genes of each parent, the number of possible combinations that can result is nearly infinite. Some progeny are likely to survive and reproduce; others die either before birth or without producing offspring.

The great variation achieved through sexual reproduction produces certain animals that satisfy the needs and desires of the breeder far more than others. On the other hand, the offspring of these outstanding animals are usually less so than their parents, although they are generally still above average.

Animal breeders have invested great effort in improving succeeding generations of domestic animals, both by limiting the differences due to the chance associated with sexual reproduction and by taking advantage of the favorable combinations that occur. Examples of these efforts include keeping records, establishing progeny testing schemes, amplifying the reproduction of outstanding individuals by AI and embryo transfer, and establishing inbred lines to capitalize on their more reliable ability to transmit characteristics to their offspring.

Because of these efforts, and because dairy cattle breeders have adopted innovative technologies through the years, far more is known about reproduction in the cow than in farm animals. The demand for milk and has provided an impetus for the speedy introduction of technologies that might prove economically advantageous.

Several observations can be made about the state of the art for 16 technologies that enhance the inherited traits of animals. (See also II-C.)

The technologies are at different search and development.

The practice of AI in dairy greatest practical impact of nologies used in the bree contrast, not a single far cessfully raised after fertilization and em fulness of several production, such and nuclear lative at this

The usef species

tional dairy herd.) And economics can also play a role; in general, the lower an animal's value, the less practical the investment in the technologies, some of which are relatively expensive.

Several technologies are critical to the introduction of others.

A methodology that could reliably induce estrus synchronization increases the economic feasibility of AI and embryo transfer. Likewise, the refinement of embryo storage and other freezing techniques would advance the development of those technologies still being developed, like sex selection and embryo transfer. Advances in in vitro fertilization will be especially useful to a better understanding of basic reproductive processes and therefore to the development and application of the more speculative technologies.

The technologies interrelate.

All the technologies combined make possible almost total control of the reproductive process of the farm animal: a cow embryo donor may be superovulated and artificially inseminated with stored, frozen sperm; the embryos may be recovered, then stored frozen or transferred directly to several recipient cows whose estrous cycles have been synchronized with that of the donor to insure continued embryonic development. Before the transfer, a few cells may be taken for identification of male or female chromosomes as a basis for sex selection. Finally, embryos may be transferred to each recipient in an effort to obtain twins. (See figure 33.)

Techniques not yet commercially applicable require embryo transfer in order to be useful include in vitro fertilization, parthenogenetic production of identical twins, cloning, chimeras, and rDNA technology.

The technologies described in this section are increase the reproductive efficiency to improve their genetic merit, general knowledge of the repro-for a variety of reasons, including specific human medical problems, fertility regulation and better fertility.

Technologies that are presently useful

SPERM STORAGE

The sperm of most cattle can be frozen to $-196°$ C, stored for an indefinite period, and then used in in vivo fertilization. Although many of the sperm are killed during freezing, success rates [or successful conceptions (table 31)] combined with other advantages of the technologies are enough to ensure widespread use of the technology. Short-term sperm storage (for one day or so) is also well-developed and widely used.

The major advantages of storing sperm are the increased use of desirable sires in breeding (see figure 34), the ease of transport and spread of desirable germplasm throughout the country and the world, and the savings from slaughtering the bull after enough sperm has been collected. The sperm can also be tested for venereal and other diseases before it is used. Therefore, the use of sperm banks is expected to increase. Little change is anticipated in semen processing, other than the continued refinement of freezing protocols, which differ for each species.

ARTIFICIAL INSEMINATION

The manual placement of sperm into the uterus has played a central role in the dissemination of valuable germplasm throughout the world's herds and flocks. Virtually all farm species can be artificially inseminated, although use of the technology varies widely for different species—e.g., 100 percent of the Nation's domestic turkeys are produced via AI compared with less than 5 percent of beef cattle. Even honey-

Table 31.—Results of Superovulation in Farm Animals

	Average number ovulations normally expected	Number of ovulations with superovulation
Cow	1	6-8
Sheep	1.5	9-11
Goat	1.5	13
Pig	13	30
Horse	1	1

SOURCE: George Seidel, Animal Reproduction Laboratory, Colorado State University, Fort Collins, Colo.

Figure 33.—The Way the Reproductive Technologies Interrelate

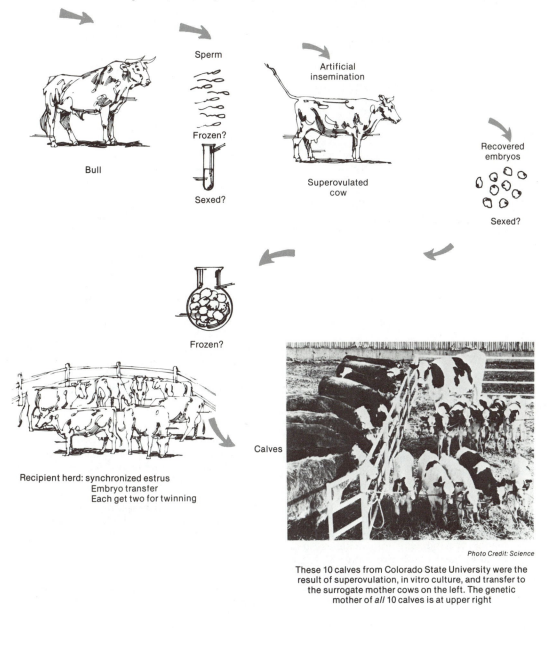

Photo Credit: Science

These 10 calves from Colorado State University were the result of superovulation, in vitro culture, and transfer to the surrogate mother cows on the left. The genetic mother of *all* 10 calves is at upper right

SOURCE: Office of Technology Assessment.

Figure 34.—Change in the *Potential* Number of Progeny per Sire per Year From 1939 to 1979

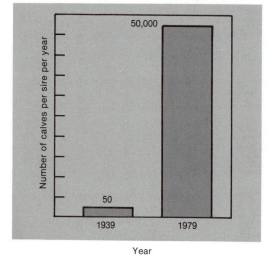

Year

SOURCE: R. H. Foote, Department of Animal Science, Cornell University, Ithaca, N.Y., unpublished data.

bees and fish can now be artificially inseminated.

It permits the widespread use of germplasm from genetically superior sires. It saves the farmer the cost of maintaining his own sires and is valuable in disease control, especially when germplasm, rather than animals, is imported or exported. An important barrier to the wider use of AI, especially in producing beef cattle, is the need for application of reliable estrus detection and estrus synchronization technologies.

An expanded role for AI in the future will depend on the availability of accurate information about the genetic value of sperm available for insemination. A nationwide information system for evaluating germplasm presently exists for only one species, dairy cattle.

ESTRUS SYNCHRONIZATION

Estrus, or "heat," is the period during which the female will allow the male to mate with her. The synchronization of estrus in a herd, using various drug treatments, greatly enhances AI and other reproduction programs.

Federal regulations that limit the use of prostaglandins or progestogens to induce synchronized estrus in horses and nonlactating cows are the major barrier to more widespread use of existing technology.

SUPEROVULATION

Superovulation is the hormonal stimulation of the female, resulting in the release from the ovary of a larger number of ova than normal. (See table 31.) Combined with AI and embryo transfer of the fertilized ova into surrogate mothers, superovulated ova can result in the production of normal offspring with the same rates of success as those following normal ovulation.

The greatest barrier to superovulation is that the degree of success cannot be predicted for an individual animal. Other barriers include widely varying quality of hormone batches for ovulation treatment, Food and Drug Administration (FDA) restrictions, and lack of data from which to judge the effects of repeated superovulation.

In the future, increased understanding of basic physiological mechanisms will facilitate efforts to improve the technology. It has additional commercial potential for sheep and cattle husbandry, and much current effort is directed towards developing and testing a commercial procedure.

EMBRYO RECOVERY

The ability to collect fertilized ova from the oviducts or uterus is a necessary step for embryo transfer or storage and for many experiments in reproductive biology. The technology is especially important for research into producing identical twins, performing embryo biopsies for sex determination, and other projects. Combining superovulation, artificial insemination, and embryo recovery makes it possible to collect embryos from a young heifer before reaching puberty. When some disorder has damaged the oviducts or uterus, embryo recovery from a valuable animal makes procreation possible.

Both surgical and nonsurgical methods are currently in use. Surgical recovery is necessary for sheep, goats, and pigs; such operations are limited by the development of scar tissue. Nonsurgical embryo recovery is preferred for the cow and the single ovulation of the horse. The approach is especially important in dairy cattle, since it can be performed on the farm without interrupting milk production.

No significant advances can be predicted for the immediate future.

EMBRYO TRANSFER

Embryos can be removed from one animal and implanted into the oviduct or uterus of another. Both surgical and nonsurgical methods are currently in use, though success rates of the latter are much lower.

The technology can obtain offspring from females unable to support a pregnancy, increasing the number of offspring from valuable females and introducing new genes into pathogen-free herds. Because more offspring can be obtained from the donor, undesirable recessive traits can be rapidly detected. The technology is also used, along with short- or long-term storage of the embryos, as a means of transporting germplasm rather than the whole animal. Current barriers to its further use are the costs in personnel and equipment, especially for surgical procedures, and the provision of suitable recipients for a successful transfer.

The use of embryo transfer should increase in the future, especially with animals of high value. Nonsurgical methods will increasingly replace surgical ones, especially for cows and horses. A role for embryo transfer can also be predicted in progeny testing of females, obtaining twins in beef cows, obtaining progeny from prepubertal females, and in combination with in vitro fertilization and a variety of manipulative treatments (production of identical twins, selfing or combining ova from the same animal, genetic engineering).

EMBRYO STORAGE

The ability to store embryos increases the advantages of embryo transfer procedures, lowers the cost of transporting animal germplasm, and reduces the need to synchronize estrus in recipients. It will also be important in the study and control of genetic drift in animals.

Adequate culture systems exist for short-term storage of embryos. They have been developed by trial-and-error and are not optimally defined for farm species at present. Nevertheless, cow embryos have been stored for 3 days in the tied oviduct of a rabbit.

Long-term storage, or freezing of embryos, exists, but protocols need to be improved. As many as two-thirds of the stored embryos die with present methods. However, for some uses embryo freezing is already profitable.

In the future, the development of precise embryo culture technology would help the development of all technologies involving the longed manipulation of gametes and emb as outside the reproductive tract. Eventually, as freezing technology improves, nearly all embryos taken from cattle in North America will be stored, rather than transferred immediately. It appears that embryos successfully stored will survive for several centuries and possibly for millenia.

SEX SELECTION

The ability to determine the sex of the unborn, or of sperm at fertilization, will have numerous practical and experimental applications. The most reliable method is karyotyping, by means of which nearly two-thirds of embryos can be sexed. Another method, which tries to identify sex-specific products of certain genes, is under development. A reliable method for separating male-producing sperm from female-producing sperm has not been achieved, though several patents are held on various tests of this type.

Before any method has any practical effect on the production of farm animals it must become simple, fast, inexpensive, reliable, and harmless to the embryo. The present state of the art is largely a consequence of research in male fertility and in sperm survival after frozen storage.

TWINNING

Twins can be artificially induced by using either embryo transfer or hormonal treatments. The first approach is more effective. Selection among female sheep for natural twin production has been very rewarding, while selection for twinning in other species has not received much attention.

Twinning in nonlitter-bearing species would greatly improve the feed conversion ratio of producing an extra offspring. The most important barriers, besides the high cost of embryo transfer techniques, include extra attention needed for the dam during gestation, parturition, and lactation.

More speculative technologies

IN VITRO FERTILIZATION

The manual joining of egg and sperm outside the reproductive tract has, for some species, been followed by successful development of the embryo through gestation to birth. The species include, at this writing, the rabbit, mouse, rat, and human. Consistent and repeatable success with in vitro fertilization in farm species has not yet been accomplished. The cases of reported success of in vitro fertilization, embryo reimplantation, and normal development in man are beginning to be documented in the scientific literature.

The in vitro work to date has attempted to develop a research tool so that the physiological and biochemical events of fertilization could be better understood. Despite the wide public attention it has received in the recent past, the technology is not perfected and will have little practical, commercial effect in producing individuals of any species in the near future.

Practical applications would include: a means of assessing the fertility of ovum and sperm; a means of overcoming female infertility by embryo transfer into a recipient animal; and, when coupled with storage and transfer, a means of facilitating the union of specific ova and sperm for production of individual animals with predicted characteristics.

Many of the practical applications should become available within the next 10 to 20 years. Further development, along with the storage of gametes, should allow fertilization of desired crosses. This technology may be combined with genetic engineering and sperm sexing in the more distant future.

PARTHENOGENESIS

Parthenogenesis, or "virgin birth," is the initiation of development in the absence of sperm. It has not been demonstrated or described for mammalian species, and the best available information indicates that the maintenance of parthenogenetic development to produce normal offspring in mammals is presently impossible.

CLONING

The possibility of producing genetically identical individuals has fascinated both scientists and the general public. As far as livestock are concerned, there are several ways to obtain genetically identical animals. The natural way is through identical twins, although these are rare in species other than cattle, sheep, and primates. For practical purposes, highly inbred lines of some mammals are already considered genetically identical; first generation crosses of these lines are also considered genetically identical and do not suffer from the depressive effect of inbreeding.

Laboratory methods for producing clones include dividing early embryos. The results of recent experiments in the production of identical offspring using these techniques are shown in table 32.

Another methodology involves the insertion of the nucleus of one cell into another, either before or after the original genetic complement of the "receiver" cell is destroyed. Researchers have found in certain amphibia that nuclear transplantation from a body cell of an embryo into a zygote can lead to the development of a sexually mature frog.

Table 32.—Experimental Production of Identical Offspring

Methodology	Result
Dividing 2-cell embryo in half	1 pair identical mouse twins
Dividing morulae[a] in half	8 pairs of identical mouse twins
Dividing 2-cell embryos in half	5 pairs of identical sheep twins
Dividing 4-cell embryos in four parts	1 set identical sheep quadruplets

[a]An embryo with 16 to 50 cells; resembles a mulberry.

SOURCE: Benjamin G. Brackett, School of Veterinary Medicine, University of Pennsylvania, Kennett Square, Pa.

The ideal technique for making genetic copies of any given adult mammal involves inserting the nucleus from a body cell (not a sex cell) from an adult individual into an ovum. Achieving this will probably take years, if indeed it is possible at all, since there is some evidence that most adult body cells are irreversibly differentiated.*

Serious technical barriers must be overcome before advantages in animal production can be foreseen.

CELL FUSION

This technology fuses two mature ova or fertilizes one ovum with another. Combining ova from the same animal is called "selfing." The combination of ova has resulted in very early development of the transferred embryo, but no further development has been reported.

Cell fusion technology may someday prove useful for transferring genetic material from a somatic cell into a fertilized single-cell embryo for the purpose of cloning. Selfing would rapidly result in pure genetic (inbred) lines for use as breeding stocks. The technique could also lead to the rapid identification of undesirable recessive traits that could be eliminated from the species.

CHIMERAS

The production of chimeras requires the fusion of two or more early embryos or the addition of extra cells to blastocysts. These genetic components may be from closely related but different species.

Live chimeras between two species of mouse have been produced. However, practical applications of chimera technology to livestock are not obvious at this stage of development. The main objective of this research is to provide a genetic tool for a better understanding of development and maternal-fetal interactions.

RECOMBINANT DNA AND GENE TRANSFER

The mechanics of directly manipulating the DNA molecules of farm animals have not yet been worked out. However, cells from mice have been mixed with pieces of chromosomal DNA, which became stably associated with the cells' own DNA. In addition, on September 3, 1980, the successful introduction of foreign DNA into mouse embryos was announced. The embryos were implanted into surrogate mothers who gave birth to mice containing altered DNA. Whether or not the DNA was *active* is unknown at this writing.

Knowledge of the genetics of farm animals must improve before rDNA or other gene transfer methods will be of practical benefit in producing meat and livestock products. Before genes can be altered they must be identified, and gene loci on chromosomes must be mapped. Work toward this goal has begun only recently and rapid progress cannot be anticipated. Multivariate genetic determinants of characteristics are anticipated to be the rule.

---

*In January 1981, it was reported that body cells from a very early embryo could act as donors of nuclei for cloned mice.

Genetics and animal breeding

Two characteristics distinguish the reproduction of farm animals from that of single-cell organisms: animal reproduction is sexual—male and female germ cells must be brought together to initiate pregnancy and produce offspring; and animal reproduction is slower (the generation interval is longer), thus the economic benefits of specific gene lines may take years to be captured. These two characteristics limit the speed and extent to which genetic improvements can be made. Reliable information about the genetic value of particular individuals is the key to overcoming limitations, for it can simplify specific breeding decisions and spread desirable genes throughout the Nations's herds and flocks.

The use of applied genetics for farm species is indirect. Breeders do not work with individual genes; rather, they must accept a genetic package that includes both beneficial and harmful traits.[11] The breeder's most important capital is embodied in the animals with which he works. To upgrade this capital, to increase the genetic value of his breeding stock, the breeder must have reliable information on the genetic value of the germplasm he is considering introducing. Since an individual farmer usually does not have the resources to collect and process data on performance of individuals outside his own herds, he must turn to outside sources of information when deciding which new germplasm to introduce.

The requirements of such an information system are extensive. In the United States today, only one such system exists. The National Cooperative Dairy Herd Improvement Program (NCDHIP) is a model program that could be adapted to other species where the benefits from advanced technologies would be enhanced by availability of populationwide data.

The National Cooperative Dairy Herd Improvement Program

Over the past 50 years, the U.S. dairy industry has used test records of individual animals to help in breeding decisions. NCDHIP is a nationwide program for collecting, analyzing, and disseminating information on the performance of dairy cattle.[12] It is the result of a memorandum of understanding among Federal and State agencies, local dairymen, and industry groups across the United States.

In NCDHIP, local Dairy Herd Improvement Association (DHIA) officials go to the dairies to collect the performance data on individual animals. These data then become part of the *Official Dairy Recordkeeping Plans*. The data are standard for all participating herds across the United States. They are sent to the Animal Improvement Programs Laboratory (AIPL) at USDA in Beltsville, Md., which analyzes them and incorporates them into the "USDA-DHIA

---

[11]Philip Handler, *Biology and the Future of Man* (New York: Oxford University Press, 1970), pp. 555-557.

[12]For a complete history of performance testing of dairy cattle in the United States, see: Gerald J. King, *The National Cooperative Dairy Herd Improvement Program*, Dairy Herd Improvement Letter 49, No. 4, July 1973, USDA, ARS.

Sire Summary List," published biannually. These summaries are public information.

In addition to the official plan, NCDHIP also includes several unofficial plans, which have less stringent regulations for data collection but which offer each dairyman a comparison of his herds with other herds across the Nation. The results of unofficial plans are not intended to be used as guidelines for selecting germplasm from outside one's herd.

The following characteristics contribute to NCDHIP's success:

- *It is a cooperative program;* no group or individual is forced to participate. Nevertheless, it has successfully brought together individuals, State and Federal agencies, breed associations, and professional and scientific societies for the pursuit of a common goal. It is almost totally financed by the dairymen themselves. In the national coordinating group, all those with an interest in the industry have a voice in formulating policy for the program.
- *It is flexible;* a dairyman can use the performance records from the unofficial plans to evaluate the animals within his herd or he can turn to the official sire summaries to make comparisons with participating herds throughout the Nation. These data are useful both for comparing the performance of one's herd and breed with others and for selecting new germplasm for introduction into the herd.
- *Its data are regarded as impartial;* disinterest on the part of the local DHIA official who collects the data and the high security surrounding the processed information are central to the program's success. AIPL's analyses and sire summaries are respected both nationally and internationally, in no small part because of freedom from commercial pressures.

Approximately 36,000 herds with almost 2.8 million cows were enrolled in the official plans of NCDHIP in 1979. In each of 18 years recorded between 1961 and 1978, cows enrolled in the Official Dairy Recordkeeping Plans in NCDHIP have outproduced cows not enrolled by over

4,000 lb of milk per lactation. In the testing year (1977-78), the superiority surpassed 5,000 lb per cow. This 5,000-lb superiority represents 52 percent more milk per lactation. The increases in production per cow result from improvement in both management techniques and genetic producing ability.

Several factors influence the rates of participation in the NCDHIP from State to State, from region to region, and from breed to breed. In some States, expansion of NCDHIP membership is not a high priority of the State Cooperative Extension Service. In some areas, the relative importance of dairying as an enterprise is low; therefore, a strong local DHIA organization does not exist. Likewise, in areas where dairying is a part-time operation, dairymen have less time and initiative for participating in the program (although many participate in NCDHIP's unofficial plans). Where dairymen rely on their own bulls and use little AI in breeding, progeny testing is extremely limited. No single factor causes dairymen in some States to take greater advantage of the superior germplasm available to them. The importance of strong national leadership cannot be overemphasized in explaining the great differences among breeds in participation rates. (See table 33.) Farsighted leadership played a large role in developing the genetic gain of Holsteins, which represent 90 percent of the U.S. dairy herd today.

The genetic gains resulting from NCDHIP are impressive, suggesting a model for spreading genetic superiority throughout the Nation's other herds. NCDHIP also shows the importance

Table 33.—National Cow-Year and Averages for All Official Herd Records, by Breed May 1, 1978–Apr. 30, 1979

Breed	Cow-years (#)	Milk (lb)	Fat (%)	Fat (lb)
Ayrshire	17,135	11,839	3.96%	469
Guernsey	57,577	10,858	4.64	504
Holstein	2,297,684	15,014	3.64	547
Jersey	89,449	10,231	4.90	501
Brown swiss.....	24,247	12,368	4.04	500
Milking shorthorn	2,130	10,451	3.65	381
Mixed and others .	83,139	13,077	3.80	497

SOURCE: U.S. Department of Agriculture, Science and Education Administration, *Dairy Herd Improvement Letter* 55, #2, December 1979, pp. 5-6.

of combining reliable evaluation of germplasm with the use of reproductive technologies. These technologies are of only academic interest when they are used alone; it is when superior germplasm can be spread throughout the Nation that the American consumer benefits.

Other species

Progeny testing schemes for other species are not as developed as they are for dairy cattle. There are several reasons for this lack of testing:

- *Difficulty in establishing a selection objective around which to design a testing program.* Milk yield and fat content were obvious traits for selection in dairy cattle. Other species have no such simple traits for selection. It has been observed that, "The lack of definition of economic selection objectives in a precise, soundly based manner is one of the serious weaknesses of much animal breeding of the past."[13]

- *Differences in management systems.* Artificial insemination is essential to the introduction of superior germplasm; where it is difficult to practice AI, elaborate testing schemes are not useful—e.g., in the Nation's beef herds, progeny testing will have to await more widespread use of AI. Though swine are increasingly raised in confined housing systems, poor fertility of boar sperm after freezing and thawing and heat detection difficulties have limited the use of AI.

- *Conflicting commercial interests.* Beef bulls, for example, continue to be sold to some extent on the basis of fancy pedigrees and lines, with relatively little objective information on their genetic merit. Although some genetic improvement programs now exist, the beef breed associations may not support interbreed comparisons because some breeds would show up poorly.

- *Conflicts between short- and long-term gains.* Cross-breeding for the benefits of hybrid-

[13]L. E. A. Rowson, "Techniques of Livestock Improvement," *Outlook on Agriculture* 6:108, 1970.

ization is particularly attractive to owners of commercial herds and flocks who constantly replace their stocks. This genetic improvement is noncumulative—the improvement does not continue from generation to generation. At present, no strong interest exists for improving the Nation's beef herd as a whole, and the individual breeder cannot effectively evaluate the germplasm available to him.

Swine.—There is no Nationwide testing program for hogs in the United States.* However, a study of needed research prepared by the USDA in 1976 noted that the production rate of approximately 13 pigs marketed per sow per year in the United States could be significantly improved. The biological potential is at least 20 to 25 pigs per year. Similarly, a successful breeding program, along with other managerial changes, could reduce the fat and increase the lean content of pork by as much as 10 to 15 lb per carcass.

The ARS study noted that ". . . an area that warrants particular attention is the development of a comprehensive national swine testing program leading to the identification, selection, and use of genetically superior boars, together with guidelines for the development and use of sow productivity and pig performance indexes."[14] In the case of swine, the increased use of intensive housing, which allows reproductive control, should increase the impetus for progeny testing. Likewise, pinpointing areas where considerable improvement remains to be made should lead to the identification of selection objectives.

Beef.—After World War II, a few breeders became increasingly interested in problems of inbreeding and the economic costs of dwarfism. By that time, some had been trained in genetics and some breed associations and State agencies initiated localized testing programs for these traits. In 1967, a "Beef Improvement Federation"

of local and breed groups was formed to try to consolidate the different systems of the State improvement programs. The Federation is now involved in:[15]

- establishing uniform, accurate records,
- assisting member organizations in developing performance programs,
- encouraging cooperation among all segments of the industry in using records,
- encouraging education by emphasizing the use of records,
- developing confidence in performance testing throughout the industry.

Despite these efforts, only about 3 percent of beef cattle nationally are recorded. This relatively low participation rate, when compared with NCDHIP, has both a technological and an institutional explanation. Under the largely extensive beef raising system in the United States, AI is difficult as long as estrus detection technologies are unavailable. Natural stud service is usually more economical. Institutional barriers also prevent the development of a strong genetic evaluation program—e.g., the breed associations are not all eager to have their breeds consistently compared with others. Likewise, some owners of bulls for stud service would lose business in a strict testing scheme.

Goats.—Though little genetic work has been done on goats in the past, the dairy goat industry has become more visible in the past few years. The desire of goat breeders to participate in NCDHIP led to the formation of a Coordinating Sub-Group for Dairy Goats. A review of the research performed indicated a great need for research in almost every area of production. As a result, AIPL developed a plan for a genetic improvement program. The leadership in the dairy goat industry was convinced that it could attain genetic improvement faster and at a lower cost via NCDHIP than it could for any other type of research.

In 1979, AIPL received a $15,000 grant from the Small Farms Research Funding to support the development of genetic evaluation proce-

---

*There are several State programs—in Indiana, North Carolina, and Tennessee. Some of these programs may test only growth and not litter size.

[14]U.S. Department of Agriculture, Agricultural Research Service, *ARS National Research Program, Swine Production*, NRP No. 20370, October 1976.

[15]R. L. Willham, "Genetic Activity in the U.S. Beef Industry," Journal Paper No. J-7923 of the Iowa Agricultural and Home Economics Experiment Station, Ames, Iowa, project No. 2,000, n.d.

dures for goats. Genetic evaluations for yield of dairy goat bucks will be available before the end of fiscal year 1980. Because limited genetic improvement for yield has occurred in dairy goats in the past, these evaluations will probably have a significant impact on the industry. AIPL can virtually guarantee beneficial results because of the data available from NCDHIP, its own expertise in genetics, statistics, and computer technology, and the decades of highly effective research on genetic improvement of dairy cattle that can be adapted for the dairy goat industry. However, funding for the goat testing program remains on a year-to-year basis.

CONCLUSION

NCDHIP has shown how important genetic information is to the production of meat and dairy products. The obstacles to such a program are also formidable, but every failure to capitalize on genetic potential is paid for by American consumers. It has also shown that where selection objectives can be identified and agreed on, and where conflicting interests can be brought together to develop a program serving all interests, genetic improvement can become a central objective in breeding programs across the country. Without reliable, evaluative data on breeding stock, the Nation's breeders will have little interest in adopting new breeding technologies as they become available.

Impacts on breeding

An improvement in germplasm, like an increase in the nutritional content of fertilizer or new and improved herbicides and pesticides, increases the quality of the physical capital used on the farm. It is likely that much improvement can still be made in the germplasm of all major farm animal species using existing technology.

Selecting for desired characteristics causes a specific qualitative change; it enhances the efficiency of the information contained within each cell. The genetic information in each cell of a farm animal is either more or less desirable or efficient than information in the cells of another animal, depending on how it performs on important traits. Superior germplasm can be used in breeding decisions to upgrade a farmer's

breeding or producing stock. (DHIA programs are the best example of how information might be distributed.)

Resources invested in genetics and in technologies related to genetics will have high payoffs—e.g., in a classic study[16] of the payoff to research in hybrid corn and in subsequent studies of other types of genetic improvement, a high cost/benefit ratio for such research was found. The original study also showed that the absolute market value of a particular product is an important factor influencing the rate of return on a given research expenditure. In general, the greater the aggregate value of the product, the greater the rate of return on a research expenditure.[17] Thus, the large expenditures for meat and animal products in the United States suggest a great payoff in applied genetic research. Beef purchases alone account for between 2 and 5 percent of the American consumer dollar, and the total market value for beef is more than twice that for corn in the United States.

DAIRY CATTLE

Total milk production has been stable for many years. While milk production per cow has gone steadily upward, the number of cows during the past 35 years has decreased proportionately. (See figure 29.) Milk production per cow should continue to increase, assuming that no radical changes in present management systems occur. The increase in production per cow could continue even if no bulls superior to those already available are found, simply as a result of more farms switching to existing technology and existing bulls. Moreover, bulls produced from this system are increasing in superiority.

The number of dairy cows calved as of January 1, 1980, was 10,810,000. It has remained relatively stable for the past year, but may de-

[16]Zvi Griliches, "Research Costs and Social Returns: Hybrid Corn and Related Innovations, "*Journal of Political Economy* 66:419, October 1958. See also R. E. Evenson, P. E. Waggoner, and V. W. Ruttan, "Economic Benefit From Research: An Example From Agriculture," *Science* 205:1101, Sept. 14, 1979.

[17]W. Peterson and Yujino Hayami, "Technical Change in Agriculture," Staff Papers series No. DP73-20, Department of Agriculture and Applied Economics, University of Minnesota, St. Paul, Minn., July 1973.

crease to around 10 million in the next decade if milk production continues to increase.

Artificial Insemination.—An example of the interaction between technologies and genetic improvement is shown in table 34. The "predicted difference" (PD) in milk production represents the ability of individual bulls to genetically transmit yield—the amount of milk above or below the genetic base that the daughters of a bull will produce on average due to the genes they receive. As indicated in table 34, the predicted difference for milk yield transferred via the bull shows an improvement from 122 to 908 lb for active AI bulls in the United States over the past 13 years.

This impressive improvement still lags behind what is theoretically possible. A hypothetical breeding program could result in an expected yearly gain of 220 lb of milk per cow, using AI; and the biological limits to this rate of gain are not known. In practice, the observed genetic trend in the U.S. national dairy herd is about 100 lb—70 lb from the PDs of bulls plus 30 lb or so from the female, most of which is actually carryover effect from the previous use of superior bulls.

AI organizations, many of which are cooperatively owned by dairymen, have not rigorously applied the principles of AI. Their efforts have been limited by reluctance to break with traditional selection practices, financial constraints for proper testing of young bulls to pro-

Table 34.—Predicted Difference (PD) of Milk Yield of Active AI Bulls

Year	PD milk (lb)
1967	122
1968	198
1969	205
1970	276
1971	301
1972	346
1973	348
1974	336
1975	425
1976	501
1977	558
1978	748
1979	908

SOURCE: Animal Improvement Programs Laboratory, Animal Science Institute, Beltsville Agricultural Research Center, USDA.

duce sires of cows, and too much emphasis on nonproductive traits of questionable economic value. The progress that has been made has resulted from the increased use of AI, the availability of data through NCDHIP, and the actual use of reliable genetic evaluations. If any of these three factors had been missing, far less improvement would have occurred.

Semen Storage.—It is doubtful that major technological changes in processing semen will occur. However, since the rate of conception is as important as the genetic merit of a sire to the economy of a dairy enterprise, more attention will be given to selecting sires of high fertility. Progress should be made in banking semen by AI studs as a hedge against costs of inflation. In the future, some of the increased costs of housing and feeding bulls will probably be offset by semen banking and earlier elimination of many bulls.

Sexed Semen.—Sexing of semen to produce heifer calves (for dairymen) or bull calves (for AI organizations) has been attempted without success for many years.

Perfect determination of the sex of progeny could practically double selection intensity in two ways—with dams to produce bulls for testing in AI and dams to produce replacements. If sexed semen is used with an AI plan, the theoretical improvement in milk yield would be 33 lb per year, with 23 lb due to selection of dams for replacements.

The value of this additional amount per year may not seem great for any individual cow, but when it is multiplied by a national herd of 7 million cows using AI and is accumulated for 10 years, the economic value, at $0.10/lb, is about $1.1 billion—an average of $110 million per year and $231 million during the 10th year. The cost of sexing semen is not known, since no one has successfully done it. If a way is found, the cost would have to be under $10 per breeding unit for the procedure to be economical.

Embryo Transfer.—The transfer of fertilized eggs from a cow to obtain progeny has been accomplished with great success. Most transfers have involved popular or exotic breeding animals with little regard for genetic potential.

Embryo transfer may never pay for itself in terms of milk production of the animals produced except indirectly through bulls. Rather, it is used mostly to produce outstanding cows for sale. Other commercial applications for cattle include obtaining progeny from otherwise infertile cows, exporting embryos instead of animals, and testing for recessive genetic traits.

Embryo transfer progeny must be worth $2,500 each to justify the costs and risks. About $1,500 of this represents costs due to embryo transfer and $1,000 the costs of producing calves normally. If genetic gain from embryo transfer comes only from dam paths, the expected gain over AI alone is 76 lb/yr. Extra gain at $0.05/lb above feed cost would have to accumulate for 79 years before added gain would equal even a $300 embryo transfer cost per pregnancy. If less semen is needed (allowing more intensive bull selection), the expected gain of 129 lb/yr must accumulate for 46 years to balance an embryo transfer cost of $300 per pregnancy.

Embryo transfer and perfect sexing of semen would combine to improve genetic gain (in milk production) slightly. The use of less semen might be possible through application of in vitro fertilization. However, feasibility based on genetic gain would still require holding all costs down to around $50 to $90 per conception. The general conclusion is that costs of embryo transfer must be greatly reduced to be economically feasible if only genetic gain is considered.

Estrus Synchronization.—The availability of an effective estrus synchronization method would provide strong impetus for increased use of AI and embryo transfer in dairy cattle. The detection of estrus is an expensive operation; effective control of estrus cycling also requires intensive management, adequate handling facilities, and close cooperation between the producer, veterinarian, and AI technician.

Summary.—

- Proper application of progeny testing with selection and AI can increase the genetic gain for milk yield more than two times faster than is occurring today. Improved evaluation of cows, proper economic emphasis on other traits, and strict adherence to selection standards are the keys. Biological limitations to this rate of genetic improvement cannot be anticipated in the foreseeable future.

- AI of dairy cattle, with the present intensity of sire selection, should increase the net worth or profit of animals (increased value minus extra costs of the AI program) about $10.00/head per year. By 1990, 8 million dairy cows in AI programs would be worth about $800 million (8 × 10^6 × $10 × 10 years) more at current market prices as a result of continued use of AI.

- Sexing of semen when used with AI may pay for itself if the cost per breeding unit can be kept between $10 and $20.

- Embryo transfer is unlikely to pay for itself genetically unless the cost is reduced to between $50 and $90 per conception. However, despite its high costs, it is used to produce animals of exceptionally high value. (See app. II-C for an explanation of reasons other than genetics why embryo transfer is used.)

- Estrus synchronization is now available for use with heifers, and should increase the use of AI and consequently the genetic improvement of dairy cattle.

- A secondary benefit of all technologies is the increased number of skilled persons who can provide technical skills as well as educate dairymen in all areas. Also, a unique pool of reproductive and genetic data has been accumulated.

BEEF CATTLE

There is no single trait of overriding importance (like milk production in dairy cows) to emphasize in the genetic improvement of beef cattle, the rate of growth is a possibility.* It is also difficult to select for several traits at once,

---

*Beef and dairy cattle are usually different breeds in the United States. In the literature and in research they are often referred to as different species. In other countries, notably in Western Europe and in Japan, so-called "dual purpose" cattle are used to produce both beef and milk. In the United States, old dairy cows usually become hamburger.

especially when some are incompatible—e.g., it is desirable to produce large animals to sell, but undesirable to have to feed large mothers to produce them. There are also other complications. Growth rate has two genetic components for which one can select—the maternal contribution (primarily milk production) and the calf's own growth potential. Other traits of interest are efficiency of growth, carcass quality traits (such as tenderness), calving ease, and reproductive traits, such as conception rate to first service with AI.

Genetic improvement programs for beef have two major advantages over those for dairy cattle. Traits such as growth rate and carcass quality can be measured in both sexes (whereas one cannot measure the milk production of bulls); and the traits are more heritable than milk production.

Artificial Insemination.—Between 3 and 5 percent of the U.S. beef herd is artificially inseminated each year. This low rate is due to several factors, including management techniques (range v. confined housing), availability of related technologies (especially, until recently, estrus synchronization), and the conflicting objectives of the individual breeders, ranchers, and breed associations.

Because little is known about the effectiveness of AI in spreading specific genes throughout the Nation's beef herds, analysts have concentrated on their reproductive performance. Calf losses are heavy throughout the Nation. The calf crop—the number of calves alive at weaning as a fraction of total number of females exposed to breeding each year—is estimated to be between 65 and 81 percent. To put these data in perspective, USDA[18] has estimated that a 5-percent increase in the national calf crop would yield a savings of $558 million per year in the supply of U.S.-grown beef. Techniques now available can produce such an increase when they are integrated into an adequate management program.

The standardized measure of weaning weight in beef cattle is the weight at 205 days, adjusted for sex of calf and age of dam. In a recent study in West Virginia—the Allegheny Highlands Project—calf weights have averaged an increase of 10 lb per year of participation in the project, via AI and crossbreeding. Estimates of increased value of calves statewide, should the same tests and AI program be expanded, add up to $3.6 million per year when calf prices average $50 per hundredweight.[19] Rapid adoption of AI could bring about this kind of increase in as little as 40 to 48 months.

The costs and returns of AI vary from farm to farm and with the number of cattle in estrus. In general, it becomes more valuable with smaller herds, more cows in estrus, higher conception rates, and better bulls. For purebred herds, even larger benefits have been estimated—e.g., in a 1969 study, the estimated increase in value per calf when AI was used was $30.02 on purebred ranches compared to $3.31 on commercial ranches in Wyoming.[20]

A major secondary, or indirect, benefit of the use of AI is feed saved for other uses. It has greatly reduced the number of sires necessary for stud service and, through radically improved milk production, the number of females as well. These reduced requirements together are equivalent to more than 1 billion bu of corn and other concentrates. This situation will be further enhanced as beef cattle AI expands.

Synchronization of Estrus.—Differences in the rates of application of AI between beef and dairy herds can be explained partly by the differing management systems for the two types of classes of cattle. Dairy herds are kept close to the barn for milking and are accustomed to being approached by humans. In contrast, beef herds may number a few thousand head on 100,000 acres of arid pasture land. The detection of estrus under these conditions is difficult.

---

[18]U.S. Department of Agriculture, Agricultural Research Service, "Beef Production," ARS National Research Program Report No. 20360 (Washington, D.C.: USDA, October 1976).

[19]B. S. Baker, M. R. Fausett, P. E. Lewis, and E. K. Inskeep, "A Program Report on the Allegheny Highlands Project" (Morgantown, W. Va.: West Virginia University, January-December 1979).

[20]D. M. Stevens and T. Mohr, "Artificial Insemination of Range Cattle in Wyoming: An Economic Analysis," Wyoming Agricultural Experiment Station Bulletin No. 496, 1969.

It has been predicted that the availability of prostaglandin agents for regulating estrus could increase the number of beef calves born from superior bulls by 10 times, and that perhaps 20 percent of the U.S. beef cow herd could receive at least one insemination artifically by 1990.[21] If this lead to a 50-lb increase in weight for 10 percent of the calves born, it should be worth $114 million to $122 million each year, assuming 80 or 85 percent net calf crop and $60 per hundredweight.

The implementation of recently developed estrus synchronization technology might increase the number of beef cows bred artificially by 4,000,000 in the United States. Such a program should be successful in advancing the calving date by one week (by decreasing the calving interval), and in increasing the quality of the calves produced. These new calves could be worth about $100 million annually, less about $50 million due to extra costs associated with the synchronization program.

Sex Control.—Sex control would have a dramatic effect on the beef industry. In 1971, it was projected that by 1980 sex control could have an annual potential benefit of $200 million based on 10 million female calves being replaced by male calves produced through the sexing of semen.[22] At the time of the prediction, the market value for steers was about $20 more than for heifers. (Steers wean heavier and gain more efficiently.) Now the margin is much greater—approximately $50. This potential method of biological control is more attractive than the use of additives like steroids or implants because of the possible hazards associated with them that preclude their use.

Embryo Transfer.—The possibilities for genetic improvement in beef cattle using embryo transfer have been analyzed. It appears that embryo transfer programs can be developed to increase the rate of genetic progress for

growth rate; but the programs are much too expensive to be used over the entire population. One problem is that the economic value of the product of a beef cow is around 25 percent (or even less) of that of a dairy cow. Nevertheless, in populations in which AI is used, embryo transfer was found to be useful for obtaining more bulls from top cows. The females produced by embryo transfer would be worth marginally more than females produced conventionally, but the costs and influence of males could spread over the population through the use of AI. The extent of this use of embryo transfer would be very small; only a few hundred bulls would be produced per year for very large populations, and over 99 percent of the population would reproduce conventionally. However, such programs could have considerable economic benefit. Care must be taken to minimize increased inbreeding of the population with such a breeding scheme.

Summary.—

- AI could substantially improve economically important traits in beef herds. However, because of the diversity of traits considered important by different breed groups and the lack of a national beef testing and recording system comparable to NCDHIP, economic estimates of its value have not been developed.
- A sexing technology to produce mostly males (they grow faster than heifers) could be of enormous potential benefit to the beef industry. However, no successful technique yet exists.
- Estrus cycle regulation could lead to a substantial increase in the number of beef cattle in AI programs. The net benefit of this technology, coupled with AI, may be as high as $50 million per year. Similarly, the availability of reliable progeny records would add to the beneficial impact of AI in beef and would probably contribute significantly to its use in beef cattle.

OTHER SPECIES

Swine.—Much progress has been made in improving the overall biological efficiency of pork production in the United States. Improved

---

[21]H. D. Hafs, "Potential Impact of Prostaglandin on Prospects for Food From Dairy Cattle," *Proc. Lutalyse Symposium*, J. W. Lauderdale and J. H. Sokolowski (eds.) (Kalamazoo, Mich.: Upjohn, 1979), pp. 9-14.

[22]R. H. Foote and P. Miller, "What Might Sex Ratio Control Mean in the Animal World," *Symposium, Am. Soc. of Animal Science*, 1971, pp. 1-10.

growth rates, feed efficiencies, carcass merit, and litter sizes have helped keep pork prices down and improve its quality in the Nation's markets. Pork today is leaner and contains more high-quality protein calories than it was just a few decades ago.

AI in swine production could expand, although it will be limited by the relatively poor ability of swine sperm to withstand freezing and by the problem of detecting estrus. It will be encouraged by the strong trend toward confinement housing and integration of all phases of hog production. The industry—especially the individual, family-farm type units—would benefit by the establishment of a progeny testing scheme to identify superior boars. Publicly available information on genetic merit would decrease dependence on a few corporate breeding organizations.

Embryo transfer in swine will be strictly limited by difficulties in developing nonsurgical methods of recovery and transfer, and by the low economic value per animal in comparison to cattle and horses. However, embryo transfer is useful in introducing new genetic material into breeding herds of specific pathogen-free swine and in transporting genetic material to various regions of the world.

Sheep.—The processes of selection and of crossing specific strains, which have been so effective in poultry and hogs, have been virtually ignored in sheep. Selection of replacement ewes from the fastest growing ewe lambs born as twins and the use of flushing to increase ovulation rates have led to annual increases of 1.8 percent in lambing; in one test the market weight of lambs was increased by 1/lb/yr of cooperation.[23]

Synchronization of estrus in ewes can be achieved with prostaglandin and many different progestogens. The technique is used extensively in many countries, but no products for this purpose are currently marketed in the United States.

AI rates abroad sometimes approach 100 percent. However, AI will not be used widely on sheep in the United States until systems for performance and progeny testing are implemented that will track the number of lambs born and their growth rate, and until routine freezing of raw semen is achieved.[24]

Goats.—The research performed on goats is largely designed for application to other animals. However, interest in goats in the United States and the demand for their products through the world is increasing.

NCDHIP has just started providing sire evaluations to goat breeders. These data, along with artifical insemination, should increase milk production. The genetic data might be of particular usefulness in the less developed countries where most goat raising occurs. Greater use of all reproductive technologies on valuable Angora goats might be expected.

Other technologies

The use of any reliable twinning or sex selection technologies will be limited until such procedures can be made simple, fast, inexpensive, and innocuous. No widespread use of these technologies should be expected within the next decade.

The more esoteric techniques for manipulating sex cells or the germplasm itself will have no impact on the production of animals or animal products within the next 20 years. In vitro manipulations, including cloning, cell fusion, the production of chimeras, and the use of rDNA techniques, will continue to be of intense interest. However, it is unlikely that they will have practical effects on farm production in the United States in this century. Each technique will require more research and refinement. Until specific genes can be identified and located, no direct gene manipulation will be practicable. A polygenic basis for most traits of importance can be expected to be the rule rather than the exception.

Should such techniques become available, limited use for producing breeding stock can be expected. Experience with early users of AI and

---

[23]E. K. Inskeep, personal communication, 1980.

[24]Ibid.

embryo transfer is strong evidence for the predicted use of the technologies, no matter what their economic justification. (See app. II-C.)

A major, secondary effect of animal research in reproductive biology is increased understanding leading to the possible solution of human problems—e.g., the concept, efficacy, and safety of the original contraceptive pill was developed and established in animals. It involves the same principle as estrous cycle regulation discussed above.

AQUACULTURE

Aquaculture is the cultivation of freshwater and marine species (the latter is often referred to as mariculture). While fish culture is about 6,000 years old, scientific understanding of its basic principles is far behind that of agriculture. Aquaculture is slowly being transformed into a modern multidisciplinary technology, especially in the industrialized countries. Increasing awareness of human nutritional needs, overfishing of natural commercial fisheries, and rising worldwide demand for fish and fish products are trends that indicate a growth in interest in aquaculture as a means to meet the food needs of the world's population.

As part of the trend toward the high technology and dense culturing of intensive aquaculture systems in the industrialized countries, problems of reproductive control, hatchery technology, feeds technology, disease control, and systems engineering are all being investigated. Reproductive control and genetic selection are important because most commercial aquaculture operations must now depend on wild seedstocks. Very little information on the animals in culture is available.

With all three of the aquaculture genera (fish, mollusks, and crustaceans), selective breeding programs have long been established, healthy gene pools are available, and advantageous hybridizations have been developed. In fish raising, culture systems often demand sterile hybrids, especially of carp and tilapia. Selective breeding of salmon has been limited by political pressures. Very little work has been conducted with catfish, the largest aquaculture industry in the United States. The use of frozen sperm, which has been successful, should increase because of the savings in transport costs. Although culture systems for mollusks are fairly well-defined, little applied genetics work has been done with these popular marine species. Some success has been reported in selection for growth rate and disease resistance of the American oyster, and selection for growth rate of the slow-growing abalone is underway. The crustaceans, of which the Louisiana crayfish is the largest and most viable industry, are the least understood. Successful hybrids of lobsters have been developed.

Aquaculture suffers from an insufficient research base on the species of interest. However, growing appreciation of and demand for marine species should result in increased support for basic and developmental work on all aspects of control, including basic reproductive biology.

POULTRY BREEDING

The quantitative breeding practices of commercial breeders have changed very little over the last 30 years. Highly heritable traits, such as growth rate, body conformation, and egg weight, are perpetuated by mass selection because little advantage is gained from hybrid vigor. Low heritable traits (egg production, fertility, and disease resistance) are perpetuated by crossbreeding and identified through progeny and family testing.

The goals of the industry are to increase egg production of the layers—both in quality and quantity—and, with broilers and turkeys, to improve growth rate, feed efficiency, and yield, as well as to reduce body fat and the incidence of defects.

The technologies of AI and semen preservation have accelerated the advances made through quantitative breeding technology. AI is widely used in commercial turkey breeding because of the inability of modern strains to mate. It makes breeding tests more efficient, steps up selection pressure on the male line, reduces the number of necessary breeder males, and increases the number of females that may be mated to one male. Semen diluents were introduced to the turkey industry about 10 years ago to lower the cost of AI. Currently, a little over half of the turkeys are inseminated with diluted semen.

Preservation of poultry semen by freezing is now practiced by several primary breeders. Although freezing chicken semen causes it to lose some potency, the practice allows increased genetic advancement and the distribution of genetic material worldwide.

The amount of genetic variation available for breeding stock is not expected to diminish in the near future. Ceilings for certain traits will eventually be reached, but certainly not in the 1980's. Advances in breeding laying chickens will be less dramatic than in the past, but efforts will continue to develop new genetic lines and to improve reserve lines and crosses to meet future needs.

The growth rate of broilers will continue to increase at 4 percent a year, which suggests that birds will be reaching 4.4 lb in 5 weeks by the 1990's. Breeding for stress resistance will be increasingly important, not only because of the increased use of intensive production systems, but also to meet the physiological stresses resulting from faster growth and greater weight.

AI will assume increasing importance. Recent advances in procedures for long-term freezing of chicken semen will allow breeders to extend the use of outstanding sires. The sale of frozen seman may eventually substitute, in part, for the sale of breeder males.

Dwarf broiler breeders will also assume increasing importance over the new few years. The dwarf breeder female is approximately 25-percent smaller than the standard female, and even though the dwarf's egg is smaller and the progeny's growth rate slightly less than that of the standard broiler, the lower cost of producing broiler chicks from the dwarf breeder more than offsets the slight loss in their growth rate. Dwarf layers and the dwarf breeder hens could reduce production costs by 20 percent and 2 percent, respectively.

There is some interest among poultry breeders in cloning, gene transfer, and sex control but progress toward successful technologies is slow.

Issue and Options for Agriculture—Animals

ISSUE: Should the United States increase support for programs in applied genetics for animals and animal products?

Advocates of a strong governmental role in support of agricultural research and development (R&D) have traditionally referred to the small size of the production unit: U.S. farms are too small to support R&D activities. Throughout this century a complicated and extensive network of Federal, State, and local agricultural support agencies has been developed to assist the farmer in applying the new knowledge produced by research institutions. This private/public sector cooperative network has produced an abundant supply of food and fiber, sometimes in excess of domestic demand. Socially oriented policies have been adopted to soften

the impacts of new technology and to rescue the marginally efficient farmer from bankruptcy.

Current projections of U.S. and world population growth show increasing demand for all food products. Other predictable trends with implications for agricultural R&D, include:

- growth in income for some populations, which will probably increase the demand for sources of meat protein;
- increasing competition among various sources of protein for the consumer's dollar;
- increasing awareness of nutrition issues among U.S. consumers;
- increasing competition for prime agricultural land among agricultural, urban, and industrial interests;
- increasing demand for U.S. food and fiber

products from abroad, leading to opportunities for increased profits for successful producers; and

- increasing demands on agricultural products for production of energy.

OPTIONS:

A. *Governmental participation in, and funding of, programs like the National Cooperative Dairy Herd Improvement Program (NCDHIP) could be increased. The efforts of the Beef Cattle Improvement Federation to standardize procedures could be actively supported, and a similar information system for swine could be established.*

The fastest, least expensive way to upgrade breeding stock in the United States is through effective use of information. Computer technology, along with a network of local representatives for data collecting, can provide the individual farmer or breeder with accurate information on the germplasm available, so that he can then make his own breeding decisions. In this way, the Nation can take advantage of population genetics and information handling capabilities to upgrade one of its most important forms of capital: poultry and livestock. Breed associations and large ranchers who sell the semen from their prize bulls based on pedigrees rather than on genetic merit may act as barriers to the effectiveness of such an objective information system.

The benefits of such programs would accrue both to U.S. consumers, in reduced real prices of meat and animal products, and to producers who participate in the programs, in increased efficiency of production. Consumers spend such a large part of their incomes on red meat that every increase in efficiency represents millions of dollars saved. Beef producers too, should welcome any assistance in upgrading their stocks. The price of semen has remained relatively stable, and semen from bulls rated highly on certain economic traits costs only a few dollars more than that from average bulls.

However, efficiency of production is not the only value to be upheld in U.S. agriculture—e.g., in milk production complex policies have been

designed to maintain constant milk supplies without large fluctuations in price.

The NCDHIP model program for dairy cattle has shown that an effective national program requires the participation by the varied interests in program policymaking in an extension network, for local collection and validation of data and for education and of expertise in data handling and analysis. Also important is a strong leadership role in establishing the program. This option implies that the Federal Government would play such a role in new programs and expand its role in existing ones.

B. *Federal funding of basic research in total animal improvement could be increased.*

The option, in contrast with option A, assumes that it is necessary to maintain or expand basic R&D to generate new knowledge that can be applied to the production of improved animals and animal products.

Information presented in this report supports the conclusion that long-term basic research on the physiological and biochemical events in animal development results in increasing the efficiency of animal production, both in total animal numbers and in quality of product. Increased understanding of the interrelationships among various systems—including reproduction, nutrition, and genetics—gradually leads to the development of superior animals that efficiently consume food not palatable to humans and are resistant to disease.

Earlier studies also support the importance of basic research—e.g., the National Research Council found in 1977 that ". . . not as much fundamental research on animal problems has been conducted in recent years . . . it should receive increased funding."[25] USDA also found, in a review of various conference proceedings, congressional hearings, special studies, and other published materials on agricultural R&D priorities, strong support for more research on the basic processes that contribute to reproduction and performance traits in farm animals:

---

[25]National Research Council, *World Food and Nutrition Study, The Potential Contributions of Research* (Washington, D. C. author, 1977), p. 97.

Specific livestock research areas identified as having signficant potential for increased production both in the United States and developing countries include: 1) control of reproductive and respiratory diseases, 2) developing genetically superior animals, 3) improving nutrition efficiency, and 4) increasing the reproductive performance of all farm animal species.[26]

---

[26]U. S. Department of Agriculture, Science and Education Administration, *Agricultural and Food Research Issues and Priorities* (Washington, D.C.: author, 1978), p. xiii.

Regardless of the effectiveness of present population control programs or of current trends in individual decisions about family size, the output of the Nation's agricultural activities must increase over the next decades if sufficient food is to be available for the world's population. Basic research is the source from which new applications to increase productivity arise.

Part III
Institutions and Society

The Question of Risk

Chapter 10

Figures

The Question of Risk

Introduction

The perception that the genetic manipulation of micro-organisms might give rise to unforeseen risks is not new. The originators of chemical mutagenesis in the 1940's were warned that harmful uncontrolled mutations might be induced by their techniques. In a letter to the Recombinant DNA Advisory Committee (RAC) of the National Institutes of Health (NIH) in December of 1979, a pioneer in genetic transformation at the Rockefeller University, wrote: ". . . I did in 1950, after some deliberation, perform the first drug resistance DNA transformations, and in 1964 and 1965 took part in early warnings against indiscriminate 'transformations' that were then being imagined."[1]

[1]Rollin D. Hotchkiss, *Recombinant DNA Research*, vol. 5, NIH publication No. 80-2130, March 1980, p. 484.

Yet none of this earlier public concern led to as great a controversy as has research with recombinant DNA (rDNA). No doubt it was encouraged because scientists themselves raised questions of potential hazard. The subsequent open debates among the scientists strengthened the public's perception that there was legitimate cause for concern. This has led to a continuing attempt to define the potential hazards and the chances that they might occur.

The initial fear of harm

For the purposes of this discussion, harm (or injury) is defined as any undesirable consequence of an act. Such a broad definition is warranted by the broad targets for hypothetical harm that genetic manipulation presents: injury to an individual's health, to animals, to the environment.

The inital concern involved injury to human health. Specifically, it was feared that combining the DNA of simian virus 40, or SV40, with an *Escherichia coli* plasmid would establish a new route for the dissemination of the virus. Although the SV40 is harmless to the monkeys from which it is obtained, it can cause cancer when injected into mice and hamsters. And while it has not been shown to cause cancer in humans, it does cause human cells to behave like cancer cells when they are grown in tissue culture. What effect such viruses might have if they were inserted into *E. coli*, a normal inhabitant of the human intestine, was unknown. This uncertainty, combined with an intuitive

judgment, led to a concern that something might go wrong. The dangerous scenario went as follows:

- SV40 causes cells in tissue culture to behave like cancer cells,
- SV40-carrying *E. coli* might be injected accidently into humans,
- humans would be exposed to SV40 in their intestines, and
- an epidemic of cancer would result.

This chain of connections, while loose, was strong enough to raise questions in at least some people's minds.

The virus SV40 has never actually been shown to cause cancer in humans; but the potential hazards led the Committee on Recombinant DNA Molecules of the National Academy of Sciences (NAS) to call in 1974 for a deferment of any experiments that attempted to join the DNA of a cancer-causing or other animal virus to vector DNA. At the same time, other experiments, that were thought to have a potential for harm —particularly those that were designed to transfer genes for potent toxins or for resistance to antibiotics into bacteria of a different species—were also deferred. Finally, one other type of experiment, in which genes from higher organisms might have been combined with vectors, was to be postponed. The fear was that latent "cancer-causing genes" might be inadvertently passed on to *E. coli*.

Throughout the moratorium, one point was certain: no evidence existed to show that harm would come from these experiments. But it was a possibility. The scientists who originally raised questions wrote in 1975: ". . . few, if any, believe that this methodology is free from risk."[2] It was recognized at that time that ". . . estimating the risks will be difficult and intuitive at first but this will improve as we acquire additional knowledge."[3] Hence two principles were to be followed: containment of the micro-organisms (see table 35, p. 213) was to be an essential part of any experiment; and the level of containment was to match the estimated risk. These principles were incorporated into the Guidelines for Research Involving Recombinant DNA Molecules, promulgated by NIH in 1976.

But the original fears surrounding rDNA research progressed beyond concern that humans might be harmed. Ecological harm to plants, animals, and the inanimate world were also considered. And other critics noted the possibility of moral and ethical harm, which might disrupt both society's structure and its system of values.

Classification of potential physical harm

Some combinations of DNA may be harmful to man or his environment—e.g., if an entire DNA copy of the poliovirus genetic material is combined with *E. coli* plasmid DNA, few would argue against the need for careful handling of this material.

For practical purposes, the potential harm associated with various micro-organisms is

shown in figure 35. Each letter (A through L) represents the consequence of a particular combination of events and micro-organisms. For example, the letters:

A,C represent the *intentional release of micro-organisms known to be harmful* to the environment or to man—e.g., in biological warfare or terrorism.

B,D represent the *inadvertent release of micro-organisms known to be harmful* to the environment or to man—e.g., in accidents at high-containment facilities where work is being carried out with dangerous micro-organisms.

E,I represent the *intentional release of micro-organisms thought to be safe* but which *prove harmful*—when the safety of organisms has been misjudged.

F,J represent the *intentional release of micro-organisms which prove safe* as expected—e.g., in oil recovery, mining, agriculture, and pollution control.

H,L represent the *inadvertent release of micro-organisms which have no harmful consequences*—e.g., in ordinary accidents with harmless micro-organisms.

G,K represent the *inadvertent release of micro-organisms thought to be safe* but which *prove harmful*—the most unlikely possible consequence, because both an accident must occur and a misjudgment about the safety must have been made.

Discussions of physical harm have recognized the possibility of intentional misuse but have minimized its likelihood. The Convention on the Prohibition of the Development, Production, and Stockpiling of Bacteriological (Biological) and Toxin Weapons and on their Destruction[4] which was ratified by both the Senate and the President in 1975,* states that the signatories will "never develop . . . biological agents or toxins . . . that have no justification for prophylactic, protective, or other peaceful purposes." Such a provision clearly includes micro-organisms carrying rDNA molecules or the toxins

---

[2]*Recombinant DNA Research,* vol. 1, DHEW publication No. (NIH) 76-1138, August 1976, p. 59.
[3]Ibid.

[4]Convention on the Prohibition of the Development, Production, and Stockpiling of Bacteriological (Biological) and Toxin Weapons and On Their Destruction. Done at Washington, London, and Moscow, Apr. 10, 1972; entered into force on Mar. 26, 1975 (26 U.S.T., 583).
*As of 1980, 80 countries have ratified the treaty; another 40 have signed but not ratified.

Figure 35.—Flow Chart of Possible Consequences of Using Genetically Engineered Micro-Organisms

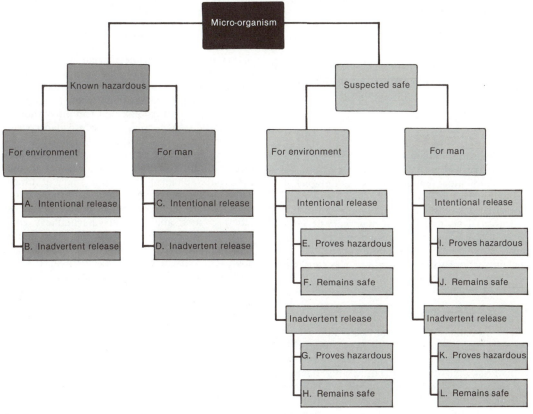

SOURCE: Office of Technology Assessment.

produced by them. It must be assumed that those who signed did so in good faith.

While there is no way to judge the likelihood of developments in this area, the problems that would accompany any attempt to use pathogenic micro-organisms in warfare—difficulties in controlling spread, protection of one's own troops and population—tend to discourage the use of genetic engineering for this purpose.* Similarly, the danger that these techniques might be used by terrorists is lessened by the scientific sophistication needed to construct a more virulent organism than those that can

already be obtained—e.g., encephalitis viruses or toxin-producing bacteria like *C. botulinum* or *C. tetani.*

Some discussions have centered around the possibility of accidents caused by a break in containment. Construction of potentially harmful micro-organisms will probably continue to be prohibited by the Guidelines; exceptions will be made only under the most extraordinary circumstances. To date, no organism known to be more harmful than the organism serving as the source of DNA has been constructed.

However, the biggest controversy has centered around unforeseen harm—that micro-organisms thought safe might prove harmful.

---

*Although stockpiling of biological warfare agents is prohibited, research into new agents is not.

Discussion of this kind of harm is hindered by the difficulty not only of quantifying the probability of an occurrence but also of predicting the type of damage that might occur. The different types of damage that can be conjured up are limited only by imagination. The scenarios have included epidemics of cancer, the spread of oil-eating bacteria, the uncontrolled proliferation of new plant life, and infection with hormone-producing bacteria.

The *risk* of harm refers to the chance of harm actually occurring. In the present controversy, it has been difficult to distinguish the possible from the probable. It is, for instance, possible that an individual will be killed by a meteor falling to the ground, but it is not probable. Analogous situations exist in genetic engineering. It is in this analysis that debate over genetic engineering has some special elements: the uncertainty of what kind of harm could occur, the uncertainty about the magnitude of risk, and the problem of the perception of risk.

Identification of possible harm

The first step in estimating risk is identifying the potential harm. It is not very meaningful to ask: How much risk does rDNA pose? The concept of risk takes on meaning only when harm is identified. The question should be: What is the likelihood that rDNA will cause a specific disease such as in a single individual or in an entire population? The magnitude of the possible harm is incorporated in the question of risk, but differs in the two cases. A statement about the risk of death to one person is different than one about the risk of death to a thousand. The right questions must be asked about a specific harm.

Since no dangerous accidents are known to have occurred, their types remain conjectural. Identifying potential harm rests on intuition and arguments based on analogy. Even a so-called risk experiment is an approximation of subsequent genetic manipulations. That is why experts disagree. No uncontestable "scientific method" dictates which analogy is useful or acceptable. By their very nature, all analogies share some characteristics with the event under consideration but differ in others. The goal is to

discover the one that is most similar and to observe it often. This process then forms the basis for extrapolation.

For example, it has been argued that ecological damage can be caused by the introduction of plants, animals, and micro-organisms into new environments. Scores of examples from history support this conclusion. The introduction to the United States of the Brazilian water hyacinth in the late 19th century has led to an infestation of the Southern waterways. Uncontrolled spread of English sparrows originally imported to control insects has made eradication programs necessary. Countless other examples are confirmation that biological organisms may, at times, cause ecological damage when introduced into a new environment. Yet there is no agreement on whether such analogies are particularly relevant to assessing potential dangers from genetically engineered organisms. It could be argued—e.g., that a genetically engineered organism (carrying less than 1 percent new genes) is still over 99 percent the same as the original, and is therefore not analogous to the "totally new" organism introduced into an ecosystem. Some experts emphasize the differences between the situations; others emphasize the similarities.

Other analogies have been raised. New strains of influenza virus arise regularly. Some can cause epidemics because the population, never before exposed to them, carries no protective antibodies. Yet can this analogy suggest that relatively harmless strains of *E. coli* might be transformed into epidemic pathogens? There is disagreement, and debates continue about what "could happen" or what is even logically possible.

Estimates of harm: risk

Assuming that agreement has been reached on the *possibility* of a specific harm, what can be done to ascertain the *probability*? What is the likelihood that damage will occur?

Damage invariably occurs as the result of a *series* of events, each of which has its own particular chance of occurring. Flow charts have been prepared to identify these steps. A typical analysis determines a probability value for each step—e.g., in figure 36 step II the probability of escape can be estimated based on the historical record of experiments with micro- organisms. Depending on the degree of containment, the probability varies. It is almost certain that experiments on an open bench top, using no precautions, will result in some escape to the surrounding environment—a much less likely event in maximum containment facilities. (See table 35.)

Two points should be noted. First, each probability can be minimized by appropriate control measures. Second, the probability that the final event will occur is equal to or less likely than the least likely link in the chain. Because the probabilities must be multiplied together, if the probability of any single step is zero, the probability of the final outcome is zero; the chain of events is broken.

THE STATUS OF THE CURRENT ASSESSMENT OF PHYSICAL RISK

A successful risk assessment should provide information about the likelihood and magnitude of damage that might occur under given circumstances. It is clear that the more types of damage that are identified, the more risk assessments must be carried out.

Figure 36.—Flow Chart to Establish Probability of Harm Caused by the Escape of a Micro-Organism Carrying Recombinant DNA

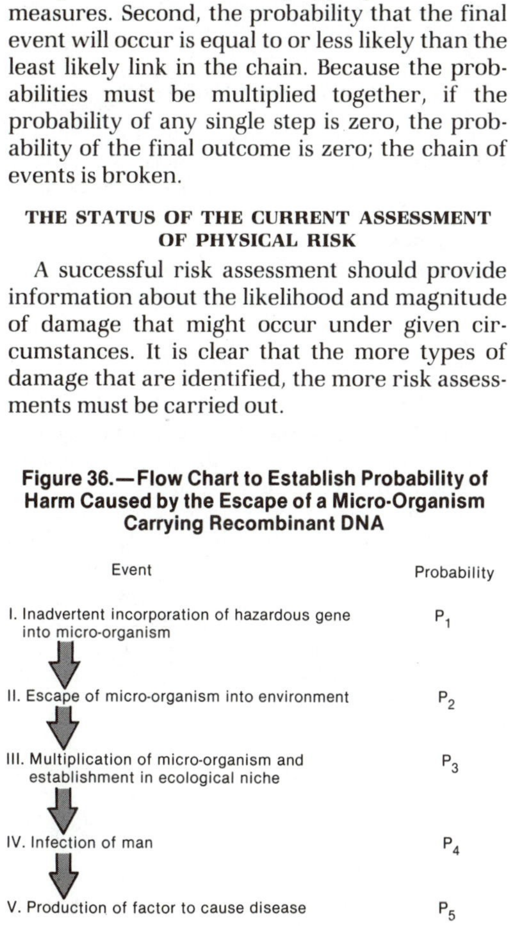

Event	Probability
I. Inadvertent incorporation of hazardous gene into micro-organism	P_1
II. Escape of micro-organism into environment	P_2
III. Multiplication of micro-organism and establishment in ecological niche	P_3
IV. Infection of man	P_4
V. Production of factor to cause disease	P_5

P_5 will always be smaller than any of the other probabilities.

SOURCE: Office of Technology Assessment.

Although the original charter of RAC underscored the importance of a risk assessment program, it was not until 1979 that the details of a formal program were published. For 5 years, risks were assessed on a case-by-case basis through: 1) experiments carried out under contract from NIH, 2) experiments that were designed for other purposes but which proved to be relevant to the question of risk, and 3) conferences at which findings were examined.

From the start, it was difficult to design experiments that could supply meaningful information—e.g., how does one test the possibility that "massive ecological disruptions might occur?" Or that a new bacterium with harmful unforseen characteristics will emerge? Still some experiments were proposed. But because these experiments had to be approximations of the actual situation, the applicability of their findings was debated. Here too, experts could and did disagree—not about the findings themselves, but about their interpretation.

For example, in an important experiment designed to test a "worst case situation," a tumor virus called polyoma was found to cause *no* tumors in test animals when incorporated into *E. coli.*[5]* Since just a few molecules of the viral DNA are known to cause tumors when injected directly into animals, it was concluded that tumor viruses are noninfectious to animals when incorporated into *E. coli.* If polyoma virus, which is the most infective tumor virus known for hamsters, cannot cause tumors in the rDNA state in *E. coli,* it is unlikely that other tumor viruses will do so. This conclusion has had widespread, but not unanimous, acceptance. It has been argued that there might be "something special" about polyoma that prevents it from causing tumors in this altered state; other tumor viruses might still be able to do so. At one meeting of RAC, in fact, it was suggested that experiments with several other viruses be carried out to confirm the generality of the finding. But how many more viruses? What is enough?

---

[5]M. A. Israel, H. W. Chan, W. P. Rowe, and M. A. Martin, "Molecular Cloning of Polyoma Virus DNA in *Escherichia Coli*: Plasmid Vector Systems," *Science* 203:883-887, 1979.

*Some combinations of free plasmid and tumor virus DNA did cause infections.

For some, one carefully planned experiment using the most sensitive tests is sufficient to allay fears. But for others, significant doubt about safety remains, regardless of how many viruses are examined. The criteria depend on an individual's *perception* of risk.

Many experiments carried out for purposes other than risk assessment have provided evidence that scenarios of doom or catastrophe are highly unlikely. This is the general consensus of specialists, not only in molecular biology, but in population genetics, microbiology, infectious diseases, epidemiology, and public health.

Experiments have revealed that the structure of genes from higher organisms (plants and animals) differ from those of bacteria. Consequently, those genes are unlikely to be expressed *accidentally* by a bacterium; the original fears of "shotgun" experiments have become less well-founded. Hence, data gathered to date have made the accidental construction of a new epidemic strain more unlikely.

Conference discussions have also contributed to a better understanding of the risks. At one such conference,[6] which was attended by 45 experts in infectious diseases and microbiology, it was concluded that:

- *E. coli* K-12 (the weakened form of *E. coli*, used in experiments) does not flourish in the intestinal tract of man;
- the type of plasmid permitted by the Guidelines has not been shown to spread from *E. coli* K-12 to other *E. coli* in the gut; and
- *E. coli* K-12 cannot be converted to a harmful strain even after known virulence factors were transferred to it using standard genetic techniques.

A workshop sponsored by NIH[7] provided a forum for scientists to discuss the risks posed by viruses in rDNA experiments. They concluded that the risks were probably *less* when a virus was placed inside a bacterium in rDNA form

than when it existed freely.* Experts in infectious disease have stressed repeatedly that the ability of a micro-organism to cause disease depends on a host of factors, all working together. Inserting a piece of DNA into a bacterium is unlikely to suddenly transform the organism into a virulent epidemic strain.

Careful calculations can also allay fears about the damage a genetically engineered micro-organism might cause. Doomsday scenarios of escaped *E. coli* that carry insulin or other hormone-producing genes were recently examined in another workshop.[8] Prior to this workshop, newspaper accounts raised the possibility that an *E. coli* carrying the gene for human insulin production might colonize humans and thus upset the hormonal balance of the body.

The participants calculated how much insulin could be produced. First, it was assumed that a series of highly unlikely events would occur—accidental release, ingestion by humans, stable colonization of the intestine by *E. coli* K-12. *E. coli* constitutes approximately 1 percent of the intestinal bacterial population, and it was assumed that all the normal *E. coli* would be replaced by the insulin-producing *E. coli*. Insulin is made in the form of a precursor molecule, proinsulin. It was assumed that 30 percent of all bacterial protein production would be devoted to this single protein, another highly unlikely situation. If so, 30 micrograms (μg)—or 0.6 units—would then be made in the intestine. Although proteins are very poorly absorbed from the intestinal cavity, it was assumed for the sake of argument that 100 percent of the proinsulin would be absorbed into the circulation. Thus, 0.6 units of insulin would be added to the normal daily human production of 25 to 30 units—an imperceptible difference.

Calculations like these have been carried out for several other hormones. Even with the most implausible series of events, leading to the greatest opportunity for hormone production,

---

[6]"Workshop on Studies for Assessment of Potential Risks Associated With Recombinant DNA Experimentation," Falmouth, Mass., June 20-21, 1977.

[7]"Workshop to Assess Risks for Recombinant DNA Experiments Involving Viral Genomes," cosponsored by the National Institutes of Health and the European Molecular Biology Organization, Ascot, England, Jan. 26-28, 1978.

*On the other hand, it has been argued that this has provided viruses with a new route for dissemination. Nevertheless, there is no evidence that viruses can readily escape from the bacteria and subsequently cause infection.

[8]"National Institute of Allergy and Infectious Diseases Workshop on Recombinant DNA Risk Assessment," Pasadena, Calif., Apr. 11-12, 1980.

the conclusion is that normal hormone levels would change by less than 10 percent. Similar conditions for interferon production could release approximately 70µg or the maximum daily dose currently used in cancer therapy. Long-term effects of such exposure are currently unknown; therefore, experiments using high-producing strains (10^6 molecules per cell or more) are likely to be monitored if such strains ever become available.

The NIH program of risk assessment, which was formally started in 1979, continues to identify possible consequences of rDNA research. Under the aegis of the National Institute of Allergy and Infectious Diseases, the program supports research studies designed to elucidate the likelihood of harm.* In addition, it collates general data from other experiments that might be relevant to risk assessment. Other risk assessments are being conducted by European organizations** and by the U.S. Environmental Protection Agency to assess the consequences of releasing micro-organisms into the environment.

Thus far, there is no compelling evidence that *E. coli* K-12 bacteria carrying rDNA will be more hazardous than any of the micro-organisms which served as the source of DNA. Nevertheless, all the experiments have dealt with one genus of bacterium. Unless the conclusions about *E. coli* can be extended to other organisms likely to be used in experiments (such as *Bacillus subtilis* and yeast), other assessments may be appropriate.

---

*Extramural efforts were first conceived in the summer of 1975 to develop and test safer host-vector systems based on *E. coli*, the interagency agreement entered into with the Naval Biosciences Laboratory tested *E. coli* systems in a series of simulated accidental spills in the laboratory. At the University of Michigan the survival of these systems was tested in mice and in cultural conditions simulating the mouse gastrointestinal tract. Tufts University tested these systems in both mice and human volunteers. Finally, the survival of host-vector systems in sewage treatment plants was tested at the University of Texas. The peak year for costs of supporting research contracts was 1978; over a half-million dollars were required. Currently, the cost of maintaining the high containment facility at Frederick, Md., is between $200,000 and $250,000 annually.

**First Report to the Committee on Genetic Experimentation , a scientific committee of the International Council of Scientific Unions, from the Working Group on Risk Assessment, July 1978.

Perception of risk

The probability of damage can be estimated for various events. The entire insurance industry is based on the fact that unfavorable events occur on a regular basis. The number of people dying annually from cancer, or automobile accidents, or homicides can be predicted fairly accurately. These estimates depend on the availability of data and the assumptions that the major determinants do not change from year to year.

But even if the probability of damage is fairly well known, a gap often exists between this "real" probability of occurrence and the "perceived" probability. Two factors that tend to affect perceptions are the magnitude of the possible damage and the lack of individual control over exposure to the risk. Both of these are significant factors in the fears associated with rDNA and the manipulation of genes. Because intuitive evaluations can contradict analytical evaluations, the question of risk cannot be resolved strictly on an analytical basis. Its resolution will have to come through the political process.

BURDEN OF PROOF

The possibility of inadvertently creating a dangerous organism does exist, but its probability is lower than was originally thought. Nevertheless, an important principle emerges from the debate. Society must decide whether the burden of proof rests with those who demand evidence of safety or with those who demand evidence of hazard. The former would halt experiments until they are proved safe. The latter would continue experiments until it is shown that they might cause harm.

A significant theoretical difference exists between the two approaches. Evidence can almost always be provided to show that something causes harm—e.g., it can be demonstrated that a poliovirus causes paralysis, that a *Pneumococcus* causes pneumonia, that a rhinovirus causes the common cold. However, it cannot be demonstrated that a poliovirus can *never* cause the common cold. It cannot be demonstrated that rDNA molecules will *never* be harmful. It can only be demonstrated that harmful events are *unlikely*. Hence, society must determine what level of uncertainty it is willing to accept.

Other concerns

Concerns raised by industrial applications

Originally concerns involved hazards that might arise in the laboratory. Now that there are industrial applications of genetic engineering, the concerns include:

- risks associated with the laboratory construction of new strains of organisms,
- risks associated with industrial production or consumer use of the new strains, and
- risks associated with the products obtained from the new strains.

Many similar considerations apply to the assessment of the first two kinds of risks. Unless the organisms used in an industrial production scheme are thoroughly characterized, conjectured fears about their ability to cause disease will continue. Even with a recombinant organism that has a well-defined sequence of DNA, a break in containment would leave its behavior in the environment questionable. Experience with substances such as asbestos gives rise to fears that exposure to the new biological systems might also cause unforeseen pathological conditions at some future time.

Hazards associated with products raise different questions. The growing consensus in Federal regulatory agencies appears to be that these products should be assessed like all others—e.g., human growth hormone (hGH) produced by genetically engineered bacteria should be tested for purity, chemical identity, and biological activity just like hGH from human pituitary glands. The possibility of product variation due to mutation of the bacteria, however, suggests that batch testing and certification might be warranted as well. (For further discussion see ch. 11.)

Concerns raised by the implications of the rDNA controversy for general microbiology

Questions about the potential harm from genetically engineered micro-organisms have led to questions about the efforts currently employed to protect the public from work being done with micro-organisms *known* to be hazardous. These viruses, bacteria, and fungi are handled daily in laboratory experiments, in the routine isolation of infectious agents from patients, and in the production of vaccines in the pharmaceutical industry.

Questions have been raised about the efficacy of regulations established for these various potentially hazardous agents. A full-scale assessment is not within the scope of this study, but it is clear that the questions are pertinent. Two conclusions have been reached.

First, there is a growing belief that the mere existence of a classification scheme for hazardous agents by the Center for Disease Control (CDC) is not enough to ensure their safe handling. The Subcommittee on Arbovirus Laboratory Safety was formed recently because of concerns expressed in academic circles. Representatives from universities, the Public Health Service, the U. S. Department of Agriculture, and the military, who constituted the subcommittee, are preparing a report based on an international survey of laboratory practices and infections. They found wide variation in the ways different agents were handled. Most of their recommendations are identical with those applicable to rDNA—that appropriate containment levels be used with different viruses, that the health of workers be monitored, and that an Institutional Biosafety Committee be appointed to serve each institution.

Second, little is known about the health record of workers involved in the fermentation and vaccine industries. For most industrial operations the evidence of harm is almost entirely anecdotal. Most industrial fermentations are regarded as harmless; representatives of industry characterize it as a "non-problem" that has never merited monitoring. Comprehensive information on the potential harmful effects associated with research using rDNA-carrying micro-organisms will not be available because the Guidelines consider it the responsibility of each institution or company to "determine, in connection with each project, the necessity for medical surveillance of recombinant-DNA research personnel." Hence some institutions might decide to keep records of some or all activities; others might not.

To be sure, some companies have exceeded the minimal medical standards set by NIH for fermentation using rDNA-carrying micro-organisms—e.g., Eli Lilly & Co. requires that *all* illnesses be reported to supervisors and that any employees who are ill for more than 5 days must report to a physician before being allowed to return to work. Any employee taking antibiotics (which might make it easier for bacteria to colonize) is restricted from areas where rDNA research is being done until 5 days after the discontinuance of the antibiotic. At Abbott Laboratories, a physician checks into the illness of any recombinant worker who is off more than 1 day—a precaution taken only after 5 days off for workers in other areas. Lilly maintains a computer listing of all workers involved in rDNA activities. Lilly, the Upjohn Co., and Merck, Sharp and Dohme have been in the process of computerizing the health records of all their employees over the past several years.

Work with rDNA has focused attention on biohazards and medical surveillance—an awareness that had arisen in the past but had not been sustained.* Consequently, several documents on the subject either have been or will be published:

- CDC is preparing a complete revision of its laboratory safety manual, which is widely used as a starting point by other laboratories.
- *The Classification of Etiologic Agents on the Basis of Hazard,* which was last revised in 1974, has been expanded by CDC in collaboration with NIH into a *Proposed Biosafety Guidelines for Microbiological and Biomedical Laboratories.* These guidelines serve the purpose fulfilled by the Dangerous Pathogens Advisory Group (DPAG) in the United Kingdom, although they lack any regulatory strength.
- A comprehensive program in safety, health, and environmental protection was developed in 1979 by and for NIH. It is administered by the Division of Safety, which includes programs in radiation safety, occupational safety and health, environmental protection, and occupational medicine.
- The Office of Biohazard Safety, National Cancer Institute has just completed a 3-year study of the medical surveillance programs of its contractors; a report is being drafted.

Although the academic, governmental, and industrial communities have shown growing interest in biosafety,* no Federal agency regulates the *possession* or *use* of micro-organisms except for those highly pathogenic to animals and for interstate transport.** Whether such regulations are necessary is an issue that extends beyond the scope of this study. Nevertheless, other countries—for instance the United Kingdom, with its DPAG—have acted on the issue. This organization functions specifically to guard against hazardous micro-organisms, by monitoring and licensing university and industrial laboratories and meting out penalties when necessary.

---

*As of September 1980, the National Institutes of Occupational Safety and Health and the Environmental Protection Agency were planning to fund assessments of the adequacy of current medical surveillance technology.

*Curiously, there is no formal society or journal, but there has been an annual Biological Safety Conference since 1955, conducted on a round-robin basis primarily by close associates of the late Arnold G. Wedum, M.D.—former Director of Industrial Health and Safety at the U.S. Army Biological Research Laboratories, Fort Detrick, Md., who is regarded as the "Father of Microbiological Safety."

**In some States and cities, licensing is required for all facilities handling pathogenic micro-organisms.

Concerns raised by the implications of the rDNA controversy for other genetic manipulation

Altering the hereditary characteristics of an organism by using rDNA is just one of the several methods of genetic engineering. The definition of rDNA refers specifically to the combination of the DNA from two organisms outside the cell. If the DNA is combined within living cells, the Guidelines do not pertain. Figure 35 shows several methods that achieve the same goal—transfering genetic material from one cell to another, bypassing the normal sexual mechanisms of mating. It is particularly significant that DNA *from different species* can be combined by all these mechanisms, only one of which is rDNA. Different species of bacteria,

Figure 37.—Alternative Methods for Transferring DNA From One Cell to Another

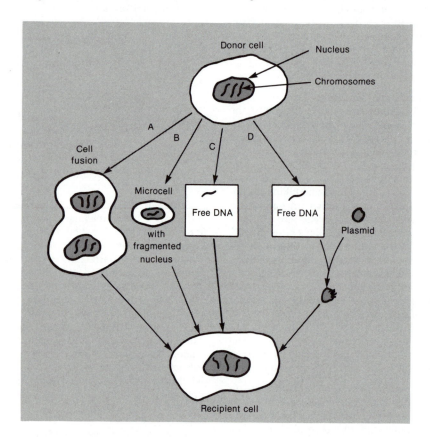

A. The two cells are fused in toto
B. A microcell with a fragmented nucleus carries the DNA
C. Free DNA can enter the recipient cell in a number of ways: by direct microinjection, through calcium-mediated transformation, or by being coated with a phospholipid membrane in order to fuse with the recipient cell
D. The free DNA can be joined to a plasmid and transferred as recombinant DNA

fungi, and higher organisms can all be fused or manipulated.*

Opponents of rDNA have stated that combining genes from different species may disturb an extremely intricate ecological interaction that is only dimly understood. Hence, such experiments, it is argued, are unpredictable and therefore hazardous. If so, all the other methods represented in figure 35 should be included in the Guidelines. Yet they are not.

The most acceptable explanation for this inconsistency is that rDNA is currently the most

---

*For example, antibiotic resistant plasmids have been transferred from *Staphylococcus aureus* to *Bacillus subtilis* across species barriers by transformation, not by rDNA. Foreign genes for the enzyme amylase have also been introduced into *B. subtilis*.

efficient and successful method of combining genes from very diverse organisms. It is reasonable to ask, however, what would happen if any of the other methods become equally successful. Will a profusion of guidelines appear? Will one committee oversee all genetic experiments

Ethical and moral concerns

The perceived risk associated with genetic engineering includes ethical and moral hazards as well as physical ones. It is important to recognize that these are part of the general topic of risk. To some, there is just as much risk to social values and structure as to human health and the environment. (For further discussion see ch. 13.)

Conclusion

Thus far, no *demonstrable* harm associated with genetic engineering, and particularly rDNA, has been found. But although demonstrable harm is based on evidence that damage *has* occurred at one time or another, it does not mean that damage *cannot* occur.

Conjectural hazards based on analogies and scenarios have been addressed and most have proved less worrisome than previously assumed. Nevertheless, there is agreement that certain experiments, such as the transfer of genes for known toxins or venoms into bacteria, should still be prohibited because of the real likelihood of danger. Still other experiments cannot clearly be shown to be hazardous or readily dismissed as harmless. Hence, a political decision is likely to be required to establish what constitutes acceptable proof and who must provide it.

Given that potential harm can be identified in some cases, its probable occurrence and magnitude quantified, and perceived risk taken into account, a decision to proceed is usually based on society's willingness to take the risk. This triad of the physical (*actual risk*), psychological (*perception of risk*), and political (*willingness to take risk*) plays a role in all decisions relating to genetic engineering.

The potential benefits must always be considered along with the risks. Decisions made by RAC have reflected this view—e.g., when it approved the cloning of the genetic material of the foot-and-mouth disease virus. The perceived benefits to millions of animals outweighed the potential hazard.

Recombinant DNA techniques represent just one of several methods to join fragments of DNA from different organisms. The current Guidelines do no extend to these other techniques, although they share some of the same uncertainties. Ignoring the consequences of the other technologies might be viewed as an inconsistency in policy.

While the initial concerns about the possibility of hazards at the laboratory level appear to have been overstated, other types of potential hazards at different stages of the technology have been identified. Emphasis has shifted somewhat from conjectured hazards that might arise from research and development to those that might be associated with production technologies. As a consequence, there is a clearer mandate for existing Federal regulatory agencies to play a role in ensuring safety in industrial settings.

Regulation of Genetic Engineering

Chapter 11

Tables

Regulation of Genetic Engineering

Introduction

Although no evidence exists that any harmful organism has been created by molecular genetic techniques, most experts believe that some risk* is associated with genetic engineering. One kind is relatively certain and quantifiable—that of working with known toxins or pathogens. Another is uncertain and hypothetical—that of the possible creation of a pathogenic or otherwise undesirable organism by reshuffling genes thought to be harmless. These may be thought of as physical risks because they concern human health or the environment.

Concern has also arisen about the possible long-range impacts of the techniques—that they may eventually be used on humans in some morally unacceptable manner or may change fundamental views of what it means to be human. These possibilities may be thought of as cultural risks, since they threaten fundamental beliefs and value systems.[1]

The issue of whether or not to regulate molecular genetic techniques—and if so, to what extent—defies a simple solution. Perceptions of the nature, magnitude, and acceptability of the risks differ drastically. Approximately 6 years ago, when the scientific community itself accepted a moratorium on certain classes of recombinant DNA (rDNA) research, some scientists considered the concern unnecessary. Today, even though the physical risks of rDNA research are generally considered to be less than originally feared—and the realization of its benefits much closer—some people would still prohibit it.

The Federal Government's approach to this issue has been the promulgation of the Guidelines for Research Involving Recombinant DNA Molecules (Guidelines), by the National Institutes of Health (NIH). (See app. III-C for information about what other countries have done with respect to guidelines for rDNA.) Three other available modes of oversight or regulation are current Federal statutes, tort law, and State and local law.

Framework for the analysis

In deciding how to address the risks posed by genetic engineering, some of the important questions that need to be examined are:

- How broadly the scope of the issue (or problem) should be defined.
  - Who identifies the risks and their magnitude?
  - Who proposes the means for addressing the problem?
- The nature of the procedural, decisionmaking mechanism.
  - Who decides?
  - Who will benefit from the proposed action and who will bear the risk?
  - Will the risk be borne voluntarily or involuntarily?
  - Who has the burden of proof?
  - Should a risk/benefit analysis, or some other approach, be used?
- The available solutions and their adequacy.
  - Should there be full regulation, no regulation, or something in-between?
  - What actions and actors should be covered?
  - What is the appropriate means for enforcing a regulatory decision?
  - Which agency or other group should do the regulating?

Underlying these questions is the proposition, widely accepted by commentators on science policy, that scientists are qualified to assess physical risk, since that involves measuring and evaluating technical data. However, a judgment of safety (the acceptability of that risk) can only be made by society through the political process, since it involves weighing and choosing among values.[2 3 4 5 6] Scientists are not nec-

---

*As used in this chapter, risk means the *possibility* of harm. The *probability* of that harm occurring may be extremely low and/or highly uncertain.

[1]H. Tristam, Engelhardt, Jr., "Taking Risks: Some Background Issues in the Debate Concerning Recombinant DNA Research," *Southern California Law Review* 51:6, pp. 1141-1151, 1978.

[2]William W. Lowrance, *Of Acceptable Risk: Science and the Determination of Safety* (Los Altos, Calif.: William Kaufmann, Inc., 1976).

essarily considered to be more qualified to make decisions concerning social values than other well-informed persons; they may in fact be less qualified when the decision involves possible restrictions on scientific research because of the

high value they place on unrestricted research and because of possible conflicts of interest. Moreover, according to this view, if society is to bear a risk, it should judge the acceptability of that risk and give its informed consent to it.[7]*

(continued from p. 211)

[3]Alvin W. Weinberg, "Science and Trans-Science," *Minerva*, 10:2, April 1972.

[4]Allan Mazur, "Disputes Between Experts," *Minerva* 11:2, April 1973.

[5]Arthur Kantrowitz, "The Science Court Experiment," *Jurimetrics Journal*, vol. 17, 1977, p. 332.

[6]David L. Bazelon, "Risk and Responsibility," *Science*, vol. 205, July 20, 1979, pp. 277-280.

[7]Engelhardt, op.cit.; Lowrance, op.cit.; and Bazelon, op. cit.

*In practice, it may often be difficult to keep the two kinds of decisions separate, since the values of individual scientists may influence their interpretation of technical data, and since policymakers may not have the technical competence to understand the risks sufficiently.[8]

[8]Weinberg, op. cit.; and Bazelon, op. cit.

Current regulation: the NIH Guidelines

The Guidelines have been developing in stages over a period of approximately 6 years as scientists and policymakers have grappled with the risks posed by rDNA techniques. (This history, discussed in app. III-A, is crucial to understanding current regulatory issues, and it serves as a basis for evaluating the Guidelines.) They represent the only Federal oversight mechanism that specifically addresses genetic engineering.

Substantive requirements

The Guidelines apply to all research involving rDNA molecules in the United States or its territories conducted at or sponsored by any institution receiving any support for rDNA research from NIH. Six types of experiments are specifically prohibited: 1) the formation of rDNA derived from certain pathogenic organisms; 2) the formation of rDNA containing genes that make vertebrate toxins; 3) the use of the rDNA techniques to create certain plant pathogens; 4) transference of drug resistance traits to micro-organisms that cause disease in humans, animals, or plants; 5) the deliberate release of any organism containing rDNA into the environment; and 6) experiments using more than 10 liters (l) of culture unless the rDNA is "rigorously characterized and the absence of harmful sequences established." A procedure is specified for obtaining exceptions

from these prohibitions. Five types of experiments are completely exempt.

Those experiments that are neither prohibited nor exempt must be carried on in accordance with physical and biological containment levels that relate to the degree of potential hazard. (See table 35.) *Physical containment* requires methods and equipment that lessen the chances that a recombinant organism might escape. Four levels, designated P1 for the least restrictive through P4 for the most, are defined. *Biological containment* requires working with weakened organisms that are unlikely to survive any escape from the laboratory. Three levels are specified. Classes of permitted experiments are assigned both physical and biological containment levels. Most experiments using *Escherichia coli* K-12, the standard laboratory bacterium used in approximately 80 percent of all experiments covered by the Guidelines, may be performed at the lowest containment levels.

ADMINISTRATION

The Guidelines provide an administrative framework for implementation that specifies the roles and responsibilities of the scientists, their institutions, and the Federal Government. The parties who are crucial to the effective operation of the system are: 1) the Director of NIH, 2) the NIH Recombinant DNA Advisory Committee (RAC), 3) the NIH Office of Recombi-

Table 35.—Containment Recommended by National Institutes of Health

Biological—Any combination of vector and host must be chosen to minimize both the survival of the system outside of the laboratory and the transmission of the vector to nonlaboratory hosts. There are three levels of biological containment:

HV1—　Requires the use of *Escherichia coli* K12 or other weakened strains of micro-organisms that are less able to live outside the laboratory.

HV2—　Requires the use of specially engineered strains that are especially sensitive to ultraviolet light, detergents, and the absence of certain uncommon chemical compounds.

HV3—　No organism has yet been developed that can qualify as HV3.

Physical—Special laboratories (P1-P4)

P1—　Good laboratory procedures, trained personnel, wastes decontaminated.

P2—　Biohazards sign, no public access, autoclave in building, hand-washing facility.

P3—　Negative pressure, filters in vacuum line, class II safety cabinets.

P4—　Monolithic construction, air locks, all air decontaminated, autoclave in room, all experiments in class III safety cabinets (glove box), shower room.

SOURCE: Office of Technology Assessment.

nant DNA Activities (ORDA), 4) the Federal Interagency Advisory Committee on Recombinant DNA Research (Interagency Committee), 5) the Institution where the research is conducted, 6) the Institutional Biosafety Committee (IBC), 7) the Principal Investigator (PI), and 8) the Biological Safety Officer.

The *Director of NIH* carries the primary burden for the Federal Government's oversight of rDNA activities, since he is responsible for implementing and interpreting the Guidelines, establishing and maintaining RAC (a technical advisory committee) and ORDA (whose functions are purely administrative), and maintaining the Interagency Committee (which coordinates all Federal activities relating to rDNA). Under this arrangement, all decisions and actions are taken by the Director or his staff. For major actions, the Director must seek the advice of RAC, and he must provide the public and other Federal agencies with at least 30 days to comment on

proposed actions. Such actions include: 1) assigning and changing containment levels for experiments, 2) certifying new host-vector systems, 3) maintaining a list of rDNA molecules exempt from the Guidelines, 4) permitting exceptions to prohibited experiments, and 5) adopting changes in the Guidelines.

For other specified actions, the Director need only inform RAC, the IBCs, and the public of his decision. The most important of these are: 1) making minor interpretive decisions on containment for certain experiments; 2) authorizing, under procedures specified by RAC, large-scale work (involving more than 10 l of culture) with rDNA that is rigorously characterized and free of harmful sequences; and 3) supporting laboratory safety training programs. Every action taken by the Director pursuant to the Guidelines must present "no significant risk to health or the environment."

RAC is an advisory committee to the Director on technical matters. It meets quarterly. Its purpose, as described in its current charter of June 26, 1980 (and unchanged since its inception in October 1974), is as follows:

> The goal of the Committee is to investigate the current state of knowledge and technology regarding DNA recombinants, their survival in nature, and transferability to other organisms; to recommend guidelines for the conduct of recombinant DNA experiments; and to recommend programs to assess the possibility of spread of specific DNA recombinants and the possible hazards to public health and to the environment. *This Committee is a technical committee, established to look at a specific problem.* (Emphasis added.)

The charter and the Guidelines also assign it certain advisory functions that have changed over time.

The RAC is composed of not more than 25 members. At least eight must specialize in molecular biology or related fields; at least six must be authorities from other scientific disciplines; and at least six must be authorities on law, public policy, the environment, public or occupational health, or related fields. In addition, representatives from various Federal agencies serve as nonvoting members.

ORDA performs administrative functions, which include reviewing and approving IBC membership and serving as a national center for information and advice on the Guidelines and rDNA activities.

The *Interagency Committee* was established in October 1976 to advise the Secretary of the then Department of Health Education and Welfare (HEW) [now Health and Human Services (DHHS)] and the Director of NIH on the coordination of all Federal activities relating to rDNA. It has thus far produced two reports. Its first, in March 1977, concluded that existing Federal law would not permit the regulation of all rDNA research in the United States to the extent considered necessary[9] and recommended new legislation, specifying the elements of that legislation.[10] The second, in November 1977, surveyed international activities on regulating the research and concluded that, while appropriate Federal agencies should continue to work closely with the various international organizations, no formal governmental action was necessary to produce international control by means of a treaty or convention.[11] It is currently considering issues arising from the large-scale industrial applications of rDNA techniques.

Under the Guidelines, essentially all the responsibility for overseeing rDNA experiments lies with those sponsoring or conducting the research. The Institution must implement general safety policies,* establish an IBC, which meets specified requirements, and appoint a Biological Safety Officer. The *Biological Safety Officer,* who is needed only if the Institution conducts experiments requiring P3 or P4 containment, (see table 35) oversees safety standards. The initial responsibility for particular experiments lies

with the PI, the scientist receiving the funding. This person is responsible for determining and implementing containment and other safeguards and training and supervising staff. In addition, the PI must also submit a registration document that contains information about the project to the IBC, and petition NIH for: 1) certification of host-vector systems, 2) exceptions or exemptions from the Guidelines, 3) and determination of containment levels for experiments not covered by the Guidelines. Furthermore, all of the above have certain reporting requirements designed so that ORDA is eventually informed of significant problems, accidents, violations, or illnesses.**

The IBC is designed to provide a quasi-independent review of rDNA work done at an institution. It is responsible for: 1) reviewing all rDNA research conducted at or sponsored by the institution and approving those projects in conformity with the Guidelines; 2) periodically reviewing ongoing projects; 3) adopting emergency plans for spills and contamination; 4) lowering containment levels for certain rDNA and recombinant organisms in which the absence of harmful sequences has been established; and 5) reporting significant problems, violations, illnesses, or accidents to ORDA within 30 days.*** The IBC must be comprised of no fewer than five members who can collectively assess the risks to health or the environment from the experiments. At least 20 percent of the membership must not be otherwise affiliated with the institution where the work is being done, and must represent the interests of the surrounding community in protecting health and the environment. Committee members cannot review a project in which they have been, or expect to be, involved or have a direct financial interest. Finally, the Guidelines suggest that IBC meetings be public; minutes of the meetings and submitted documents must be available to the public on request.

---

[9]*Interim Report of the Federal Interagency Committee on Recombinant DNA Research: Suggested Elements for Legislation,* Mar. 15, 1977, pp. 9-10.

[10]Ibid., pp. 11-15.

[11]*Report of the Federal Interagency Committee on Recombinant DNA Research: International Activities,* November 1977, pp. 13-15.

*These include conducting any health surveillance that it determines to be necessary and ensuring appropriate training for the IBC, Biological Safety Officers, Principal Investigators, and laboratory staff.

**The PI is required to report this information within 30 days to ORDA and his IBC. The Biological Safety Officer must report the same to the Institution and the IBC unless the PI has done so. The Institution must report within 30 days to ORDA unless the PI or IBC has done so.

***It does not have to report if the PI has done so.

The requirements imposed on an institution and its scientists are enforced by the authority of NIH to suspend, terminate, or place other conditions on its funding of the offending projects or all projects at the institution. Compliance is monitored through the requirements for notification mentioned above.

PROVISIONS FOR VOLUNTARY COMPLIANCE

Organizations or individuals who do not receive any NIH funds for rDNA research are not covered by the Guidelines. These include other Federal agencies, institutions and individuals funded by those agencies, and corporations.

Federal agencies other than NIH that conduct or fund rDNA research have proclaimed their voluntary compliance with the Guidelines.* Staff scientists have been so informed by memoranda. As for outside investigators, this policy has been implemented through the grant application process. Instructions in grants applications contain policy statements regarding compliance with the Guidelines, and applicants are sometimes contacted to ascertain their knowledge of the Guidelines. Information has been requested for certain experiments, and IBC membership has been reviewed. From time to time, the agencies have consulted with NIH on matters that need interpretation.

Part VI of the Guidelines is designed to encourage voluntary compliance by industry. It creates a parallel system of project review and IBC approval analogous to that required for NIH-funded projects, modified to alleviate industry's concerns about protection of proprietary information.

The Freedom of Information Act requires Federal agencies, with certain exceptions, to make their records available to the public on request. One of the exceptions is for trade secrets and proprietary information obtained from others. Part VI contains several provisions for protecting this information. Perhaps the most important is a process whereby a corporation may request a presubmission review of the records needed to register its projects with NIH. The DHHS Freedom of Information Officer makes an informal determination of whether the records would have to be released. If they are determined to be releasable, the records are returned to the submitting company. The Guidelines also require that NIH consult with any institution applying for an exemption, exception, or other approval about the content of any public notice to be issued when the application involves proprietary information. As a matter of practice, such applications are also considered by RAC in nonpublic sessions.

Large-scale experiments (more than 10 l of culture) with rDNA molecules are prohibited unless the rDNA is "rigorously characterized and the absence of harmful sequences established." Such experiments are actually scale-ups of potential industrial processes. Those meeting this standard may be approved by the Director of NIH under procedures specified by RAC.* At its September 1979 meeting, RAC adopted procedures for review that require the applicant to submit information on its laboratory practices and containment equipment. Subsequently, recommendations were developed for large-scale uses of organisms containing rDNA. These were published in the Federal Register on April 11, 1980. Besides setting large-scale containment levels, they require the institution to appoint a Biological Safety Officer with specified duties, and to establish a worker health surveillance program for work requiring P3 containment. At its September 1980 meeting, RAC modified its review procedures so that the application need only specify the large-scale containment level at which the work would be done, without providing details on containment equipment. RAC will continue to review the biological aspects of the applications in order to determine that rDNA is rigorously characterized, that the absence of harmful sequences is established, and that the proposed containment is at the appropriate level.

---

*These agencies are the National Science Foundation, the Department of Agriculture, the Department of Energy, the Veterans Administration, and the Center for Disease Control. Two other agencies, which have expressed interest in this research but are not currently sponsoring any projects, are the Department of Defense and the National Aeronautics and Space Administration.

---

*It is NIH, not the company proposing the scale-up, that determines if the rDNA to be used is "rigorously characterized and the absence of harmful sequences established.".[12]

[12]Guidelines for Research Involving Recombinant DNA Molecules, sec. IV-E-1-b-(3)-(d).

Evaluation of the Guidelines

Two basic issues must be addressed. The first is how well the Guidelines confront the risks from genetic engineering, which may not have a definitive answer in view of the uncertainty associated with most of the risks. Consequently, it is also necessary to consider a second issue—whether confidence is warranted in the decisionmaking process responsible for the Guidelines.

THE PROBLEM OF RISK

The Guidelines are designed to address the risks to public health and the environment from either rDNA molecules or organisms and viruses containing them. The underlying premise is that research should not be unreasonably restricted. This is essentially a risk-benefit approach; at the time that the original Guidelines were drafted, it represented a compromise between the extremes of no regulation and of no research without proof of safety. Physical and biological containment levels were established for various experiments based on estimated degrees of risk. The administrative mechanism created by the Guidelines is that of a Federal agency—NIH—advised by a diverse body of experts—RAC. Scientific advice on the technical aspects of risk assessment is provided by technical experts on RAC; public input is provided by experts in nontechnical subjects and by the right of the public to comment on major actions, which are published in the Federal Register. Compliance is accomplished by a combination of local self-regulation and limited Federal oversight, with the ultimate enforcement resting in the Federal funding power.

Since their initial appearance, the Guidelines have evolved. As scientists learned more about rDNA and molecular genetics, two trends occurred. First, containment levels were progressively lowered. Major revisions were made in 1978 and 1980; minor revisions were often made quarterly, as proposals were submitted to the RAC at its quarterly meetings, recommended by RAC, and accepted by the Director. By now, approximately 85 percent of the permitted experiments can be done at the lowest physical and biological containment levels. Second, the degree of centralized Federal over-

sight has been substantially reduced to the point where almost none remains. Under the 1976 Guidelines, all permitted experiments ultimately had to be reviewed by the IBC and ORDA before they could be started; the 1978 Guidelines no longer required preinitiation review of most experiments by ORDA, although ORDA continued to maintain a registry of experiments and to review IBC decisions. Under the November 1980 revision to the Guidelines, there will be no Federal registration or review of experiments for which containment levels are specified in the Guidelines. About 97 percent of the permitted experiments fall into this category.

Preinitiation review of experiments by RAC has been an important part of the oversight mechanism. Expert review encourages experimental design to be well thought out and provides a means for catching potential problems, e.g., one application reviewed by RAC never mentioned that the species to be used as a DNA donor was capable of manufacturing a potent neurotoxin; it was turned down after a RAC member familiar with the species brought this fact to the Committee's attention.[13]

The burdens imposed on rDNA activities by the Guidelines appear to be reasonable in view of continuing concerns about risk. Less than 15 percent of permitted experiments require preinitiation approval by the local IBC's, which usually meet monthly. Preinitiation approval of experiments by NIH is required only for: 1) experiments that have not been assigned containment levels by the Guidelines; 2) experiments using new host-vector systems, which must be certified by NIH; 3) certain experiments requiring case-by-case approval; and 4) requests for exceptions from Guideline requirements. The lowest containment levels place minimal burdens on the experimenter. (see table 35). For industrial applications, NIH approval must be received not only when the project is scaled-up beyond the 10-l limit, but also for each additional scale-up of the same project. Many representatives of industry consider these subse-

---

[13]R. M. Henig, "Trouble on the RAC—Committee Splits Over Downgrading of *E. coli* Containment," *BioScience*, vol. 29, pp. 759, 762, December 1979.

quent approvals to be unnecessary and burdensome.

Information about whether the Guidelines have been a disadvantage for U.S. companies in international competition is scanty. Examples include the approximately 1-year headstart two European groups were given while the cloning of hepatitis B virus was prohibited, the advantage some European companies had in using certain species of bacteria for cloning under conditions that were prohibited in the United States, and the delays some pharmaceutical companies faced because they had to build better containment facilities.

The present Guidelines are a comprehensive, flexible, and nonburdensome way of dealing with the physical risks associated with rDNA research while permitting the work to go forward. That is all they were ever intended to do.

The Scope of the Guidelines.—In many respects, the Guidelines do not address the full scope of the risks of genetic engineering. They cover one technique, albeit the most important; they do not address the admittedly uncertain, long-term cultural risks; they are not legally binding on researchers receiving funds from agencies other than NIH; and they are not binding on industry.

Other genetic techniques present risks similar to those posed by rDNA, but to a lesser degree. Recombinant DNA is the most versatile and efficient technique; it uses the greatest variety of genetic material from the widest number of sources with reasonable assurance of expression by the host cell. Cell fusion of micro-organisms, which also involves the uncertain risk of recombining the genetic material of different species, is significantly less versatile and efficient than rDNA but mixes more genetic material. In addition, the parental cells may contain partial viral genomes that could combine to form a complete genome when the cells are fused. Transformation, a technique known for decades, similarly involves moving pieces of DNA between different cells. However, it is significantly less versatile and efficient than cell fusion, and it is generally considered to be virtually risk-free. Thus, cell fusion is in a gray area

between the other two techniques; yet no risk assessment has been done, and no Federal oversight exists.

Another limitation in the scope of the guidelines—and in the process by which they were formulated—is that long-range cultural risks (as distinguished from policy issues related to safety) were never addressed. As noted by the Director of NIH:[14]

> . . . NIH has been addressing the policy questions involving the safety of this research, not the 'potential future application . . . to the altering of the genetic character of higher forms of life, including man' . . .

Perhaps it was inappropriate to do more. Such ethical issues might be considered premature in view of the level of the development of the technology. The desire among many molecular biologists to move ahead with the research meant that experiments were being done; therefore the immediate potential for harm was to health and the environment. Thus, it was arguably necessary to develop a framework to deal with the risks based on what was known at the time. On the other hand, the broader questions of where the research might eventually lead and whether it should be done at all have been raised in the public debate. They have not been formally considered by the Federal Government.

Another limitation in the scope of the Guidelines is their nonapplicability to research funded or performed by other Federal agencies. However, agencies supporting such research are complying with the Guidelines as a matter of policy. There appears to be little reason for questioning these declarations of general policy. In practice, problems might arise if a mission is perceived to be at odds with the Guidelines or because of simple bureaucratic defense of territory—e.g., when the 1976 Guidelines were promulgated, two agencies—the Department of Defense (DOD) and the National Science Foundation (NSF)—reserved the right to deviate for reasons of national security or differing interpreta-

[14] 43 F.R. 60103, Dec. 22, 1978, citing 43 F.R. 33067, July 28, 1978.

tions, respectively.* DOD no longer claims an exception for national security.[15] NSF took its position when it approved funding for an experiment using a particular species of yeast that had not been certified by NIH, relying on an ambiguously worded section[16] in the Guidelines to assert that it could certify the host. Subsequent revisions explicitly stated that these hosts had to be certified by the Director of NIH[17] [18] and removed many similar ambiguities.

In the final analysis, NIH has indirect leverage over the actions of other agencies through its funding. All non-NIH funded rDNA projects at an institution which also receives NIH funds for rDNA work must comply with the Guidelines; otherwise NIH funds may be suspended or terminated.

While the procedures of other agencies for administering compliance are significantly less formal than those created by the Guidelines for NIH, they do rely heavily on NIH for help and advice, and they coordinate their efforts through the Interagency Committee and their nonvoting membership on RAC. So far, this voluntary compliance by the agencies appears to be working fairly well.

The most significant limitation in the scope of the Guidelines is their nonapplicability to industrial research or production on other than a voluntary basis. This lack of legal authority raises concerns not only about compliance but also about NIH's ability to implement a voluntary program effectively.

Whether every company working with rDNA will view voluntary compliance to be in its best interest depends on a number of factors. In the past, certain short-sighted actions by even a few companies in a given industry has led to

well-documented abuses and a host of Federal laws to curtail them. However, at least two constraints are operating in the case of the biotechnology industry. First, the possibility of tort lawsuits is an inducement to comply with the Guidelines, which would probably be accepted as the standard of care against which alleged negligence would be evaluated. (This concept is discussed in greater detail in the section on Tort Law and Workman's Compensation.) Second, the threat of statutory regulation, which the companies have sought to avoid, always exists. Other factors are also at work. Except for the 10-l limitation, for which case-by-case exceptions must be sought, the large-scale containment recommendations of April 11, 1980, are not excessively burdensome, at least for pharmaceutical companies. The requirements are similar to measures that must currently be taken to prevent product contamination. In addition, the public debate should have made each company aware of the problems and the need for voluntary compliance before it invested substantially in biotechnology; expensive controls will not have to be retrofitted. However, one definite concern is that new companies attracted to the field will perceive their interests differently. Because they did not actually experience the period when legislation seemed inevitable and because they will be late entries in the race, they may be inclined to take shortcuts.

Besides the concern about whether industry has sufficient incentive to comply, there are a number of other reasons for questioning the effectiveness of the voluntary program. First,until very recently no member of RAC was an expert in industrial fermentation technology—yet the Committee has been considering applications from industry for large-scale production since September 1979.* This drawback was demonstrated at its March 1980 meeting, when the Committee expressed uncertainty over what Federal or State safety regulations presently cover standard fermentation technology em-

---

*For a statement of the DOD position, see the minutes of the November 23, 1976, meeting of the Federal Interagency Committee. At that time, DOD had no active or planned rDNA projects. NSF's statement of its intention to "preserve some level of independence of decision" was expressed in an internal NIH memorandum dated February 24, 1978, from the Deputy Director for Science, NIH, to the Director, NIH.

[15]Dr. John H. Moxley, III., Assistant Secretary of Defense for Health Affairs, personal communication, Nov. 18, 1980.

[16]"Fungal or Similar Lower Eukaryotic Host-Vector Systems," 41 F.R. 27902, 27920, July 7, 1976.

[17]43 F.R. 60108, Dec. 22, 1978, sec. III-C-5 of the 1978 Guidelines.
[18]45 F.R. 6724, Jan. 29, 1980, sec. III-C-5 of the 1980 Guidelines.

---

*At its September 1980 meeting, RAC passed the following resolution, which has been accepted by the Director of NIH:[19]

Members should be chosen to provide expertise in fermentation technology, engineering, and other aspects of large-scale production.

A fermentation technology expert was appointed in January 1981.

[19]45 F.R. 77373, Nov. 21, 1980.

ployed by the drug industry. Various members expressed concern in the March and June 1980 meetings about the Committee's continuance to make recommendations on the applications without a firm knowledge of large-scale production.

Second, the provisions in part VI of the Guidelines, which allow prior review of submitted information by the DHHS Freedom of Information Act Officer, give an industrial applicant the option of withholding potentially important information on the grounds of trade secrecy, even when DHHS disagrees. Third, because some RAC members have been opposed to discussing industrial applications in closed session (needed to protect proprietary information), they have chosen not to participate in those sessions. Thus, some diversity of opinion and expertise has been lost. Fourth, monitoring for compliance after the scale-up applications are granted is limited. Some early applications were granted on the condition that NIH could inspect facilities, and at least one inspection was made. Under procedures adopted at the September 1980 meeting, a company's IBC will be responsible for determining whether the facilities meet the standards for the large-scale containment level assigned by RAC. A working group of RAC may visit the companies and their IBCs from time-to-time but only for information gathering purposes, rather than for regulatory actions. Fifth, even if noncompliance were found, no penalties can be imposed.

The members of RAC, acutely aware of the problems with voluntary compliance by industry, have been deliberating about them for almost 2 years. At a meeting in May 1979, they decided, by a vote of nine to six with six abstentions, to support the principle of mandatory compliance with the Guidelines by non-NIH funded institutions. However, the Secretary of HEW (Joseph Califano) decided to continue with the development of voluntary compliance provisions[20] which were adopted as Part VI of the Guidelines in January 1980. Actual RAC review of submissions from the private sector for large-scale work began in September 1979. At a meeting in June 1980, RAC debated the effectiveness

of NIH's quasi-regulation of industry. A primary concern was whether the RAC would be viewed as giving a "stamp of approval" to industrial projects, when, in fact, it has neither the authority nor the ability to do so. One member, lawyer Patricia King, stated:[21]

> Voluntary compliance is the worst of all possible worlds . . . .You achieve none of the objectives of regulation and none of the benefits of being unregulated. All you're saying is 'I give a stamp of approval to what I see here before me without any authority to do anything.'

Most of the speakers expressed the desire that the various agencies in the Interagency Committee be responsible for such regulation. However, the Interagency Committee, which has been studying the problem since January 1980, has yet to decide what it can do. Thus, many of its members see RAC as filling a regulatory void until the traditional agencies take action.

Some regulatory agencies have begun to deal with specific problems within their areas of interest. The Occupational Safety and Health Administration will decide its regulatory policy on the basis of a study of potential risks to workers posed by the industrial use of rDNA techniques being conducted by the National Institute of Occupational Safety and Health (NIOSH). In a letter to the Director of NIH dated September 24, 1980, Dr. Eula Bingham, then Assistant Secretary for Occupational Safety and Health of the Department of Labor, estimated this process would take approximately 2 years. The Environmental Protection Agency (EPA) has awarded several contracts and grants to assess the risks of intentional release of genetically engineered micro-organisms and plants into the environment. And the Food and Drug Administration (FDA) has begun to develop policy with respect to products made by processes using genetically engineered micro-organisms. (Further details on agency actions are discussed in the section, *Federal Statutes.)*

Compliance.—The primary mechanism in the Guidelines for enforcing compliance is local self-regulation, with very limited Federal over-

---

[20]RAC minutes of Sept. 6-7, 1979, p. 16, in *Recombinant DNA Research,* vol. 5, (Wash., D.C.: HEW, 1980), p. 165.

[21]Susan Wright, "Recombinant DNA Policy: Controlling Large-Scale Processing," *Environment,* vol. 22, September 1980, pp. 29,32.

sight. Penalties are based on NIH's power to restrict or terminate its funding.

The initial responsibility for compliance lies with the scientist doing the experiments. A researcher's attitude toward the risks of rDNA techniques and the necessity for the Guidelines appear to be an influential factor in the degree of compliance. A science writer who worked for 3 months in a university lab in 1976 noted sloppy procedures and a cavalier attitude, stating: "Among the young graduate students and post-doctorates it seemed almost chic not to know the NIH rules."[22] On the other hand, in the case of a recent violation of the Guidelines, it appears as if the investigator's graduate students were the first to raise questions.[23][24] Competitiveness is another important factor. Novice scientists must establish reputations, secure tenure in a tight job market, and obtain scarce research funds; established researchers still compete for grants and certainly for peer recognition. This competitive pressure could provide strong incentives to bend the Guidelines; on the other hand, it might be channeled to encourage compliance if it is believed that NIH will in fact penalize violations by restricting or terminating funding.

The first level of actual oversight occurs at the institution. An argument can be made that reliance on the PI and an IBC (that might be composed mostly of the PI's colleagues) provides too great an opportunity for lax enforcement or coverups. On the other hand, spreading responsibility among the institution, the PI, the IBC, and, in the case of more hazardous experiments, the Biological Safety Officer might reduce the chance of violations being overlooked or condoned. This responsibility is enhanced by the reporting requirements borne by each of these parties, designed so that ORDA learns of "significant" problems, accidents, violations, and illnesses. What is "significant" is not defined.

Public involvement at the local level acts as an additional safeguard. Twenty percent of the

IBCs members must be unaffiliated with the institution. IBC documents, including minutes of meetings, are publicly available, but meetings are not required to be held in public. On the other hand, the probable inability of the members who represent the public to understand the technical matters might limit their effectiveness.

How successful has compliance been? Three known violations have occurred. In each, no threat to health and the environment existed. In each, there was some confusion as to why the violations occurred. NIH is presently investigating the third violation. For the first two, it accepted explanations of misunderstandings and misinterpretations of the Guidelines. However, a Senate oversight report concluded:[25]

> While undoubtedly most researchers have observed the guidelines conscientiously, it is equally clear that others have substituted their own judgments of safety for those of NIH.

No firm conclusions can be drawn on the question of compliance. The reporting of only a few violations could be evidence that the compliance mechanism embodied in the Guidelines has been working well. Or it could mean that some violations are not being discovered or reported.

The November 1980 amendments to the Guidelines substantially changed procedures designed to monitor compliance by abolishing a document called a Memorandum of Understanding and Agreement (MUA). It had been required for 15 to 20 percent of all experiments, those thought to be potentially most risky. The MUA, which was to be filed with ORDA by an institution, provided information about each experiment, and it was the institution's certification to NIH that the experiment complied with the Guidelines. By having the MUAs, ORDA could monitor for inconsistencies in interpreting the Guidelines, actual noncompliance, and the consistency and quality with which IBCs functioned nationwide. The amendments continued a trend begun in January 1980, when approximately 80 percent of the experiments,

[22]Janet L. Hopson, "Recombinant Lab for DNA and My 95 Days in It," *Smithsonian*, vol. 8, June 1977, p. 62.

[23]D. Dickson, "Another Violation of NIH Guidelines," *Nature* vol. 286, Aug. 14, 1980, p. 649.

[24]D. Dickson, "DNA Recombination Forces Resignation," *Nature* vol. 287, Sept. 18, 1980, p. 179.

[25]"Recombinant DNA Research and Its Applications," *Oversight Report*, Subcommittee on Science, Technology, and Space of the Senate Committee on Commerce, Science and Transportation, Aug. 1978, p. 17.

those done with *E. coli* K-12, were exempted from the MUA requirement.

The abolition of the MUA essentially abolished centralized Federal monitoring of rDNA experiments. The only current Guideline provision that serves this kind of monitoring function is the requirement that the institution, the IBC, or the PI notify ORDA of any significant violations, accidents, or problems with interpretation. Limited monitoring of large-scale activities continues. Under NIH procedures (which are not part of the Guidelines) for reviewing applications for exemptions from the 10-l limit, the application must include a copy of the registration document filed with the IBC. The manufacturing facilities may also be inspected by NIH, not for regulatory purposes, but to gather information for updating its recommended large-scale containment levels. The abolition of the MUA is consistent with traditional views that Government should not interfere with basic scientific research. Whether or not it will reduce either the incentive to comply with the Guidelines or the likelihood of discovering violations remains to be seen.

THE DECISIONMAKING PROCESS

Another way to evaluate the Guidelines besides considering their substantive requirements is to look at the process by which they were formulated. In a situation where there is uncertainty and even strong disagreement about the nature, scope, and magnitude of the risks, it is difficult to judge whether or not a proposed solution to a problem will be a good one. Society's confidence in the decisionmaking process and in the decisionmakers then becomes the issue. As David L. Bazelon, Chief Judge of the U. S. Court of Appeals for the District of Columbia, has stated:[26]

> When the issues are controversial, any decision may fail to satisfy large portions of the community. But those who are dissatisfied with a particular decision will be more likely to acquiesce in it if they perceive that their views and interests were given a fair hearing. If the decision-maker has frankly laid the competing considerations on the table, so that the public knows the worst as well as the best, he is unlike-

ly to find himself accused of high-handedness, deceit, or cover-up. We simply cannot afford to deal with these vital issues in a manner that invites public cynicism and distrust.

The manner in which the Guidelines themselves evolved has been controversial. (For a detailed discussion see app. III-A.) Initially, the scope and nature of the problem was defined by the scientific community; NIH organized RAC along the lines suggested by the NAS committee letter referred to in app. III-A. One of the goals of RAC was to recommend guidelines for rDNA experiments; it was not charged with considering broader ethical or policy issues or the fundamental question of whether the research should have been permitted at all. The original Guidelines were produced by a committee having only one nonscientist.

In late 1978, the Secretary of HEW significantly restructured RAC and modified the Guidelines in order to increase the system's accountability to the public, to "provide the opportunity for those concerned to raise any ethical issues posed by recombinant DNA research", and to make RAC "the principal advisory body . . . on recombinant DNA policy."[27] However, it has remained in large part a technically oriented body. Its charter was not changed in this respect; the Guidelines themselves state that its advice is "primarily scientific and technical," and matters presented for its consideration have continued to be mostly technical. One area where RAC has played a significant policy role, however, is in dealing with the issue of voluntary compliance by industry.

It could be argued that the system did provide for sufficient public input into the formulation of the problem* and that no other formulation was realistic. The two meetings in 1976 and 1977 of the NIH Director's Advisory Committee and the hearing chaired by the general counsel of HEW in the fall of 1978 provided the opportunity for public comment on the overall Fed-

---

[26]D. L. Bazelon, "Coping With Technology Through the Legal Process," 62 *Cornell Law Review* 817,825, June 1977.

[27]Joseph A. Califano, "Notice of Revised Guidelines—Recombinant DNA Research," 43 F.R. 60080-60081, Dec. 22, 1978.

*The problem was conceived in terms of how to permit the research to be done while limiting the physical risks to an acceptable level. Other formulations were possible, the broadest being how to limit all risks, including cultural ones, to an acceptable level. Such a formulation could have resulted in a prohibition of the research.

eral approach to the controversy, including whether or not the problem had been too narrowly phrased. Similarly, Congress had the opportunity in 1977 to reevaluate the entire institutional response, taking into account any moral objections to the research in addition to those concerning safety. Yet the principal bills were based on the proposition that the research continue in a regulated fashion.

A related issue is the one of burden of proof. Should the proponents of a potentially beneficial technology be required to demonstrate minimal or acceptable risk even if that risk is uncertain or even hypothetical? Or should its opponents be required to demonstrate unacceptable risk? If the proposition is accepted that those who bear the risks, in this case the public as well as the scientists, must judge their acceptability, then the burden must be on the proponents. The scientific community clearly accepted this burden. The moratorium proposed by the NAS committee in July 1974 called for a suspension of certain types of rDNA experiments until the risks could be evaluated and procedures for adequately dealing with those risks could be developed. The Guidelines prohibited some experiments, specified containment levels for others, and required certification of host-vector systems. All actions approved by the Director of NIH, including the lessening of the restrictions imposed by the original Guidelines, have had to meet the requirement of presenting "no significant risk to health or the environment."

Two other criticisms have been directed against RAC, particularly in its early days. The first concerned inherent conflicts of interest. RAC's members were drawn from molecular biology and related fields. One of the early drafts of the Guidelines was criticized as being "tailored to fit particular experiments that are already on the drawing boards."[28] However, only a few of the members were actually working with rDNA.[29] A more serious criticism was the lack of a broad range of expertise. Although

the risks had been expressed in terms of potential hazards to human health and the environment, the original RAC had no experts in the areas of epidemiology, infectious diseases, botany or plant pathology, or occupational health. It did have one expert in enteric organisms, *E. coli* in particular.

These shortcomings were eventually remedied by expanding RAC's membership to allow the appointment of other experts, including some from nontechnical fields such as law and ethics. In addition to providing knowledge of other fields, these members served as disinterested advisors, since they had no direct interest in expediting the research. Thus, the Government dealt with the problem of conflicts of interest by offsetting the interested group with other groups. In view of the need for the technical expertise of the molecular biologists, this approach seems reasonable; nevertheless the matter could probably have been handled more expeditiously. Although the April 1975 amendment to the RAC charter added experts from such fields as epidemiology and infectious diseases, the charter did not require plant experts until September 1976 (shortly after the passage of the original Guidelines) and occupational health specialists until December 1978. In addition, while two nontechnical members were added in 1976 (one before and one after passage of the Guidelines), their number was not increased until Secretary Califano reconstituted the Committee in late 1978.

The present makeup of RAC is fairly diverse. As of September 1980, nine of its members specialized in molecular biology or related fields, seven were from other scientific disciplines, and eight were from the areas of law, public policy, the environment, and public or occupational health.[30] Moreover, since December 1978, representatives of the interested Federal agencies have been sitting as nonvoting members. In January 1981, an expert on fermentation was added.

[28]N. Wade, "Recombinant DNA: NIH Sets Strict Rules to Launch New Technology," 190 *Science* 1175,1179, 1975.

[29]Dr. Elizabeth Kutter, a member of RAC at that time, personal communication, Sept. 11, 1980.

[30]Dr. Bernard Talbot, Special Assistant to the Director, NIH, personal comunication, Sept. 18, 1980.

One conflict of interest not solved by expanding the diversity of the RAC's membership is institutional in nature. NIH, the agency having primary responsibility for developing and administering the Guidelines, views its mission as one of promoting biomedical research. Although the Guidelines are not regulations, they contain many of the elements of regulations. They set standards, offer a limited means to monitor for compliance, and provide for enforcement, at least for institutions receiving NIH grants to do rDNA work; thus, they may be considered quasi-regulatory. Regulation is not only foreign but antithetical to NIH's mission. The current Director stated publicly at the June 1980 RAC meeting that the role of NIH is not one of a regulator, a role that must be avoided. Under these circumstances, perhaps another agency, or another part of DHHS, might be more appropriate for overseeing the Guidelines, since the attitudes and priorities of promoters are usually quite different from that of regulators.

If RAC has always been essentially a technical advisory body, who then has made the value decisions concerning the acceptability of the risks presented by rDNA and the means for dealing with them? The final decisionmaker has been the Director of NIH, with the notable exception in the case of the 1978 Guidelines, which contained the significant procedural revisions needed to meet Secretary Califano's approval.[31] The Director did have access to diverse points of view through the Director's Advisory Committee meetings and the public hearings held before the 1978 Guidelines. (See app. III-A.) In addition, major actions were always accompanied by a statement discussing the relevant issues and explaining the basis for the decisions; after the 1978 revisions, major actions had to be proposed for public comment before decisions were made. In theory, it may have been preferable for the public to have been substantially involved in the actual formation of the original Guidelines rather than simply to have reacted to a finished product. However, this probably would have slowed the process at a time when the strong desire of the molecular biologists to

use the rDNA techniques could have threatened the notion of self-regulation. Today, there appears to be reasonable opportunity for public input through the process of commenting on proposed actions.

Conclusion

The Guidelines are the result of an extraordinary, conscientious effort by a combination of scientists, the public, and the Federal Government, all operating in an unfamiliar realm. They appear to be a reasonable solution to the problem of how to minimize the risks to health and the environment posed by rDNA research in an academic setting, while permitting as much of that research as possible to proceed. They do not in any way deal with other molecular genetic techniques or with the long-term social or philosophical issues that may be associated with genetic engineering.

The Guidelines have been an evolving document. As more has been learned about rDNA and molecular genetics, containment levels have been significantly lowered. Also, the degree of Federal oversight has been substantially lessened. Under the November 1980 Guidelines, virtually all responsibility for monitoring compliance is placed on the IBCs. NIH's role will involve primarily: 1) continuing interpretation of the Guidelines, 2) certifying new host-vector systems, 3) serving as a clearinghouse of information, 4) continuing risk assessment experiments, and 5) coordinating Federal and local activities.

The most significant short-term limitation of the Guidelines is the way they deal with commercial applications and products of rDNA techniques. Although large-scale containment levels and related administrative procedures exist, there are several reasons for questioning the effectiveness of the voluntary compliance concept. The most serious problem has been the lack of expertise in fermentation technology on RAC. In addition, since the Guidelines are not legally binding upon industry, the NIH lacks enforcement authority, although there has been no evidence of industrial noncompliance. Finally, because of its role as a promoter of bio-

---

[31]Califano, op. cit.

medical research, NIH cannot be expected to act aggressively to fill this regulatory void.

As a model for societal decisionmaking on technological risks, the system created by the Guidelines could serve as a valuable precedent. It does a reasonable job of combining substantive scientific evaluation of technical issues with procedural safeguards designed to accommodate social values and to limit conflicts of interest. The only major criticism is that procedural safeguards and public input were not significant factors when the rDNA problem was first addressed.

Other means of regulation

There are three other means available for regulating molecular genetic techniques and their products—current Federal statutes, tort law and workmen's compensation, and State and local laws. These all may be used to remedy some of the limitations of the Guidelines.

Federal statutes

The question of whether existing Federal statutes provide adequate regulatory authority first arose with respect to rDNA research. In March 1977, the Interagency Committee concluded that while a number of statutes* could provide authority to regulate specific phases of work with rDNA, no single one or combination would clearly reach all rDNA research to the extent deemed necessary by the Committee. Furthermore, while some could be broadly interpreted, the Committee believed that regulatory action taken on the basis of those interpretations would be subject to legal challenge.[32] This was the basis for their conclusion that specific legislation was needed and was one of the reasons behind the legislative effort discussed in app. III-A.

With respect to commercial uses and products of rDNA and other genetic techniques, a much more certain basis for regulation exists. Many of the Federal environmental, product safety, and public health laws are directed toward industrial processes and products. To a large extent, the genetic technologies will produce chemicals, foods, and drugs—as well as pollutant byproducts—that will clearly come within the scope of these laws. However, there may be limitations in these laws and questions of their interpretation that may arise with respect to the manufacturing process, which employs large quantities of organisms, and when there is an intentional release of microorganisms into the environment—e.g., for cleaning up pollution. For a list of pertinent laws, see table 36.)

The Federal Food, Drug, and Cosmetic Act (FFDCA) and section 351 of the Public Health Service Act (42 U.S.C. 262) give FDA authority over foods, drugs, biological products (such as vaccines), medical devices, and veterinary medicines. This authority will also apply to those products when they are made by genetic engi-

Table 36.—Statutes That Will Be Most Applicable to Commercial Genetic Engineering

1. Federal Food, Drug, and Cosmetic Act (21 U.S.C. §301 et. seq.)
2. Occupational Safety and Health Act (29 U.S.C. §651 et. seq.)
3. Toxic Substances Control Act (15 U.S.C. §2601 et. seq.)
4. Marine Protection, Research, and Sanctuaries Act (33 U.S.C. §1401 et. seq.)
5. Federal Water Pollution Control Act, as amended by the Clean Water Act of 1977 (33 U.S.C. §1251 et. seq.)
6. The Clean Air Act (42 U.S.C. §7401 et. seq.)
7. Hazardous Materials Transportation Act (49 U.S.C. §1801 et. seq.)
8. Solid Waste Disposal Act, as amended by the Resource Conservation and Recovery Act of 1976 (42 U.S.C. §6901 et. seq.)
9. Public Health Service Act (42 U.S.C. §201 et. seq.)
10. Federal Insecticide, Fungicide, and Rodenticide Act (7 U.S.C. §136 et. seq.)

SOURCE: Office of Technology Assessment.

---

*The Committee concentrated on the following statutes: 1) the Occupational Safety and Health Act (29 U.S.C. §651 et. seq.); 2) the Toxic Substances Control Act (15 U.S.C. §2601 et. seq.); 3) the Hazardous Materials Transportation Act (49 U.S.C. §1801 et. seq.); and 4) sec. 361 of the Public Health Service Act (42 U.S.C. §264).

[32]*Interim Report of the Federal Interagency Committee on Recombinant DNA Research: Suggested Elements for Legislation*, op. cit.

neering methods. However, interpretive questions arising out of the unique nature of the technologies—such as the type of data necessary to show the safety and efficacy of a new drug produced by rDNA techniques—will have to be resolved by the administrative process on a case-by-case basis.

FDA has not published any statements of official policy toward products made by genetic engineering. Since it has different statutory authority for different types of products, it is likely that regulation will be on a product-by-product basis through the appropriate FDA bureau. Substances produced by genetic engineering will generally be treated as analogous products produced by conventional techniques with respect to standards for chemistry, pharmacology, and clinical protocols; however, quality controls may have to be modified to assure continuous control of product purity and identity. In addition, for the time being, the Bureau of Drugs and the Bureau of Biologics will require a new Notice of Claimed Investigational Exemption for a New Drug and a new New Drug Application for products made by rDNA technology, even if identity with the natural substance or with a previously approved drug is shown. This policy is based on the position that drugs or biologics made by rDNA techniques have not become generally recognized by experts as safe and effective and therefore meet the statutory definition of a "new drug."[33]*

FFDCA also permits regulation of drug, food, and device manufacturing. Certain FDA regulations, called Good Manufacturing Practices, are designed to assure the quality of these products. FDA may have to revise these to accommodate genetic technologies; it has the authority to do so. It probably does not have the authority to use these regulations to address any risks to workers, the public, or the environment, since FFDCA is designed to protect the consumer of the regulated product.

The statute most applicable to worker health and safety is the Occupational Safety and Health Act, which grants the Secretary of Labor broad power to require employers to provide a safe workplace for their employees. This power includes the ability to require an employer to modify work practices and to install control technology. The statute creates a general duty on employers to furnish their employees with a workplace "free from recognized hazards that are causing or are likely to cause death or serious physical harm," and it requires employers to comply with occupational safety and health standards set by the Secretary of Labor. According to a recent Supreme Court case, a standard may be promulgated only on a determination that it is "reasonably necessary and appropriate to remedy a significant risk of material health impairment."[34] Because these fairly stringent requirements limit the Act's applicability to recognized hazards or significant risks, the statute could not be used to control manufacturing where the genetic techniques presented only hypothetical risks. However, it should be applicable to large-scale processes using known human toxins, pathogens, or their DNA.

The Secretary of Labor is also directed to account for the "urgency of the need" in establishing regulatory priorities. How the Department of Labor will view genetic technologies within its scale of priorities remains to be seen. NIOSH, the research organization created by this statute, has been studying rDNA production methods to determine what risks, if any, are being faced by workers. It has conducted fact-finding inspections of several manufacturers, and it is planning a joint project with EPA to assess the adequacy of current control technology. In addition, a group established by the Center for Disease Control (CDC) together with NIOSH will be making recommendations on: 1) the medical surveillance of potentially exposed workers, 2) the central collection and analysis of medical data for epidemiological purposes, and 3) the establishment of an emergency response team.[35]

[33]Minutes of the Industrial Practices Subcommittee of the Federal Interagency Advisory Committee on Recombinant DNA Research, Dec. 16, 1980, p. 3.
*Sec. 201(p) of the FFDCA (21 U.S.C. §321(p)) defines a new drug as "any drug . . . the composition of which is such that such drug is not generally recognized, among experts qualified by scientific training and experience . . . as safe and effective . . . ."

[34]*Industrial Union Department, AFL-CIO* v. *American Petroleum Institute,* 100 S.Ct. 2844,2863, 1980.
[35]Minutes of the Industrial Practices Subcomittee of the Federal Interagency Advisory Committee on Recombinant DNA Research, Dec. 16, 1980, op. cit., p. 6.

The Toxic Substances Control Act (TSCA) was intended by Congress to fill in the gaps in the other environmental laws. It authorizes EPA to acquire information on "chemical substances" in order to identify and evaluate potential hazards and then to regulate the production, use, distribution, and disposal of those substances.

A "chemical substance" is defined under section 3(2) of this Act as "any organic or inorganic substance of a particular molecular identity," including "any combination of such substances occurring in whole or in part as a result of a chemical reaction or occurring in nature."* This would include DNA molecules; however, it is unclear if the definition would encompass genetically engineered organisms. In promulgating its Inventory Reporting Regulations under TSCA on December 23, 1977, EPA took the following position in response to a comment that commercial biological preparations such as yeasts, bacteria, and fungi should not be considered chemical substances:[36]

> The Administrator disagrees with this comment . . . . This definition [of chemical substance] does not exclude life forms which may be manufactured for commercial purposes and nothing in the legislative history would suggest otherwise.

However, in a December 9, 1977, letter responding to a Senate inquiry, EPA Administrator Douglas M. Costle stated:[37]

[A]lthough there is a general consensus that recombinant DNA molecules are "chemical substances" within the meaning of section 3 of TSCA, it is not at all clear whether a host organism containing recombined DNA molecules fits—or was intended to fit—that definition . . . . If such organisms are subject to TSCA on the grounds that they are a "combination of . . . substances occurring in whole or in part as a result of a chemical reaction," the Agency might logically have to include all living things in the definition of "chemical substance"—an inter-

pretation which I am confident the Congress neither contemplated nor intended.

If EPA were to take the broader interpretation, and if that were to survive any legal challenge, TSCA would have great potential for regulating commercial genetic engineering by regulating the organisms. Under section 4 of this Act, EPA can adopt rules requiring the testing of chemical substances that "may present an unreasonable risk"* to health or the environment when existing data are insufficient to make a determination. Under section 5, the manufacturer of a new chemical substance is required to notify EPA 90 days before beginning production and to submit any test data available on the chemical's health or environmental effects. If EPA decides that the data are insufficient for evaluating the chemical's effects and that it "may present an unreasonable risk" or will be produced in substantial quantities, the chemical substance's manufacture or use can be restricted or prohibited. Under section 6, EPA can prohibit or regulate the manufacture or use of any chemical substance that "presents, or will present an unreasonable risk of injury to health or the environment."

As with the Occupational Safety and Health Act, the scientific evidence probably does not support a finding that most genetically engineered molecules or organisms present an unreasonable risk. On the other hand, the standard in section 5—may present an unreasonable risk—and the requirement for a premanufacturing notice would permit EPA to evaluate cases where genetically engineered micro-organisms were proposed to be released into the environment.

Several other environmental statutes will apply, mainly with respect to pollutants, wastes, or hazardous materials.** The Marine Protec-

---

*Substances subject solely to FFDCA or the Federal Insecticide, Fungicide, and Rodenticide Act are excluded from this definition.
[36]42 F.R., 64572, 64584, Dec. 23, 1977.
[37]Letter to Adlai E. Stevenson, Chairman, Subcommittee on Science, Technology, and Space, U.S. Senate Committee on Commerce, Science, and Transportation, in *Oversight Report, Recombinant DNA Research and Its Applications*, 95th Cong., 2d sess., August 1978, p.88.

*The term "unreasonable risk" is not defined in the statute. However, the legislative history indicates that its determination involves balancing the probability that harm will occur and the magnitude and severity of that harm, against the effect of the proposed regulatory action and the availability to society of the benefits of the substance.[38]
[38]H. Rept. 94-1341, 94th Cong., 2d sess. 1976, pp. 13-15.
**Two consumer protection statutes were considered but were determined to be virtually inapplicable These were: the Federal Hazardous Substances Act (15 U.S.C. §1261 et. seq.); and the Consumer Product Safety Act (15 U.S.C. §2051 et. seq.).

tion, Research, and Sanctuaries Act prohibits ocean dumping without an EPA permit of any material that would "unreasonably degrade or endanger human health, welfare, or amenities, or the marine environment, ecological systems, or economic potentialities."[39] "Material" is defined as "matter of any kind or description, including . . . biological and laboratory waste . . . and industrial . . . and other waste."[40] The Federal Water Pollution Control Act regulates the discharge of pollutants (which include biological materials) into U.S. waters, and the Solid Waste Disposal Act regulates hazardous wastes. The Clean Air Act regulates the discharge of air pollutants, which includes biological materials. Especially applicable is section 112 (42 U.S.C. § 7412), which allows EPA to set emission standards for hazardous air pollutants—those for which standards have not been set under other sections of the Act and which "may reasonably be anticipated to result in an increase in mortality or an increase in serious irreversible, or incapacitating reversible, illness." The Hazardous Materials Transportation Act covers the interstate transportation of dangerous articles, including etiologic (disease-causing) agents. The Secretary of Transportation may designate as hazardous any material that he finds "may pose an unreasonable risk to health and safety or property" when transported in commerce in a particular quantity and form.[41]

Section 361 of the Public Health Service Act (42 U.S.C. §264) authorizes the Secretary of HEW (now DHHS) to ". . . make and enforce such regulations as in his judgment are necessary to prevent the introduction, transmission, or spread of communicable diseases . . . ." Because of the broad discretion given to the Secretary, it has been argued that this section provides sufficient authority to control all rDNA activities.* Others have argued that its purpose is to protect only human health; for regulations to be valid, there would have to be a supportable finding of a connection between rDNA and human disease. In any event, HEW declined to promulgate any regulations.

The following conclusions can therefore be made on the applicability of existing statutes. First, the products of genetic technologies—such as drugs, chemicals, pesticides,** and foods—would clearly be covered by statutes already covering these generic categories of materials. Second, uncertainty exists for regulating either production methods using engineered micro-organisms or their intentional release into the environment, when risk has not been clearly demonstrated. Third, the regulatory agencies have begun to study the situation but have not promulgated specific regulations. Fourth, since regulation will be dispersed throughout several agencies, there may be conflicting interpretations unless active efforts are made by the Federal Interagency Committee to develop a comprehensive, coordinated approach.

Tort law and workmen's compensation

Statutes and regulations are usually directed at preventing certain types of conduct. While tort law strives for the same goal, its primary purpose is to compensate injuries. (A tort is a civil wrong, other than breach of contract, for which a court awards damages or other relief.) By its nature, tort law is quite flexible, since it has been developed primarily by the courts on a case-by-case basis. Its basic principles can easily be applied to cases where injuries have been caused by a genetically engineered organism, product, or process. It therefore can be applied to cases involving genetic technologies as a means of compensating injuries and as an incentive for safety-conscious conduct. The most applicable concepts of tort law are negligence and strict liability. (A related body of law—workmen's compensation—is also pertinent.)

Negligence is defined as conduct (an act or an omission) that involves an unreasonable risk of harm to another person. For the injured party to be compensated, he must prove in court that: 1) the defendant's conduct was negligent, 2) the

---

[39] 33 U.S.C. § 1412.
[40] 33 U.S.C. § 1402(c).
[41] 49 U.S.C. § 1803.
*On Nov. 11, 1976, the Natural Resources Defense Council and the Environmental Defense Fund petitioned the Secretary of HEW to promulgate regulations concerning rDNA under this Act.

**Pesticides are subject to the Federal Insecticide, Fungicide, and Rodenticide Act, 7 U.S.C. §136 et. seq..

defendant's actions in fact caused the injury, and 3) the injury was not one for which compensation should be denied or limited because of overriding policy reasons.

Because of the newness of genetic technology, legal standards of conduct (e.g., what constitutes unreasonable risk) have not been articulated by the courts. If a case were to arise, a court would undoubtedly look first to the Guidelines. Even if a technique other than rDNA were involved, they would provide a general conceptual framework for good laboratory and industrial techniques. Other sources for standards of conduct include: 1) CDC's guidelines for working with hazardous agents; 2) specific Federal laws or regulations, such as those under the Public Health Service Act covering the interstate transportation of biologic products and etiologic agents; and 3) industrial or professional codes or customary practices, such as generally accepted containment practices in the pharmaceutical industry or in a microbiology laboratory. Compliance with these standards, however, does not foreclose a finding of negligence, since the courts make the ultimate judgment of what constitutes proper conduct. In several cases, courts have decided that an entire industry or profession has lagged behind the level of safe practices demanded by society.* Conversely, noncompliance with existing standards almost surely will result in a finding of negligence, if the other elements are also present.

Causation may be difficult to prove in a case involving a genetically engineered product or organism. In the case of injury caused by a pathogenic micro-organism—e.g., it may be difficult to isolate and identify the micro-organism and virtually impossible to trace its origin, especially if it had only established a transitory ecological niche. In addition, it might be difficult to reconstruct the original situation to determine if the micro-organism simply escaped despite

precautions or if culpable human action was involved. On the other hand, if a micro-organism or toxin is identified, it may be so unique because of its engineering that it can be readily associated with a company known to produce it or with a scientist known to be working with it.**

The law recognizes that not every negligent act or omission that causes harm should result in liability and compensation—e.g., the concept of "foreseeable" harm serves to limit a defendant's liability. The underlying social policy is that the defendant should not be liable for injuries so random or unlikely as to be not reasonably foreseeable. This determination is made by the court. In the case of a genetically engineered organism, extensive harm would probably be foreseeable because of the organism's ability to reproduce; how that harm could occur might not be foreseeable.

Unlike negligence, strict liability does not require a finding that the defendant breached some duty of care owed to the injured person; the fact that the injury was caused by the defendant's conduct is enough to impose liability regardless of how carefully the activity was done. For this doctrine to apply, the activity must be characterized as "abnormally dangerous." To determine this, a court would look at the following six factors, no one of which is determinative:[42]

1. existence of a high risk of harm,
2. great gravity of the harm if it occurs,
3. inability to eliminate the risk by exercising reasonable care,

---

*For example, see: The T. J. Hooper, 60 F. 2d 737 (2d Cir. 1932), concerning tugboats; and Helling v. Carey, 519 P. 2d 981 (1974), where the court held that the general practice among ophthalmologists of not performing glaucoma tests on asymptomatic patients under 40 (because they had only a one in 25,000 chance of having the disease) would not prevent a finding of negligence when such a patient developed the disease.

**If several companies were working with the micro-organism, it could be impossible to prove which company produced the particular ones that caused the harm. A recent California Supreme Court case, Sindell v. Abbott Laboratories, 26 Cal.3d 588, 1980, could provide a way around this problem if the new theory of liability that it establishes becomes widely accepted by courts in other jurisdictions. The Court ruled that women whose mothers had taken diethylstilbestrol, a drug that allegedly caused cancer in their daughters, could proceed to trial against manufacturers of the drug, even though most of the plaintiffs would not be able to show which particular manufacturers produced the drug. The Court said that when the defendant manufacturers had a substantial share of the product market, liability, if found, would be apportioned among the defendants on the basis of their market share. A particular defendant could escape liability only by proving it could not have made the drug.

[42]Restatement (Second) of Torts §520 (1976).

4. extent to which the activity is not common,
5. inappropriateness of the activity to the place where it is done, and
6. the activity's value to the community.

Given the current consensus about the risks of genetic techniques, it would be difficult to argue that the doctrine of strict liability should apply. However, in the extremely unlikely event that a serious, widespread injury does occur, that alone would probably support a court's determination that the activity was abnormally dangerous, regardless of its probability. In such cases, the courts have generally relied on the principle of "enterprise liability"—that those engaged in an enterprise should bear its costs, including the costs of injuries to others.[43]

For either negligence or strict liability, the person causing the harm is liable. Under the legal principle of *respondeat superior,* liability is also imputed from the original actor to people or entities who have a special relationship with him—e.g., employers. Thus, a corporation can be liable for the torts of its scientists or production workers. Similarly, a university, an IBC, a Biological Safety Officer, and a PI would probably be liable for the torts of scientists and students under their direction.

Another body of law designed to compensate injuries deserves brief mention. Workmen's compensation is a statutory scheme adopted by the States and—for specific occupations or circumstances—by the Federal Government to compensate injuries without a need for showing fault. The employee need only show that the injury was job-related. He is then compensated by the employer or the employer's insurance company. It would clearly apply to genetic engineering.

Tort law and workmen's compensation will be available to compensate any injuries resulting from the use of molecular genetic techniques, especially from their commercial application. Tort law may also indirectly prevent potentially hazardous actions, although the de-

terrent effect of compensation is less efficient than direct regulation—e.g., the threat of lawsuits will not necessarily discourage high-risk activities where problems of proof make recovery unlikely, where the harm may be small and widespread (as with mild illness suffered by a large number of people), or where profits are less than the cost of prevention but greater than expected damage awards and legal costs.

Tort law has two other limitations. First, tort litigation involves high costs to the plaintiff, and indirectly to society. Second, it cannot adequately compensate the victims of a catastrophic situation where liability would bankrupt the defendant.

State and local law

Under the 10th amendment to the Constitution, all powers not delegated to the Federal Government are reserved for the States or the people. One of those is the power of the States and municipalities to protect the health, safety, and welfare of their citizens. Thus, they can regulate genetic engineering.

The reasons espoused in favor of local regulation are based on the traditional concept of local autonomy; those most likely to suffer any adverse affects of genetic engineering should control it. Also, local and State governments are usually more accessible to public input than the Federal Government. Consequently, judgments on the acceptability of the risks will more precisely reflect the will of the segment of the public most directly affected.

A number of arguments have been made against local as opposed to Federal regulation. The primary one is that regulation by States and communities would give rise to a random patchwork of confusing and conflicting controls. In addition, States and especially localities may not have the same access as the Federal Government to the expertise that should be used in the formulation of rational controls. Finally, any risks associated with rDNA or other techniques are not limited by geographic boundaries; therefore, they ought to be dealt with nationally. The above arguments reflect the position that regulation of genetic technologies is a na-

---

[43]R. Dworkin, "Biocatastrophe and the Law: Legal Aspects of Recombinant DNA Research," in *The Recombinant DNA Debate,* Jackson and Stitch (eds.) (Englewood Cliffs, N.J.: Prentice-Hall, Inc. 1979), pp. 219, 223.

tional issue that can be handled most effectively at the Federal level.

A few jurisdictions have used their authority in the case of rDNA.* The most comprehensive regulation was created by the States of Mary-

*Cambridge, Mass., established a citizens' study group that recommended that researchers be subject to some additional restrictions beyond those of the Guidelines. These were embodied in an ordinance passed by the City Council on Feb. 7, 1977. Berkeley, Calif., passed an ordinance requiring private research to conform to the Guidelines. Similar ordinances or resolutions were passed by Princeton, N.J., Amherst, Mass., and Emeryville, Calif.

land and New York.[44] [45] Currently, there is little, if any, effort on the State or local level to pass laws or ordinances covering rDNA or similar genetic techniques, and there is little activity under the existing laws.

[44]*Annotated Code of Maryland*, art. 43 §§ 898-910 (supp. 1978).
[45]*McKinney's Consolidated Laws of New York*, Public Health Law, art. 32-A §§3220-3223 (supp. 1980)

Conclusion

The initial question with respect to regulating genetic engineering is how to define the scope of the problem. This will depend largely on what groups are involved in that process and how they view the nature, magnitude, and acceptability of the risks. Similarly, the means of addressing the problem will be determined by how it is defined and who is involved in the actual decisionmaking process. For these reasons, it is important that regulatory mechanisms combine scientific expertise with procedures to accommodate the values of those bearing the risk so that society may have confidence in those mechanisms.

Currently, genetic techniques and their products are regulated by a combination of the Guidelines, Federal statutes protecting health and the environment, some State or local laws, and the judicially created law of torts, which is available to compensate injuries after they occur. In most cases, this system appears adequate to deal with the risks to health and the environment. However, there is some concern regarding commercial applications for the following reasons: 1) the voluntary applicability of the Guidelines to industry, 2) RAC's insufficient expertise in fermentation technology, 3) the potential interpretive problems in applying existing law to the workplace and to situations where micro-organisms are intentionally released into the enviornment, and 4) the absence of a definitive regulatory posture by the agencies.

Issue and Options

ISSUE: **How could Congress address the risks associated with genetic engineering?**

A number of options are available, ranging from deregulation through comprehensive new regulation. An underlying issue for most of these options is: What are the constitutional constraints placed on congressional regulation of molecular genetic techniques, particularly when they are used in research? (This is discussed in app. III-B.)

OPTIONS:

A: *Congress could maintain the status quo by letting NIH and the regulatory agencies set the Federal policy.*

This option requires Congress to determine that legislation to remedy the limitations in current Federal oversight would result in unnecessary and burdensome regulation. No known harm to health or the environment has occurred under the current system, and the agencies generally have significant legal authority and expertise that should permit them to adapt to most new problems posed by genetic engineering. The agencies have been consulting with each other through the Interagency Committee, and the three agencies that will play the most important role in regulating large-scale commercial activities—FDA, OSHA, and EPA—have been studying the situation.

The disadvantages of this option are the lack of a centralized, uniform Federal response to the problem, and the possibility that risks associated with commercial applications will not be adequately addressed. Certain applications, such as the use of micro-organisms for oil recovery are not unequivocally regulated by current statutes; broad interpretations of statutory language in order to reach these situations may be overturned in court. Conflicting or redundant regulations of different agencies would result in unnecessary burdens on those regulated. In addition, some commercial activity is now at the pilot plant stage, but the responsible agencies have yet to establish official policy and to devise a coordinated plan of action.

B: Congress could require that the Federal Interagency Advisory Committee on Recombinant DNA Research prepare a comprehensive report on its members' collective authority to regulate rDNA and their regulatory intentions.

The Industrial Practices Subcommittee of this Committee has been studying agency authority over commercial rDNA activities. Presently, there is little official guidance on regulatory requirements for companies that may soon market products made by rDNA methods.—e.g., companies are building fermentation plants without knowing what design or other requirements OSHA may mandate for worker safety. As was stated by former OSHA head, Dr. Eula Bingham, it will take at least 2 years for OSHA to set standards, if the current NIOSH study shows a need for them.[46]

A congressionally mandated report would assure full consideration of these issues by the agencies and expedite the process. It could in-

---

[46]Letter from Dr. Eula Bingham, Assistant Secretary for Occupational Safety and Health, to Dr. Donald Fredrickson, Director, NIH, Sept. 24, 1980.

clude the following: 1) a section prepared by each agency that assesses its statutory authority and articulates what activities and products will be considered to come within its jurisdiction, 2) a summary section that evaluates the adequacy of existing Federal statutes and regulations as a whole with respect to commercial genetic engineering, and 3) a section proposing any specific legislation considered to be necessary.

The principal disadvantages of this option are that it may be unnecessary and impractical. The agencies are studying the situation, which must be done before they can act. Also, it is often easier and more efficient to act on each case as it arises, rather than on a hypothetical basis before the fact.

C: Congress could require Federal monitoring of all rDNA activity for a limited number of years.

This option represents a "wait and see" position by Congress and the middle ground between the status quo and full regulation. It recognizes and balances the following factors: 1) the absence of demonstrated harm to human health or the environment from genetic engineering; 2) the continuing concern that genetic engineering presents risks; 3) the lack of sufficient knowledge from which to make a final judgment; 4) the existence of an oversight mechanism that seems to be working well, but that has clear limitations with respect to commercial activities; 5) the virtual abolition of Federal monitoring of rDNA activities by the recent amendments to the Guidelines; and 6) the expected increase in commercial genetic engineering activities.

Monitoring involves the collection and evaluation of information about an activity in order to know what is occurring, to determine the need for other action, and to be able to act if necessary. More specifically, this option would provide a data base that could be used for: 1) determining the effectiveness of voluntary compliance with the Guidelines by industry and mandatory compliance by Federal grantees, 2) determining the quality and consistency of IBC decisions and other actions, 3) continuing a formal risk assessment program, 4) identifying vague or conflicting provisions of the Guidelines for revision, 5) identifying emerging trends or problems, and 6) tracing any long-term adverse impacts on health or the environment back to their sources.

The obvious disadvantages of this option are the increased paperwork and effort by scientists, universities, corporations, and the Federal Government. Those working with rDNA would have to gather the required information periodically and prepare reports, which would be filed by the sponsoring institution with a designated existing Federal agency. A wide-range of information would be required for each project. The agency would have to process the reports and take other actions, such as preparing an annual report to Congress, to implement the underlying purposes of this option. Additional manpower would most likely be needed by that agency.

A statute implementing this option could include the following elements: 1) periodic collection of information in the form of reports from all institutions in the United States that sponsor any work with rDNA, 2) active evaluation of that information by the collecting agency, 3) annual reports to Congress, and 4) a sunset clause. Important information would include: 1) the sponsoring institution's name; 2) all places where it sponsors the research; and 3) a tabular or other summary that discloses for each project continuing or completed during the reporting period: the culture volume, the source and identity of the DNA and the host-vector system, the containment levels, and other information deemed necessary to effect the purposes of the act. The statute could also require employers to institute and report on a worker health surveillance program.

For this option to work, the monitoring agency would have to take an active role in evaluating the data. It should have the authority to require amendments to the reports when any part is vague, incomplete, or inconsistent with another part. It could also be required to notify the appropriate Federal funding agency of apparent cases of noncompliance with the Guidelines by their grantees. Finally, it should pre-

pare an annual report to Congress on the effectiveness of Federal oversight.

The choice of an agency to administer the statute would be important. The selection of NIH would permit the use of an existing administrative structure and body of expertise and experience. On the other hand, one of the regulatory agencies may take a more active monitoring role and be more experienced with handling proprietary information.

This approach is similar to a bill introduced in the 96th Congress, S. 2234, but broader in scope. The latter covered only institutions not funded by NIH, and did not contain provisions for requiring amendments to the reports or for notifying other agencies of possible noncompliance. The bill was broader in one respect because it would have required information about prospective experiments. This provision had been criticized because of the difficulty of projecting in advance the course that scientific inquiry will take. The goals of a monitoring program can be substantially reached by monitoring ongoing and completed work.

D. *Congress could make the NIH Guidelines applicable to all rDNA work done in the United States.*

The purpose of this option is to alleviate any concerns about the effectiveness of voluntary compliance. RAC itself has gone on record as supporting mandatory compliance with the Guidelines by non-NIH funded instituions, including private companies.

This option has the advantages of using an existing oversight mechanism, which would simply be extended to industry and to academic research funded by agencies other than NIH. Specific requirements on technical questions such as containment levels, host-vector systems, and laboratory practices would continue to be set by NIH in order to accommodate new information expeditiously; the statute would simply codify the responsibilities and procedures of the current system. There would be few transitional administrative problems, since the expertise and experience already exist at NIH. However, it would be necessary to appoint several experts in fermentation and other industrial technologies to RAC if production, as well as research, is to be adequately covered. In addition, the recommendations for large-scale containment procedures would have to be made part of the Guidelines.

The major changes would have to be made with respect to enforcement. Present penalties for noncompliance—suspension or termination of research funds—are obviously inapplicable to industry. In addition, procedures for monitoring compliance could be strengthened. Some of the elements of option C could be used. An added or alternative approach would be to inspect facilities.

The main disadvantage of this option is that NIH is not a regulatory agency. Since NIH has traditionally viewed its mission as promoting biomedical research, it would have a conflict of interest between regulation and promotion. One of the regulatory agencies could be given the authority to enforce the Guidelines and to adopt changes therein. NIH could then continue in a scientific advisory role.

E. *Congress could require an environmental impact statement and agency approval before any genetically engineered organism is intentionally released into the environment.*

There have been numerous cases where an animal or plant species has been introduced into a new environment and has spread in an uncontrolled and undersirable fashion. One of the early fears about rDNA was that a new pathogenic or otherwise undesirable micro-organism could establish an environmental niche. Yet in pollution control, mineral leaching, and enhanced oil recovery, it might be desirable to release large numbers of engineered micro-organisms into the environment.

The Guidelines currently prohibit deliberate release of any organism containing rDNA without approval by the Director of NIH on advice of RAC. The obvious disadvantage of this prohibition is that it lacks the force of law. The release of such an organism without NIH approval would be a prima facie case of negligence, if the organism caused harm. However, it may be more desirable social policy to attempt to prevent this type of harm through regulation rather than to compensate for injuries through lawsuits. Another possible disadvantage of the present system is that approval may be granted on a finding that the release would present "no significant risk to health or the environment;" a tougher or more specific standard than this may be desirable.

A required study of the possible consequences following the release of a genetically engineered organism, especially a micro-organism, would be an important step in ensuring safety. This option could be implemented by requiring those proposing to release the organism to file an impact statement with an agency such as NIH or EPA, which would then grant or deny permission to release the organism. A disadvantage of this option is that companies and individuals might be discouraged from developing useful organisms if this process became too burdensome and costly.

F. *Congress could pass legislation regulating all types and phases of genetic engineering, from research through commercial production.*

The main advantage of this option would be to deal comprehensively and directly with the risks of novel molecular genetic techniques, rather than relying on the current patchwork system. A specific statute would eliminate the uncertainties over the extent to which present law covers particular applications of genetic engineering, such as pollution control, and any concerns about the effectiveness of voluntary compliance with the Guidelines.

Other molecular genetic techniques, while not as widely used and effective as rDNA, raise similar concerns. Of the current techniques, cell fusion is the prime candidate for being treated like rDNA in any regulatory framework. It permits the recombination of chromosomes of species that do not recombine naturally, and it may permit the DNA of latent viruses in the cells to recombine into harmful viruses. No risk assessment of this technique has been done, and no Federal oversight exists.

The principal arguments against this option are that the current system appears to be working fairly well, and that the limited risks of the techniques may not warrant the significantly increased regulatory burden and costs that would result from such legislation. Congress will have to decide if that system will remain adequate as commercial activity grows.

If Congress were to decide on this option, the legislation could incorporate some or all of options C, D, and E. The present mechanism created by the Guidelines could be appropriately modified to provide the regulatory framework. The modifications could include a registration and licensing system to provide information on what work was actually being done and a means for continuous oversight. One important type of information would be health and safety statistics gathered by monitoring workers involved in the production of products from genetically engineered organisms. Another modification could be a sliding scale of penalties for violations, ranging from monetary fines through revocation of operating licenses to criminal penalties for extreme cases.

It would not be necessary to create a new agency, which would duplicate some of the responsibilties of existing agencies. Instead, Congress could give these agencies clear regulatory authority by amending the appropriate statutes. Designating a lead agency would assure a more uniform interpretation and application of the laws.

G. Congress could require NIH to rescind the Guidelines.

This option requires Congress to determine that the risks of rDNA techniques are so insignificant that no control or oversight is necessary. Deregulation would have the advantage of allowing funds and personnel currently involved in implementing the Guidelines at the Federal and local levels to be used for other purposes. In fiscal year 1980, NIH spent approximately $500,000 in administering the Guidelines; figures are not available for the analogous cost to academia and industry. Personnel hours

spent have not been estimated. Very few people work full-time on administering or complying with the Guidelines. NIH employs only six people full-time for this purpose, and some institutions employ full-time biological safety personnel. However, over 1,000 people nationally devote some effort to implementing the Guidelines—members of the IBCs and the scientists conducting the rDNA experiments who must take necessary steps to comply.

There are several reasons for retaining the Guidelines. First, sufficient scientific concern about risks exists for the Guidelines to prohibit certain experiments and require containment for others. Second, they are not particularly burdensome, since an estimated 80 to 85 percent of all experiments can be done at the lowest containment levels and an estimated 97 percent will not require NIH approval. Third, NIH will continue to serve an important role in continuing risk assessments, in evaluating new host-vector systems, in collecting and dispersing information, and in interpreting the Guidelines. Fourth, if the Guidelines were abolished, regulatory activity at the State and local levels could again become active. Finally, the oversight system has been flexible enough in the past to liberalize restrictions as evidence indicated lower risk.

H. Congress could consider the need for regulating work with all hazardous micro-organisms and viruses, whether or not they are genetically engineered.

Micro-organisms carrying rDNA, according to an increasingly accepted view, represent just a subset of micro-organisms and viruses, which, in general, pose risks. CDC has published guidelines for working with hazardous agents such as polio virus. However, such work is not currently subject to legally enforceable Federal regulations. It was not within the scope of this study to examine this issue, but it is an emerging one that Congress may wish to consider.

Patenting Living Organisms

A landmark decision

In a 5 to 4 decision (*Diamond* v. *Chakrabarty,* June 16, 1980), the Supreme Court ruled that a manmade mico-organism is patentable under the current patent statutes. This decision was alternately hailed as having "assured this country's technology future"[1] and denounced as creating "the Brave New World that Aldous Huxley warned of."[2] However, the Court clearly stated that it was undertaking only the narrow task of determining whether or not Congress, in enacting the patent statutes, had intended a manmade micro-organism to be excluded from patentability solely because it was alive. Moreover, the opinion invited Congress to overturn the decision if it disagreed with the Court's interpretation.

---

[1]Prepared Statement of Genentech, Inc., cited in "Science May Patent New Forms of Life, Justices Rule, 5 to 4," *The New York Times,* June 17, 1980, p. 1.

[2]Prepared statement of the Peoples' Business Commission, cited in "Science May Patent New Forms of Life, Justices Rule, 5 to 4," *The New York Times,* June 17, 1980, p. 1.

Congress may want to reconsider the issue of whether and to what extent it should specifically provide for or prohibit the patentability of living organisms. While the judiciary operates on a case-by-case basis, Congress can consider all the issues related to patentability at the same time, gathering all relevant data and taking testimony from the interested parties. The issues involved go beyond the narrow ones of scientific capabilities and the legal interpretations of statutory wording. They require broader decisions based on public policy and social values; Congress has the constitutional authority to make those decisions for society. It can act to resolve the questions left unanswered by the Court, overrule the decision, or develop a comprehensive statutory approach, if necessary. Most importantly, Congress can draw lines; it can specifically decide which organisms, if any, should be patentable.

Legal protection of inventions

The inherent "right" of the originator of a new idea to that idea is generally recognized, at least to the extent of deserving credit for it when used by others. At the same time, it is also believed that worthwhile ideas benefit society when they are widely available. Similarly, when an idea is embodied in a tangible form, such as in a machine or industrial process, the inventor has the "right" to its exclusive posession and use simply by keeping it secret. However, if he may be induced to disclose the invention's details, society benefits from the new ideas embodied therein, since others may build upon the new knowledge. The legal system has long recognized the competing interests of the inventor and the public, and has attempted to protect both. The separate laws covering trade secrets and patents are the mean by which this is done.

Trade secrets

The body of law governing trade secrets recognizes that harm has been done to one person if another improperly obtains a trade secret and then uses it personally or discloses it to others. A trade secret is anything—device, formula, or information—which when used in a business provides an advantage over competitors ignorant of it—e.g., improper acquisition includes a breach of confidence, a breach of a specific promise not to disclose, or an outright theft. Trade secrecy is derived from the common law, as opposed to being specifically created by statute; the State courts recognize and protect it as a form of property. The underlying policy is one of preventing unfair competition or unjust benefits. The protection lasts indefinitely. Two well-known examples of long-time trade secrets are the formulas for Coca Cola and for Smith Brothers' black cough drops; the latter is supposedly over 100 years old.

A company relying on trade secrecy to protect an important invention must take several steps to effect that protection. These include: permitting only key personnel to have access, requiring such people to sign complex contracts involving limitations on subsequent employment, and monitoring employees and competitors for possible breaches of security. Even so, there are practical limitations to what can be done and what can be proved to the satisfaction of a court. Moreover, independent discovery of the secret by a competitor is not improper, including the discovery of a secret process by an examination of the commercially marketed product. Most importantly, once a trade secret becomes public through whatever means, it can never be recaptured. Thus, reliance on trade secrecy for protecting inventions can be risky.

Patents

In contrast to the common law development of trade secrecy, patent law is a creation of Congress. The Federal patent statutes (title 35 of the United States Code) are derived from article I, section 8, of the Constitution, which states:

> The Congress shall have Power . . . To promote the Progress of Science and useful Arts, by securing for limited Times to Authors and Inventors the exclusive Right to their respective Writings and Discoveries.

This clause grants Congress the power to create a Federal statutory body of law designed to encourage invention by granting inventors a lawful monopoly for a limited period of time. Under the current statutory arrangement, which is conceptually similar to the first patent statutes promulgated in 1790, a patent gives the inventor the right to *exclude* all others from making, using, or selling his invention within the United States without his consent for 17 years. In return, the inventor must make full public disclosure of his invention. The policy behind the law is twofold. First, by rewarding successful efforts, a patent provides the inventor and those who support him with the incentive to risk time and money in research and development. Second, and more importantly, the patent system encourages public disclosure of technical information, which may otherwise have remained secret, so others may use the knowledge. The inducement in both cases is the potential for economic gain through exploitation of the limited monopoly. Of course, there are many reasons why this potential may not be realized, including the existence of competing products.

To qualify for patent protection, an invention must meet three statutory requirements: it must be capable of being classified as a process, machine, manufacture, or composition of matter; it must be new, useful, and not obvious; and it must be disclosed to the public in sufficient detail to enable a person skilled in the same or the most closely related area of technology to construct and operate it. Plants that reproduce asexually may also be patented, but slightly different criteria are used.

Although the categories in the first requirement are quite broad, they are not unlimited. In fact, the courts have held such things as scientific principles, mathematical formulas, and products of nature to be unpatentable on the grounds that they are only discoveries of pre-existing things—not the result of the inventive, creative action of man, which is what the patent laws are designed to encourage. This concept was reaffirmed in the *Chakrabarty* opinion.

The requirement that an invention be useful, new, and not obvious further narrows the range of patentable inventions. Utility exists if the invention works and would have some benefit to society; the degree is not important. Novelty signifies that the invention must differ from the "prior art" (publicly known inventions or knowledge). Novelty is not considered to exist, —e.g., if: 1) the applicant for a patent is not the inventor, 2) the invention was previously known or used publicly by others in the United States, or 3) the invention was previously described in a U.S. or foreign patent or publication. The inability to meet the novelty requirement is another reason why products of nature are unpatentable. Nonobviousness refers to the degree of difference between the invention and the prior art. If the invention would have been obvious at the time it was made to a person with ordinary skill in that field of technology, then it is not patentable. The policy behind the dual criteria of novelty and nonobviousness is that a patent should not take from the public something which it already enjoys or potentially enjoys as an obvious extension of current knowledge.

The final requirement—for adequate public disclosure of an invention—is known as the enablement requirement. It is designed to ensure that the public receives the full benefit of the new knowledge in return for granting a limited monopoly. As a public document, the patent must contain a sufficiently detailed description of the invention so that others in that field of technology can build and use it. At the end of this description are the claims, which define the boundaries of the invention protected by the patent.

The differences between trade secrets and patents, therefore, center on the categories of inventions protected, the term and degree of protection, and the disclosure required. Only those inventions meeting the statutory requirements outlined above qualify for patents and then only for a limited time, whereas anything giving an advantage over business competitors qualifies as a trade secret for an unlimited time. A patent requires full public disclosure, while trade secrecy requires an explicit and often costly effort to withhold information. The patent law provides rights of exclusion against everyone, even subsequent independent inventors, while the trade secrecy law protects only against wrongful appropriation of the secret.

Living organisms

Although the law for protecting inventions is usually thought of as applying to inanimate objects, it also applies to certain living organisms.

Any organism that both meets the broad definition of a trade secret and may be lawfully owned by a private person or entity can be protected by that body of law, including microorganisms, plants, animals, and insects. In addition, plants are covered specifically by two Federal statutes, the Plant Patent Act of 1930 and the Plant Variety Protection Act of 1970. Furthermore, the Supreme Court has now ruled that manmade micro-organisms are covered by the patent statutes. Its determination of congressional intent in the *Chakrabarty* case was based significantly on an analysis of the two plant protection statutes.

Patent protection for plants was not available until Congress passed the Plant Patent Act of 1930, recognizing that not all plants were products of nature because new varieties could be created by man. This Act covered new and distinct asexually reproduced varieties other than tuber-propagated plants or those found in nature.* The requirement for asexual reproduction was based on the belief that sexually reproduced varieties could not be reproduced true-to-type and that it would be senseless to try to protect a variety that would change in the next generation. To deal with the fact that organisms reproduce, the Act conferred the right to exclude others from asexually reproducing the plant or from using or selling any plants so reproduced. It also liberalized the description requirement for plants. Because of the impossibility of describing plants with the same degree of specificity as machines, their description need only be as complete as is "reasonably possible."

By 1970, plant breeding technology had advanced to where new, stable, and uniform varieties could be sexually reproduced. As a result, Congress provided patent-like protection to novel varieties of plants that reproduced sexually by passing the Plant Variety Protection Act of 1970. Fungi, bacteria, and first-generation hybrids were excluded.** Hybrids have a built-

---

*Approximately 4,500 plant patents have been issued to date, most for roses, apples, peaches, and chrysanthemums.

**Originally, six vegetables—okra, celery, peppers, tomatoes, carrots, and cucumbers—were also excluded. On Dec. 22, 1980, President Carter signed legislation (H.R. 999) amending the Plant Variety Protection Act to include these vegetables, to extend the term of protection to 18 years, and to make certain technical changes.

in protection, since the breeder can control the inbred, parental stocks and the same hybrid cannot be reproduced from hybrid seed.

The 1970 Act, administered by the Office of Plant Variety Protection within the U.S. Department of Agriculture (USDA), parallels the patent statutes to a large degree. Certificates of Plant Variety Protection allow the breeder to exclude others from selling, offering for sale, reproducing (sexually or asexually), importing, or exporting the protected variety. In addition, others cannot use it to produce a hybrid or a different variety for sale. However, saving seed for crop production and for the use and reproduction of protected varieties for research is expressly permitted. The term of protection is 18 years.

The *Chakrabarty* case

In 1972, Ananda M. Chakrabarty, then a research scientist for the General Electric Co., developed a strain of bacteria that would degrade four of the major components of crude oil. He did this by taking plasmids from several different strains, each of which gave the original strain a natural ability to degrade one of the crude oil components, and putting them into a single strain. The new bacterium was designed to be placed on an oil spill to break down the oil into harmless products by using it for food, and then to disappear when the oil was gone. Because anyone could take and reproduce the microbe once it was used, Chakrabarty applied for a patent on his invention. The U.S. Patent and Trademark Office granted a patent on the process by which the bacterium was developed and on a combination of a carrier (such as straw) and the bacteria. It refused to grant patent protection on the bacterium itself, contending that living organisms other than plant were not patentable under existing law. On appeal, the Court of Customs and Patent Appeals held that the inventor of a genetically engineered microorganism whose invention otherwise met the legal requirements for obtaining a patent could not be denied a patent solely because the invention was alive. The Supreme Court affirmed.

The majority opinion characterized the issue as follows:[3]

The question before us in this case is a narrow one of statutory interpretation requiring us to construe 35 U.S.C. §101, which provides:

"Whoever invents or discovers any new and useful process, machine, manufacture, or composition of matter, or any new and useful improvement thereof, may obtain a patent therefor, subject to the conditions and requirements of this title."

Specifically, we must determine whether respondent's micro-organism constitutes a "manufacture" or "composition of matter" within the meaning of the statute.

After evaluating the words of the statute, the policy behind the patent laws, and the legislative history of section 101 of the patent statutes and of the two plant protection Acts, the Court ruled that Congress had not intended to distinguish between unpatentable and patentable subject matter on the basis of living versus nonliving, but on the basis of "products of nature, whether living or not, and human-made inventions."[4] Therefore, the majority ruled, "[t]he patentee has produced a new bacterium with markedly different characteristics from any found in nature and one having potential for significant utility. His discovery is not nature's handiwork, but his own; accordingly it is patentable subject matter under §101."[5] The majority did not see their decision as extending the limits of patentability beyond those set by Congress.

The Court found that, in choosing such expansive terms as "manufacture" and "composition of matter"—words that have been in every patent statute since 1793—Congress plainly intended the patent laws to have a wide

[3]*Diamond* v. *Chakrabarty*, 100 S.Ct. 2204, 2207 (1980).

[4]Ibid, p. 2,210.
[5]Ibid, p. 2,208.

scope. Moreover, when these laws were last re-codified in 1952, the congressional committee reports affirmed the intent of Congress that patentable subject matter "include anything under the sun that is made by man."[6] The Court acknowledged that not everything is patentable; laws of nature, physical phenomena, and abstract ideas are not.

The Court found the Government's arguments unpersuasive. Specifically, that passing the Plant Patent Act of 1930 and the Plant Variety Protection Act of 1970, which excluded bacteria, was evidence of congressional understanding that section 101 did not apply to living organisms; otherwise; these statutes would have been unnecessary. In disagreeing, the Court stated that the 1930 Act was necessary to overcome the belief that even artificially bred plants were unpatentable products of nature and to relax the written description requirement, permitting a description as complete as is "reasonably possible." As for the 1970 Act, the Court stated that it had been passed to extend patent-like protection to new sexually reproducing varieties, which, in 1930, were believed to be incapable of reproducing in a stable, uniform manner. The 1970 Act's exclusion of bacteria, which indicated to the Government that Congress had not intended bacteria to be patentable, was considered insignificant for a number of reasons.

The Government had also argued that Congress could not have intended section 101 to cover genetically engineered micro-organisms, since the technology was unforeseen at the time. The majority responded that the very purpose of the patent law was to encourage new, unforeseen inventions, which was why section 101 was so broadly worded. Furthermore, as for the "gruesome parade of horribles"[7] that might possibly be associated with genetic engineering, the Court stated that the denial of a patent on a micro-organism might slow the scientific work but certainly would not stop it; and the consideration of such issues involves policy judgments that the legislative and executive

branches of Government, and not the courts, are competent to make. It further recognized that Congress could amend section 101 to specifically exclude genetically engineered organisms or could write a statute specifically designed for them.

The dissenting Justices agreed that the issue was one of statutory interpretation, but interpreted section 101 differently. They saw the two plant protection Acts as strong evidence of congressional intent that section 101 not cover living organisms. In view of this, the dissenters maintained that the majority opinion was actually extending the scope of the patent laws beyond the limit set by Congress.

The stated narrowness of the Court's decision may limit its impact as precedent in subsequent cases that raise similar issues, although not necessarily. Certainly, the decision applies to any genetically engineered micro-organism. It is a technical distinction without legal significance that most of the work being done on such organisms involves recombinant DNA (rDNA) techniques, which Chakrabarty did not use. The real question is whether or not it would permit the patenting of other genetically engineered organisms, such as plants, animals, and insects. Any fears that the decision might serve as a legal precedent for the patenting of human beings in the distant future are totally groundless. Under our legal system, the ownership of humans is absolutely prohibited by the 13th amendment to the Constitution.

Although the *Chakrabarty* case involved a micro-organism, there is no reason that its rationale could not be applied to other organisms. In the majority's view, the crucial test for patentability concerned whether or not the micro-organism was manmade. Conceptually, there is nothing in this test that limits it to micro-organisms. The operative distinction is between humanmade and naturally occurring "things," regardless of what they are. Thus, the *Chakrabarty* opinion could be read as precedent for including any genetically engineered organism (except humans) within the scope of section 101. Whether a court in a subsequent case will interpret *Chakrabarty* broadly or narrowly cannot be predicted.

[6]S. Rept. No. 1979, 82d Cong., 2d sess., p. 5, 1952; H.R. Rept. No. 1923, 82d Cong., 2d sess., p. 6, 1952, cited in *Diamond* v. *Chakrabarty*, 100 S.Ct. 2204, 2207 (1980).

[7]*Diamond* v. *Chakrabarty*, 100 S.Ct. 2204, 2211 (1980).

Even if section 101 were interpreted as covering other genetically engineered organisms, they probably could not be patented for failure to meet another requirement of the patent laws—the enablement requirement. It is generally impossible to describe a living organism in writing with enough detail so that it can be made on the basis of that description. Relaxing this requirement for plants was one reason behind the Plant Patent Act of 1930. For microorganisms, the problem is solved by depositing a publicly available culture with a recognized national repository and referring to the accession number in the patent.* While such an approach

*This procedure was accepted by the Court of Customs and Patent Appeals (CCPA) in upholding a patent on a *process* using microorganisms. *Application of Argoudelis*, 434 F.2d 1390 (CCPA 1970). This procedure should also be acceptable for patents on microorganisms themselves.

may be theoretically possible for animals and insects, it may be logistically impractical. However, if tissue culture techniques advance to the point where genetically engineered organisms can be made from single cells and stored indefinitely in that form, there appears to be no reason to treat them any differently than microorganisms, in the absence of a specific statute prohibiting their patentability.

Potential impacts of the decision and related policy issues

During the 8-year history of the *Chakrabarty* case and the surrounding public debate, numerous assertions were made about the potential impacts of permitting patents on genetically engineered organisms. They ranged from more immediate effects on the biotechnology industry, the patent system, and academic research to the long-term impacts on genetic diversity and the food supply. In addition, two major policy issues that have been raised are the morality of patenting living organisms; and the propriety of permitting private ownership of inventions from publicly funded research.

Impacts on industry

The basic question for industry is the extent to which permitting patents on genetically engineered organisms will stimulate both their development and the growth of the industries employing them. To ascertain this requires first an examination of the theory and social policies underlying the patent system.

THE RELATIONSHIP BETWEEN PATENTS AND INNOVATION

The patent system is supposed to stimulate innovation—the process by which an invention is brought into commercial use—because the inventor does not receive financial rewards until the invention is used commercially. The Constitution itself presumes this, as do the statutes enacted pursuant to the patent clause in article I, section 8. Attempts have been made to subject this presumption to empirical analysis; but innovation is extraordinarily complex and involves interacting factors that are difficult to separate. In addition, the existence of patents and trade secrets as alternative means for protection makes it almost impossible to study the effects of patents alone on invention and innovation.*

*A major reason for the lack of empirical studies has been the lack of appropriate data. The information available on the number of patents applied for and issued does not indicate the importance, economic benefits, or economic costs of inventions (whether patented or unpatented) that may not have existed at all or may have been created more slowly if not for the patent system.[8] In Presi-

Several reasonable arguments have been presented to support the presumption that the patent system stimulates innovation. First, the potential for the exclusive commercialization of a new product or process creates the incentive to undertake the long, risky, and expensive process from research through development to marketing. At every stage of innovation—from defining priorities and making initial estimates of an invention's value to advertising the finished product—the inventor and his backers must spend time, money, and effort, not only to develop a product but to convince others of its worth. Only a small percentage of new ideas or inventions survive. If a competitor, particularly a larger firm with a well-developed marketing capability, were free to copy a product at this point, smaller firms would have little incentive to undertake the process of innovation.

Second, the information and new knowledge disclosed by the patent allows others to develop competing, and presumably better, products by improving on the patented product or "inventing around" it. Third, patents may reduce unnecessary costs to individual firms, thereby freeing resources for further innovation. Once a patent is issued, competitors can redirect research and development (R&D) funds into other areas. For the firm holding the patent, maintaining control over the technology is theoretically less expensive, since the costs of trade secret protection are no longer required.**

Anecdotal accounts support the proposition that patents stimulate innovation; probably the best known is the story of penicillin. Although

Sir Alexander Fleming had discovered a promising weapon against bacterial infection, it took him over 10 years to get the money and facilities he needed to purify and produce penicillin in bulk. Only World War II and an international effort finally accomplished that task. Sir Howard Florey, who shared the Nobel prize with Fleming for developing penicillin, attributed the delay to their not having patented the drug, which he termed "a cardinal error."[10]

Some have claimed that the monopoly power of a patent can be used to retard innovation. A corporation can legally refuse to license a patent on a basic invention to holders of patents on improvements, thus protecting its product from becoming less attractive or obsolete. On the other hand, unless the corporation can satisfy the market for its product, it is usually in its economic interest to engage in cross-licensing arrangements with holders of improvement patents; it receives royalties and all parties can market the improved product. Cross-licensing has been misused several times by a few dominant firms in an attempt to exclude innovative new firms from their markets. Such arrangements violate the antitrust laws. Whether or not that body of law adequately prevents patent misuse is beyond the scope of this report.

THE ADVANTAGES OF PATENTING LIVING ORGANISMS

Given the presumed connection between patents and innovation, the next question is whether patenting a living organism would add significant protection for the patent holder, or whether alternative approaches would be sufficient. In this context, it is necessary to focus on the present industrial applications—which involve only micro-organisms—to examine alternative forms of patent coverage and to compare the protection offered by trade secrecy with that offered by patents.

Opinions vary widely among spokesmen for the genetic engineering companies on the value of patenting micro-organisms.[11] Spokesmen for Genentech, Inc., have stated numerous times

dent Carter's recent report on industrial innovation, the patent policy committee, composed of industry representatives having long experience with the patent system, recommended ways of enhancing innovation by improving the patent system, including the patenting of industrially important living organisms. However, they provided no hard economic data to support their recommendation.[9]

[8]Carole Kitti, and Charles L. Trozzo, *The Effects of Patent and Antitrust Laws, Regulations, and Practices on Innovation*, vol. II (Arlington, Va., Institute for Defense Analyses, 1976), pp. 2,9.

[9]U.S. Department of Commerce, *Advisory Committee on Industrial Innovation: Final Report*, September 1979, pp. 148-149.

**Patent rights can be very expensive to enforce against an infringer, however, should litigation be necessary.

[10]Ibid, pp. 170-171.

[11]D. Dickson, "Patenting Living Organisms: How to Beat the Bug-Rustlers," *Nature*, vol. 283, Jan. 10, 1980, pp. 128-129.

that such patents are crucial to the development of the industry, while others have stated their preference for trade secrecy.

Genentech's friend-of-the-court brief filed in the *Chakrabarty* case stated, "The patent incentive did, and doubtless elsewhere it will, prove to be an important if not indispensible factor in attracting private support for life-giving research."[12] Genentech has also supported increased patent protection because, to attract top scientists to the company, it had to give assurances that they would be able to publish freely.[13] This severely curtails any reliance on trade secrets.

The rationale behind the contrary position is based on the belief that the industry is moving so quickly that today's frontrunner is not necessarily tomorrow's, and that unique knowledge translates into competitive advantage. Thus, in a strategy similar to that of the advanced microelectronics industry, firms may prefer to rely on trade secrets even for patentable inventions, coupled with an intense marketing effort once an invention has reached the commercial stage. The idea is to get the jump on competitors and to stay in front.[14]

The uncertainty about whether micro-organisms could be patented before the Supreme Court's decision does not appear to have hindered the development of the industry. Clearly, companies did not have any difficulty raising capital—e.g., before the decision, Cetus Corp. had a paper value of $250 million without holding a single patent on a genetically engineered organism. Moreover, products such as insulin, human growth hormone, and interferon were being made, albeit in small quantities, by unpatented, genetically modified organisms. (See ch. 4.)

Before the decision, companies relied either entirely on trade secrecy for protection, or on a combination of patents on the microbiological process and the product and trade secret protection of the mico-organism itself. Considering the existence of such protection, the question is what the actual advantages are to patenting the micro-organisms as well.

One advantage results from the ability of a living organism to reproduce itself. Developing a new microbe for a specific purpose, such as the production of human insulin, can be a long, difficult, and costly procedure. Yet once it is developed, it reproduces endlessly, and anybody acquiring a culture would have the benefit of the development process at little or no cost unless the organism were patented.

Often, a company is able to keep the microbe a trade secret, since only the product is sold. However, where the microbe is the product—such as with Chakrabarty's oil-eating bacterium—patenting the organism is the best means of protection. Moreover, even when a microbe itself can be kept under lock and key, a company desiring to patent the process in which it is used must place a sample culture in a public repository to meet the enablement requirement.

A competitor could legally obtain the micro-organism. If the competitor were to use it to make the product for commercial purposes, the company might suspect infringement but have difficulty proving it, especially when the product is not patented. The infringing activity would take place entirely within the confines of the competitor's plant. Mere suspicion is not sufficient legal grounds for inspecting the competitor's plant for evidence of infringement when the unpatented product could theoretically be made by many different methods besides the one patented.*

A second, but less certain, advantage provided by patenting the micro-organism is that even uses and products of the organism not discovered by the inventor would be protected indirectly. That is, while new uses and products could be patented by their inventors, those patents would be "dominated" by the micro-organism patent. Royalties would have to be paid

[12]Brief for Genentech as Amicus Curiae, p. 3.
[13]Thomas Kiley, Vice President and General Counsel for Genentech, personal communication, Apr. 15, 1980.
[14]Dickson, op cit., p. 128.

*Some would answer this assertion by saying that a lawsuit could be started even on the basis of little evidence; the suing company would rely on the discovery process, which is liberal and wide-ranging, to provide any existing evidence of infringement.

whenever the micro-organism was used for commercial purposes. Whether this would be a significant advantage in practice is uncertain. Usually, only one product is optimally produced by a given micro-organism and only one micro-organism is best for a given process. Presumably, the micro-organism's inventor would also have discovered and patented its best use and product.

Another alternative to patenting a man-made micro-organism, besides trade secrecy, is to patent its manmade components. Examples of these include a plasmid containing the cloned gene, a sequence of DNA, or a synthetic gene made by the reverse transcriptase process. These components, which are nothing more than strings of inanimate chemicals, would not be unpatentable products of nature if they were made in the laboratory and were not identical to the natural material. Patenting them would not be equivalent to patenting the entire organism, since their function would be affected in varying degrees by the internal environment of their host. Nevertheless, the inventor of a particularly useful component, such as an efficient and stable plasmid, might want to patent it regardless of whether or not the organism could be patented, since it could be used in an indefinite number of different micro-organisms.

Thus, if Congress were to prohibit patenting of micro-organisms because they are alive, industry could compensate to a large degree by patenting inanimate components. On the other hand, if Congress allows the Supreme Court's decision to stand, certain components will undoubtedly still be patented. In fact, such patents may become more important than patents for micro-organisms, since the components are the critical elements of genetic engineering.

PATENT V. TRADE SECRET PROTECTION

Even with the advantages provided by patenting a micro-organism, a company could still decide to rely on trade secrecy. In choosing between these two options, it would evaluate the following factors:[15]

[15]R. Saliwanchik, "Microbiological Inventions: Protect by Patenting or Maintain as a Trade Secret?" *Developments in Industrial Microbiology*, vol. 19, 1978, pp. 273, 277.

- whether the organism itself or the substance that it makes will be the commercial product,
- whether there is any significant doubt of its meeting the legal requirements for patenting,
- whether there is the likelihood of others discovering it independently,
- whether it is a pioneer invention,
- what its projected commercial life is and how readily others could improve on it if it were disclosed in a patent,
- whether there are any plans for scientific publication, and
- what the costs of patenting are versus reliance on trade secrecy.

The first two factors make the decision easy. Obviously, an organism like Chakrabarty's can best be protected by a patent. In most instances, the substance made by the organism is the commercial product. In that case, if there are significant doubts that the organism can meet all the legal requirements for patentability, the company would probably decide to rely on trade secrecy.

The next three factors require difficult decisions to be made on the basis of the characteristics of the new organism, its product, and the competitive environment. If research to develop a particular product is widespread and intense (as is the case with interferon), the risk of a competitor developing the invention independently provides a significant incentive for patenting. On the other hand, reverse engineering (examination of a product by experts to discover the process by which it was made) by competitors is virtually impossible for products of micro-organisms because of the variability and biochemical complexity of microbiological processes.

Thus, greater protection may often lie in keeping a process secret, even if the microbe and the process could be patented. This is especially true for a process that is only a minor improvement in the state of the art or that produces an unpatentable product already made by many competitors. The commercial life of the process might be limited if it were patented because infringement would be difficult to detect and not worth the time and money to prosecute. Reliance on trade secrecy might then extend its commercial life.

Most companies would patent truly pioneer inventions, which often provide the opportunity for developing large markets. Moreover, patents of this sort tend to have long commercial lives, since it is difficult to circumvent a pioneer invention and since any improvements are still subject to the pioneer patent. Furthermore, infringement is easy to detect because of the invention's trailblazing nature.

The last two factors involve considerations secondary to a product and its market. Obviously, any publication of the experiments leading to an invention foreclose the option of trade secrecy. Also, company must evaluate the options of protection via either patenting or trade secrecy in terms of their respective cost effectiveness.

IMPACT OF THE COURT'S DECISION ON THE BIOTECHNOLOGY INDUSTRY

The *Chakrabarty* decision will add some protection for microbiological inventions by providing companies with an additional incentive for the commercial development of their inventions, particularly in marginal cases, by lowering uncertainty and risk. A greater effect will result from the new information disclosed in patents on inventions that otherwise might have been kept secret indefinitely. Competitors and academicians will gain new knowledge as well as a new organism upon which to build. The Patent Office had deferred action on about 150 applications, while awaiting the Court's decision; as of December 1980, it was processing approximately 200 applications on micro-organisms.[16*]

Depending on the eventual number and importance of patented inventions that would have otherwise been kept as trade secrets, the ultimate effect of the decision on innovation in the biotechnology industry could be significant.

Conversely, if the Court had reached the opposite decision, the industry would have been held back only moderately because of reasonably effective alternative means of protection.

Impacts of the Court's decision on the patent law and the Patent and Trademark Office

The key rationale supporting the Court's holding Chakrabarty's microbe to be patentable was the fact that it was manmade; its status as a living organism was irrelevant. The Patent Office interprets this decision as also permitting patents on micro-organisms found in nature but whose useful properties depend on human intervention other than genetic engineering,[17] e.g., if the isolation of a pure culture of a microbial strain induces it to produce an antibiotic, that pure culture would be patentable subject matter.

Because of the complexity, reproducibility, and mutability of living organisms, the decision may cause some problems for a body of law designed more for inanimate objects than for living organisms. It raises questions about the proper interpretation and application of the requirements for novelty, nonobviousness, and enablement. In addition, it raises questions about how broad the scope of patent coverage on important micro-organisms should be and about the continuing need for the two plant protection Acts. These uncertainties could result in increased litigation, making it more difficult and costly for owners of patents on living organisms to enforce their rights.

The complexity of living matter will make it difficult for anyone examining the invention to determine if it meets the requirements for novelty, nonobviousness, and enablement. Micro-organisms can have different characteristics in different environments. Moreover, microbial taxonomists often differ on the precise classification of microbial strains. Even after expensive tests, uncertainty may still exist about whether a specific micro-organism is distinct from other known strains; scientists do not have complete

---

[16]Rene Tegtmeyer, Assistant Commissioner for Patents, U.S. Patent and Trademark Office, personal communication, Jan. 8, 1981.

*These applications include about 100 on genetically engineered microbes and about 100 on cultures of strains isolated from nature.

[17]Ibid, Jan. 7, 1981.

knowledge of any single organism's biophysical and biochemical mechanisms. Consequently, there may be cases where it is difficult to know the prior art precisely enough to make a determination of novelty.

Similarly, microbial complexity raises problems in determining nonobviousness because there are so many different ways of engineering a new organism with a desired trait—e.g., a gene could be inserted into a given plasmid at several different positions. If a microbe with the gene at one position in the plasmid were patented, could a patent be denied to an otherwise structurally identical organism with the gene at a different position because the second was obvious? Perhaps not. The second organism would probably not be an obvious invention if it provided significantly more of the product, a better quality product under similar fermenting conditions, or the same product under cheaper operating conditions.

As to enablement, the major problem has been discussed previously; placing a culture of the micro-organism into a repository is the accepted solution. One problem with repositories, however, is their potential misuse. In a case involving alleged price fixing and unfair competition—e.g., the Federal Trade Commission found that micro-organisms placed in a public repository pursuant to process and product patents on the antibiotic Aureomycin did not produce the antibiotic in commercially significant amounts; in actual practice, other strains were being used for production, and the company involved was able to benefit from a patent, while, in effect, retaining the crucial micro-organism as a trade secret.[18]*

Complexity also raises questions about the appropriate scope of patent coverage. In a patent, the inventor is permitted to claim his invention as broadly as possible, so long as the claims

---

[18]*American Cyanamid Co., et. al.,* 63 FTC 1747, 1905 n. 14 (1963), *vacated and remanded,* 363 F.2d 757 (6th Cir. 1966), *readopted* 72 FTC 623 (1967), *affirmed* 401 F.2d 574 (6th Cir. 1968), *cert. denied,* 394 U.S. 920 (1969).

*The company had maintained that sec. 112 simply required it to deposit a strain that conformed to the description of the one found in the patent application. However, it is often the case with bacteria that many strains of a species will conform to even the most precisely written description.

made do not overlap with any "prior art" or obvious extensions thereof—e.g., a person who developed a particular strain of *Escherichia coli* that produced human insulin through a genetically modified plasmid could be entitled to a patent covering *all* strains of *E. coli* that produce the insulin in the same way. Chakrabarty's patent application—e.g., claimed "a bacterium from the genus *Pseudomonas* containing therein at least two stable energy generating plasmids, each of said plasmids providing a separate hydrocarbon degradative pathway." Several species and hundreds of strains of *Pseudomonas* fit this description. A patent limited to a particular microbial strain is not particularly valuable because it can easily be circumvented by applying the inventive concept to a sister strain; on the other hand, a patent covering a whole genus of micro-organism (or several) may retard competition. This problem will probably be resolved by the Patent Office and the courts on a case-by-case basis.

Another aspect of the same problem is whether a patent on an organism would cover mutants. It would not if the mutation occurred spontaneously and sufficiently altered the claimed properties. However, if a new organism were made in a laboratory with a patented organism as a starting point, the situation would be analogous to one where an inventor can patent an improved version of a machine but must come to terms with the holder of the "dominant" patent before marketing it.

The *Chakrabarty* decision also raises questions about the scope of section 101 and its relation to the plant protection Acts—e.g., plant tissue culture is, in effect, a collection of micro-organisms; should it be viewed as coming under section 101 instead of either of the plant protection Acts? Could plants excluded under these Acts—such as tuber-propagated plants or first-generation hybrids—be patented under section 101? Could any plants or seeds be patented under section 101, and if so, is there still a need for the plant protection Acts? If there is a need, would the Acts be administered better by only one agency? The Senate Committee on Appropriations has directed the Departments of Commerce and Agriculture to submit a report within 120 days of the *Chakrabarty* decision on the advisability of shifting the examining function to USDA.[19] As of December 1980, this issue was still under study. These questions could be resolved by the courts, but they are probably more amenable to a statutory solution.

Another effect of the decision could be on patent enforcement. The various uncertainties discussed above may have to be resolved through costly litigation. Moreover, in specific cases, the problems associated with describing a micro-organism in sufficient detail may increase the chances that a patent will be declared invalid. In any event, litigation costs would probably increase as more expert testimony is needed.

The fact that organisms mutate might introduce still another complication into infringement actions. A deposited micro-organism is the standard by which possible infringement would be judged. If it has mutated with respect to one of its significant characteristics, a patent holder who is seeking to prove infringement may have no case. While this problem does not appear to be amenable to a statutory solution, the risk of such a mutation is actually quite small.*

Because a living invention reproduces itself, the statutory definition of infringement may have to be changed. Presently, infringement consists of making, using, or selling a patented invention without the permission of the patent holder. Theoretically, someone could take part of a publicly available micro-organism culture, reproduce it, and give it away. Arguably, this is not "making" the invention, and the patent holder would have the burdensome and expensive task of going after each user. The two plant protection statutes deal with this problem by specifically prohibiting unauthorized reproduction of the protected plant. This approach may be necessary for other living inventions.

How all of these uncertainties will affect the Patent Office's processing of applications cannot be predicted. Currently, the average processing time for all applications is 22 months; separate

information on genetic engineering applications is not available.[20] It may take examiners longer to process applications on micro-organisms than for those covering only microbiological processes or products because of the interpretive problems mentioned. Moreover, the Patent Office will have to develop greater expertise in molecular genetics—a frontier scientific field that has only recently been the subject of patent applications. On the other hand, the Office generally faces this problem for any new area of technology.

In terms of increased numbers of applications, the decision is not expected to have a significant effect on the Patent Office operations in the next few years. The Office receives approximately 100,000 applications a year, and it has about 900 examiners, each processing an average of about 100 applications per year. Figures on the number of applications on genetically engineered organisms vary, depending on how the category is defined, and precise information has not been tabulated by the Patent Office. Rough estimates indicate that in February 1980 about 50 applications were pending, and by December 1980, that number had increased to about 100. Applications are being filed at the rate of about 5 per month. Also, just over 100 are pending on microbes that have been isolated and purified from natural sources, but have not been genetically engineered. Four examiners are working on both categories as well as others. Thus, in view of the total operations of the Office, these applications require only a small part of its resources. Over the next few years, the number is expected to increase because of the decision and developments in the field but not to a point where more than a few additional examiners will be needed.[21]

Impact of the Court's decision on academic research

Many academicians have voiced concerns about the effects on research of the *Chakrabarty* decision and the commercialization of molecular biology in general. They claim that the re-

---

[19]S. Rept. No. 96-251, 96th Cong. 1st sess., 1979, p. 46.
*Most micro-organisms can be stored in a freeze-dried form, which entails virtually no risk of mutation.

[20]Rene Tegtmeyer, personal communication, Dec. 15, 1980.
[21]Ibid., Dec. 15, 1980, and Jan. 8, 1981.

sults of rDNA research are not being published while patent applications are pending, discussion at scientific meetings is being curtailed, and novel organisms are less likely to be freely exchanged. A related concern is that scientific papers may not be citing the work of other scientists to avoid casting doubt on the novelty or inventiveness of the author's work, should he decide to apply for a patent. Finally, there is concern that the granting of patents on basic scientific processes used in the research laboratory will directly impede basic research—e.g., two scientists have recently been granted a patent on the most fundamental process of molecular genetic technology—the transfer of a gene in a plasmid using rDNA techniques.[22] The patent has been transferred to the universities where they did their work—Stanford and the University of California at San Francisco (UCSF). Although both universities have stated they would grant low-royalty licenses to anyone who complied with the National Institutes of Health (NIH) Guidelines, subsequent owners of fundamental process patents may not be so altruistic.

There are several reasons for believing that these concerns, although genuinely held, are somewhat overstated. First, patents on fundamental scientific processes or organisms should not directly hinder research. The courts have interpreted patent coverage as not applying to research; in other words, the patent covers only the commercial use of the invention.[23] Also, it would be difficult and prohibitively expensive for a patent holder to bring infringement actions against a large number of geographically separated scientists. Second, patents ultimately result in full disclosure. If patents were not available, trade secrecy could be relied on, with the result that important information might never become publicly available. Third, although delays occur while a patent application is pending, they often happen anyway while experiments are being conducted or while articles

are being prepared for publication because of the competitive nature of modern science.

Essentially, the issue is the effect of the commercialization of research results on the research process itself. Even if patents were not available for biological inventions, the inventor would simply keep his results secret if he were interested in commercialization. Viewed from this perspective, it is difficult to see why the availability of patents should affect the exchange of scientific information in genetic research any more than it does in any other field of research with commercial potential. The *Chakrabarty* decision may inhibit the dissemination of information only if it creates an atmosphere that stimulates academic scientists to commercialize their findings. However, if it encourages them to rely on patents rather than on trade secrets, it will ultimately enhance the dissemination of information.

Impacts of the Court's decision on genetic diversity and the food supply

Some public interest groups have claimed that patenting genetically modified organisms will adversely affect genetic diversity and the food supply. The claim is based on an analogy to a situation alleged to exist for plants. Briefly, the groups claim that patenting micro-organisms will irrevocably lead to patents on animals, which will have the same deleterious effects on the animal gene pool and the livestock industry as the two plant protection Acts have had on the plant gene pool and the plant breeding industry. The alleged effects are: loss of germplasm resources as a result of the elimination of thousands of varieties of plants; the increased risk of widespread crop damage from pests and diseases because of the genetic uniformity resulting from using a single variety; and the increasing concentration of control of the world's food supply in a few multinational corporations through their control of plant breeding companies.[24]

Only limited evidence is available, but no conclusive connection has been demonstrated be-

[22]U.S. Patent No. 4,237,224, issued Dec. 2, 1980.

[23]*Kaz Manufacturing Co.* v. *Chesebrough-Ponds, Inc.,* 211 F. Supp. 815 (S.D.N.Y. 1962) (dictum), *affirmed* 317 F.2d 679 (2d Cir. 1963); *Chesterfield* v. *United States,* 159 F. Supp. 371 (Ct. Cl. 1958); *Dugan* v. *Lear Avia,* 55 F. Supp. 223 (S.D.N.Y. 1944) (dictum); *Akro Agate Co.* v. *Master Marble Co.,* 18 F. Supp. 305 (N.D.W.Va. 1937).

[24]Brief for the Peoples' Business Commission as Amicus Curiae, pp. 6-13, *Diamond* v. *Chakrabarty,* 100 S. Ct. 2204 (1980).

tween the plant protection laws and the loss of genetic diversity, the encouragement of using a single variety, and any increased control by a few corporations of the food supply. (For a detailed discussion, see ch. 8.) Therefore, any connection between patenting micro-organisms and potential detrimental impacts on the livestock industry appears tenuous at best. The assumptions that the *Chakrabarty* decision will inevitably lead to patenting animals, and that the consequences will be the same as those claimed to result from granting limited ownership rights to varieties of plants, are speculative.

The morality of patenting living organisms

The moral issue is difficult to analyze because it embodies at least three overlapping questions: whether it is moral to grant exclusive rights of ownership to a living species; whether patents on lower forms of life will inevitably lead to genetic engineering of humans; and whether patenting organisms undermines the generally held belief in the uniqueness and sanctity of life, especially human life.

It is difficult to assess the extent of the belief that patenting living organisms is intrinsically immoral, and no such assessment has been done. Its extent and intensity will probably be directly correlated with the complexity of the organism involved. Fewer people will be disturbed about patenting micro-organisms than about patenting cattle. A belief in the immorality of patenting a living organism is a value judgment to which Congress may wish to give some consideration.

The second aspect of the moral issue revolves around the well-known metaphor of the "slippery slope"—the fear that the first steps along the path of genetic engineering may irrevocably lead to man. Technology, at times, appears to have its own momentum; the aphorism "what can be done, will be done" has been true in the past. Thus, some people fear that patenting micro-organisms may indeed set a dangerous precedent and encourage the technology to progress to the point of the ultimate dehumaniza-

tion—the engineering of people as an industrial enterprise.[25]

The *Chakrabarty* opinion was written in narrow terms. But while its reasoning might be applied to a future case involving an animal or insect, it simply could not be used to justify the patenting of human beings because of the 13th amendment to the Constitution, which prohibits the ownership of humans.

One way to negotiate the slippery slope is to deal directly with the adverse aspects of the technology. Barriers can be erected along the slope; the Constitution already protects humans. Congress can erect other barriers by statute, specifically drawing lines as to which organisms can or cannot be patented.

The third part of the issue is religous or philosophical in nature. For many, the patenting of a living organism undermines the awe and deep respect they hold for the unique nature of life. Moreover, it raises apprehensions of an ultimate threat to concepts of the nature of humanity and its place in the universe. To these people, if life can be engineered and patented, perhaps it is not special or sacred. If this is true of lower organisms, why would human beings be different? (This and other aspects of the morality issue are discussed in greater detail in ch. 13.)

Private ownership of inventions from publicly funded research

Much of the basic research in molecular genetics has been funded by Federal grants. Most of the work leading to the development of rDNA techniques—e.g., was performed at Stanford University and UCSF under NIH grants. The scientists involved have received a patent on that fundamental scientific process. Some opponents of patenting organisms have argued that private parties should not be permitted to own inventions resulting from federally funded R&D; and in any event, there is something special about molecular genetics that requires the Federal Government to retain ownership of

[25]Ibid., p. 25.

federally funded inventions and to make them generally available through nonexclusive licenses.

Until recently, there had been no comprehensive, governmentwide policy regarding ownership of patents on federally funded inventions. Some agencies, such as the Department of Health and Human Services (DHHS), permitted nonprofit institutional grantees to own patents on inventions (subject to conditions deemed necessary to protect the public interest) if they had formal procedures for administering them. However, most agencies generally retained title to such patents, making them available to anyone in the private sector for development and possible commercialization through nonexclusive licenses.

The rationale behind the policy was simply that inventions developed by public money should be available to all—including private industry—on a nonexclusive basis. This arrangement had been criticized as not providing sufficient incentive for industry to take the risks to develop the inventions. Of the more than 28,000 patents owned by the Government, less than 4 percent have been successfully licensed; on the other hand, universities, which do grant exclusive licenses on patents that they own, have been able to license 33 percent of their patents.[26]

On December 12, 1980, President Carter signed the Government Patent Policy Act of 1980. The Act sets forth congressional policy that the patent system be used to promote the utilization of inventions developed under federally supported R&D projects by nonprofit organizations and small businesses. To this end, the organization or firm may elect to retain title to those inventions, subject to various conditions designed to protect the public interest. Such conditions include retention by the funding agency of a nonexclusive, irrevocable, paid-up license to use the invention, and the right of the Government to act where efforts are not being made to commercialize the invention, in cases of health or safety needs, or when the use of the invention is required by Federal regulations.

There is still the question of whether patents on molecular techniques or genetically engineered micro-organisms are sufficiently different to merit exception from any general patent policy decided on by Congress. For some, the molecular genetic techniques are unique because they are powerful scientific tools that can manipulate the life processes as never before. However, in a November 1977 report, NIH took the following position with regard to patents on rDNA inventions developed under DHHS-NIH support:[27]*

> There are no compelling economic, social, or moral reasons to distinguish these inventions from others involving biological substances or processes that have been patented, even when partially or wholly developed with public funds.

The report was prompted by the Stanford-UCSF patent application. Even though the application was in accord with the funding agreements between the institutions and NIH, the universities requested a formal NIH opinion on the issue in view of the intense public interest in rDNA research. NIH solicited comments from a group of approximately 67 individuals, ranging from academic and industrial scientists to students, lawyers, and philosophers.[28] The review and analysis of the responses were referred to the Federal Interagency Committee on rDNA Research, the Public Health Service, and the Office of the General Counsel of the Department of Health, Education, and Welfare (now DHHS). A fairly uniform consensus on the above-quoted finding developed in this process; the one significant dissenter, the Department of Justice, contended that the Government should retain ownership of any invention resulting from federally funded rDNA research because of the great public interest in that research.

---

[26]S. Rept. No. 96-480, 96th Cong. 1st sess, 1979, p. 2.

[27]The Patenting of Recombinant DNA Research Inventions Developed under DHEW Support: An Analysis by the Director, National Institutes of Health, November 1977, p. 16.

*The report further concluded that no change was necessary in the basic NIH policy permitting nonprofit organizations to own patents on inventions developed under contracts or grants from the Department of Health, Education, and Welfare (now DHHS), subject to several conditions to protect the public interest. The only recommended change was that the formal agreements between NIH and the institutions be amended to require that any licensees of institutional patent holders comply with the containment standards of the NIH Guidelines in any production or use of rDNA molecules under the license agreement.

[28]Ibid., app. I, pp. 5-8.

Issue and Options

ISSUE: To what extent could Congress provide for or prohibit the patentability of living organisms?

In its *Chakrabarty* opinion, the Supreme Court stated that it was undertaking only the narrow task of determining whether or not Congress, in enacting the patent statutes, had intended a manmade micro-organism to be excluded from patentability solely because it was alive. Moreover, the opinion specifically invited Congress to overrule the decision if it disagreed with the Court's interpretation.

Congress has several options. It can act to resolve the questions left unanswered by the Court, overrule the decision, or develop a comprehensive statutory approach. Most importantly, Congress can draw lines; it can decide which organisms, if any, should be patentable.

OPTIONS

A: Congress could maintain the status quo.

Congress could choose not to address the issue of patentability and allow the law to be developed by the courts. The advantage of this option is that issues will be addressed as they arise in the context of a tangible, nonhypothetical case. Some of the issues raised in the debate on patenting may turn out to be irrelevant as the technology and the law develop. Moreover, many of the uncertainties raised by the *Chakrabarty* decision regarding provisions of the patent law other than section 101 may be incapable of statutory resolution. The complexity of living organisms and the increase in knowledge of molecular genetics will raise such broad and varied questions that legal interpretations of whether a particular biological invention meets the requirements of novelty, nonobviousness, and enablement will best be done on a case-by-case basis by the Patent Office and the Federal courts.

There are two disadvantages to this option. First, a uniform body of law may take time to develop, since judicial decisions about new legal questions by different Federal courts may initially conflict. Second, the Federal judiciary is not designed to take sufficient account of the broader political and social interests involved.

B: Congress could pass legislation dealing with the specific legal issues raised by the Court's decision.

Many of the legal questions do not readily lend themselves to statutory resolution. However, three questions are fairly narrow and well-defined and may therefore be better resolved by statute: 1) Is there a continuing need for the plant protection Acts if plants can be patented under section 101? 2) If there is a continuing need for these Acts, could they be administered better by one agency? 3) Should the definition of infringement be clarified by amending section 271 of the Federal Patent Statutes (title 35 U.S.C.) to include reproduction of a patented organism for the purpose of selling it?

Congressional action to clarify these issues would provide direction for industry and the Patent Office, and it would obviate the need for a resolution through costly, time-consuming litigation. Lessening the chances of litigation or the chances of a patent being declared invalid will provide some stimulation for innovation by lessening the risks in commercial development. In addition, Congress could determine that the plant protection Acts could be better administered by one agency or should be incorporated under the more general provisions of the patent law; if so, some administrative expenses probably could be saved.

C: Congress could mandate a study of the plant protection Acts.

Two statutes, the Plant Patent Act of 1930 and the Plant Variety Protection Act of 1970, grant ownership rights to plant breeders who develop new and distinct varieties of plants. They could serve as a model for studying the broader, long-term potential impacts of patenting living organisms. An empirical study of the impacts of the plant protection laws has not been done. Such a study would be timely, not only because of the *Chakrabarty* decision, but also because of allegations that the Acts have encouraged the planting of uniform varieties, loss of germplasm resources, and increased concentration in the plant breeding industry. In addition, information about the Acts' affect on innovation and competition in the breeding industry would be relevant to this aspect of the biotechnology industry. However, it may be extremely difficult to isolate the effects of these laws from the effects of other factors.

D: Congress could prohibit patents on any living organism or on organisms other than those already subject to the plant protection Acts.

By prohibiting patents on any living organisms, Congress would be accepting the arguments of those who consider ownership rights in living organisms to be immoral, or who are concerned about other potentially adverse impacts of such patents. Some of the claimed impacts are: 1) patents would stimulate the development of molecular genetic techniques, which will eventually lead to human genetic engineering; 2) patents contribute to an atmosphere of increasing interest in commercialization, which will discourage the open exchange of information crucial to scientific research; and 3) plant patents and protection certificates have encouraged the planting of uniform varieties, loss of germplasm resources, and increasing concentration in the plant breeding industry. Also, by repealing the plant Acts, Congress would be reversing the policy determination it made in 1930 and in 1970 that ownership rights in novel varieties of plants would stimulate plant breeding and agricultural innovation.

A prohibitory statute would have to deal with those organisms at the edge of life, such as viruses. Although there are uncertainties and disagreements in classifying some entities as living or nonliving, Congress could be arbitrary in its inclusions and exclusions, so long as it clearly dealt with all of the difficult cases.

This statute by itself would slow but not stop the development of molecular genetic techniques and the biotechnology industry because there are several good alternatives for maintaining exclusive control of biological inventions:

maintaining organisms as trade secrets; patenting microbiological processes and their products; and patenting the inanimate components of a genetically engineered micro-organism, such as plasmids, which are the crucial elements of the technique anyway. The development would be slowed primarily because information that might otherwise become public would be kept as trade secrets. A major consequence would be that desirable products would take longer to reach the market. Also, certain organisms or products that might be marginally profitable yet beneficial to society, such as some vaccines, would be less likely to be developed. In such cases, the recovery of development costs would be less likely without a patent to assure exclusive marketing rights.

Alternatively, Congress could overrule the *Chakrabarty* decision by amending the patent law to prohibit patents on organisms other than the plants covered by the two statutes mentioned in option C. This would demonstrate congressional intent that living organisms could be patented only by specific statute and alleviate concerns of those who fear the "slippery slope."

E: Congress could pass a comprehensive law covering any or all organisms (except humans).

This option recognizes the fact that Congress can draw lines where it sees fit in this area. It could specifically limit patenting to micro-organisms or encourage the breeding of agriculturally important animals by granting patent rights to breeders of new and distinct breeds. Any fears that such patents would eventually lead to patents on human beings would be unfounded, since the 13th amendment to the Constitution, which abolished slavery, prohibits ownership of human life.

The statute would have to define included or excluded species with precision. Although there are taxonomic uncertainties in classifying organisms, Congress could arbitrarily include or exclude borderline cases.

A statute that permitted patents on several types of organisms could be modeled after the Plant Variety Protection Act—e.g., it should cover organisms that are novel, distinct, and uniform in reproduction; such terms would have to be defined. Infringement should include the unauthorized reproduction of the organism—although reproduction for research should be excluded to allow the development of new varieties. In fact, consideration should be given to covering in one statute plants and all other organisms that Congress desires to be patentable. This would provide the advantage of comprehensiveness and uniform treatment; it could also address the problems discussed under option B.

The impact of this law cannot be assessed precisely. A comprehensive statute would stimulate the development of new organisms and their products and would encourage dis-semination of technical information; however, such a statute is not essential to the development of the biotechnology industry, since incentives and alternative means for protection already exist. The secondary impacts on society of the legislation are even harder to assess because of the scarcity of data from which to draw conclusions. The policy judgments will have to be made by Congress after it weighs the opinions of the various interest groups. Through legislation, Congress has the chance to balance competing views on this controversial issue and, if necessary, to alleviate the primary concerns about the long-term impacts of the decision—that higher organisms will inevitably be patented.

Chapter 13
Genetics and Society

Chapter 13

Genetics and Society

Genetics and modern science

In 1979, the Organization for Economic Cooperation and Development (OECD)* published a survey of mechanisms for settling issues involving science and technology in its member countries.[1] The OECD report noted that:[2]

> Science and technology . . . have a number of distinguishing characteristics which cause special problems or complications. One is ubiquity: they are everywhere. They are at the forefront of social change. They not only serve as agents of change, but provide the tools for analyzing social change. They pose, therefore, special challenges to any society seeking to shape its own future and not just to react to change or to the sometimes undesired effects of change.

After surveying member countries, OECD identified six factors that distinguish issues in science and technology from other public controversies.

1. *The rapidity of change in science and technology* often leads to concern. The *science* of genetics is one of the most rapidly expanding areas of human knowledge in the world today. And the *technology* of genetics is causing quick and fundamental changes on a variety of fronts. The news media have consistently reported developments in genetics, often with front-page stories. Consequently, the public has become increasingly aware of developments in genetics and genetic technologies and the speed with which knowledge in the field is gathered and applied.

2. *Many issues in today's science and technology are entirely new.* Protoplast fusion, recombinant DNA (rDNA), gene synthesis, chimeras, fertilization of mammalian embryos in vitro, and the successful introduction of foreign genes into mammals were the subjects of science fiction until a few years ago. Now they appear in newspapers and popular magazines. Yet the general public's understanding of these phenomena is limited. It is difficult for people to evaluate competing claims about the dangers and benefits of this new technology.

3. *The scale, complexity, and interdependence among the technologies* are greater than people suspect. As in other fields, applications of biological technology often depend on parallel developments in areas that provide critical support systems. Breakdowns in these systems are often as limiting as failures in the new technology itself. In other parts of this report for example, sophisticated breeding systems in farm animals and large-scale fermentation processes for single-cell cultures are described. Besides the biological technology required to support these systems, precise computerized operations are required to ensure purity, safety, and process control in fermentation and to provide the population statistics necessary for breeding decisions.

4. Some scientific and technological achievements may be *irreversible in their effects.* Because living organisms reproduce, some fear that it will be impossible to contain and control a genetically altered organism that finds its way into the environment and produces undesirable effects. Scenarios of escaping organisms, pandemics, and careless researchers are often drawn by critics of today's genetics research. The intentional release of recombinant organisms into the environment is a related issue that will need to be resolved in the future.

Another example of irreversibility, brought about by the demands placed on

---

*The members of OECD are: Australia, Austria, Belgium, Canada, Denmark, Finland, France, West Germany, Greece, Iceland, Ireland, Italy, Japan, Luxembourg, the Netherlands, New Zealand, Norway, Portugal, Spain, Sweden, Switzerland, Turkey, the United Kingdom, and the United States.

[1] Guild K. Nichols, *Technology on Trial: Public Participation in Decision-Making Related to Science and Technology* (Paris: Organization for Economic Cooperation and Development, 1979).

[2] Ibid., p. 16.

world resources, is the accelerating loss of plant and animal species. Concern over this depletion of the world's germplasm arises because genetic traits that might meet as yet unknown needs are being lost.

5. There exist strong *public sensibilities about real or imagined threats to human health.* Mistrust of experts has been stimulated by such events as the accident at the Three-Mile Island nuclear plant and the burial of toxic chemical wastes in the Love Canal. Regardless of the real dangers involved, the public's perception of danger can be a significant factor in decisionmaking. At present, some perceive genetic technologies as dangerous.

6. *A challenge to deeply held social values is being raised by scientific and technological issues.* The increasing control over the inherited characteristics of living things causes concern in the minds of some as to how widely that control should be exercised and who should be deciding about the kinds of changes that are made. Furthermore, because genetics is basic to all living organisms, technologies applicable to lower forms of life are theoretically applicable to higher forms as well, including human beings. Some wish to discourage applications in lower animals because they fear that the use of the technologies will progress in increments, with more and more complex organisms being altered, until human beings themselves become the object of genetic manipulation.

Special problems posed by genetics

Genetics is just one among several disciplines of the biological sciences in which major advances are being made. Other areas, such as neurobiology, behavior modification, and sociobiology, arouse similar concerns.

Genetics differs from the physical sciences and engineering because of its intimate association with people. The increasing control over the characteristics of organisms and the potential for altering inheritance in a directed fashion is causing many to reevaluate themselves and their role in the world. For some, this degree of control is a challenge, for others, a threat, and for still others, it causes a vague unease. Different groups have different reasons for embracing or fearing the new genetic technologies. Religious, political, and ethical reasons have been advanced to support different viewpoints.

The idea that research in genetics may lead some day to the ability to direct human evolution has caused particularly strong reactions. One reason is that such capability brings with it responsibility for retaining the genetic integrity of people and of the species as a whole, a responsibility formerly entrusted to forces other than man.

Others find the idea of directing evolution exciting. They view the development of genetics technologies in a positive light, and see opportunities to improve humanity's condition. They argue that the capability to change things is, in fact, a part of evolution.

Religious arguments on both sides of this challenge have been made. Pope John Paul II has decried genetic engineering as running counter to natural law. On the other hand, one Catholic philosopher has written:[3]

... We have always said, often without real belief, that we were and are created by God in His own image and likeness. "Let us make man in our image, after our likeness" logically means that man is by nature a creator, like his Creator. Or at least a cocreator in a very real, awesome manner. Not mere collaborator, nor administrator, nor caretaker. By divine command we are creators. Why, then, should we be shocked today to learn that we can now or soon will be able to create the man of the future? Why should we be horrified and denounce the sci-

---

[3]Robert T. Francoeur, "We Can—We Must: Reflections on the Technological Imperative," *Theological Studies* 33:3, September 1972, p. 429 and at footnote 2.

entist or physician for daring to "play God?" Is it because we have forgotten the Semitic (biblical) conception of creation as God's ongoing collaboration with man? Creation is our God-given role, and our task is the ongoing creation of the yet unfinished, still evolving nature of man.

Man has played God in the past, creating a whole new artificial world for his comfort and enjoyment. Obviously we have not always displayed the necessary wisdom and foresight in that creation; so it seems to me a waste of time and energy for scientists, ethicists, and laymen alike to beat their breasts today, continually pleading the question of whether or not we have the wisdom to play God with human nature and our future. It is obvious we do not, and never will, have all the foresight and prudence we need for our task. But I am also convinced that a good deal of the wisdom we lack could have been in our hands if we had taken seriously our human vocation as transcendent crea-

tures, creatures oriented toward the future (here and hereafter), a future in which we are cocreators.

Genetics thus poses social dilemmas that most other technologies based in the physical sciences do not. Issues such as sex selection, the abortion of a genetically defective fetus, and in vitro fertilization raise conflicts between individual rights and social responsibility, and they challenge the religious or moral beliefs of many. Furthermore, people sense that genetics will pose even more difficult dilemmas in the future. Although many cannot fully articulate the basis for their concern, considerations such as those discussed in this section are cited. The strong emotions aroused by genetics and by the questions of how much and what kind of research should be done are at least partly rooted in deeply held human values.

Science and society

The public's increasing concern about the effects of science and technology has led to demands for greater participation in decisions on scientific and technological issues, not only in the United States but throughout the world. The demands imply new challenges to systems of representative government; in every Western country, new mechanisms have been devised for increasing citizen participation. An increasingly informed population, skilled at exerting influence over policymakers, seems to be a strong trend for the future. The media has played an important role in this development, reporting both on breakthroughs in science and technology and on accidents, pollution, and the side-effects of some technologies.

One result has been the growing politicization of science and technology. While perhaps misunderstanding the nature of science as a process, the public has become disenchanted by recent accidents associated with technology, by experts who openly disagree with one another, and by the selective use of information by some scientific supporters to obtain a political objective. The public has seen that technology affects

the distribution of benefits in society; it can have unequal impacts, and those who pay or who are most in need are not necessarily always those who benefit.

A national opinion survey of a random sample of 1,679 U.S. adults conducted for the National Commission for the Protection of Human Subjects of Biomedical and Behavioral Research[4] made clear that there is public doubt concerning equity. Sixty percent of those polled felt that new tests and treatments deriving from medical research are not equally accessible to all Americans. Seventy percent felt that those most likely to benefit from a new test or treatment of limited availability were those who could pay for it or who knew an important doctor. This should be compared with the 85 percent who felt that a new test or treatment should be available to those who apply first or who are most in need.

4"Special Study, Implications of Advances in Biomedical and Behavioral Research," Report and Recommendations of the National Commission for the Protection of Human Subjects of Biomedical and Behavioral Research, DHEW publication No. (OS) 78-0015.

Public concern and demand for involvement in the policy process is illustrated by the response of communities to plans for laboratories that would conduct rDNA research. Perhaps the best known example is Cambridge, Mass., where plans were announced for construction of a moderate containment laboratory at Harvard University. Concern over this facility led to the formation of the Cambridge Experimentation Review Board (CERB). Composed of nine citizens—all laymen with respect to rDNA research—the CERB spent 6 months studying the subject and listening to testimony from scientists with opposing points of view. Their final recommendations did not differ substantially from the NIH Guidelines; but the process was crucial. CERB demonstrated that citizens could acquire enough knowledge about a highly technical subject to develop realistic criteria and apply them. Similar responses to proposed laboratories have occurred in a number of other American communities, including Ann Arbor, Mich., and Princeton, N.J.[5]

These reactions, and similar phenomena surrounding controversies like nuclear power, indicate that the desire for citizen participation is strong and widespread. Recognizing this, each Federal agency has its own rules and mechanisms for citizen input. Special ad hoc commissions are sometimes formed to collect information from private citizens before decisions are made on particular projects. Congressional hearings held around the country and in Washington, D.C., are perhaps the best known of these inquiries. While these mechanisms sometimes slow the decisionmaking process, they help legitimize some decisions, and their role will probably expand in the future.

In corporate science and technology, public demands are being felt as well. Present regulations for environmental protection and worker and product safety have significantly altered corporate research and development efforts. The public is also becoming more involved in corporate decisionmaking—e.g., through "public accountability" campaigns by stockholders to influence company policies.

With the politicization of science, the process of research itself is coming under increasing public scrutiny—most recently in cases of possible biohazards, research with human subjects, and research on fetuses. Some efforts are underway to require better treatment of research animals as well.

The relationship between science and society, between human beings and their tools, is a constantly evolving one. The process that has been called the "dialogue within science and the dialogue between the scientific community and the general public"[6] will continue to search for standards of responsibility. It is likely that as long as science remains as dependent on public funds as it has over the past 40 years, it will be held accountable to public values. As has been noted:[7]

> The technologies of war, industrialization, medicine, environmental protection, etc., appear less as the demonstrations of superior claims of knowledge and more and more as the symbols of the ethical and political choices underlying the distribution of the power of scientific knowledge among competing social values . . . . This cultural shift of emphasis from the role of science in the intellectual construction of reality to the role of science in the ethical construction of society may indicate a profound transformation in the parameters of the social assessment of science and its relations to the political order.

---

[5]Richard Hutton, *Bio-Revolution: DNA and the Ethics of Man-Made Life* (New York: New American Library (Mentor), 1978).

[6]Daniel Callahan, "Ethical Responsibility in Science in the Face of Uncertain Consequences," *Ethical and Scientific Issues Posed by Human Uses of Molecular Genetics*, Marc Lappe and Robert S. Morison (eds.), Annals of the New York Academy of Sciences 265, Jan. 23, 1976, p. 10.

[7]Yaron Ezrahi, "The Politics of the Social Assessment of Science" in *The Social Assessment of Science*, E. Mendelson, D. Nelkin, P. Weingart (eds.), Conference Proceedings (Bielefeld, West Germany: A&W Opitz, 1978), p. 181.

The "public" and "public participation"

These are terms with vastly different meanings to different people. Some take "the public" to mean an organized public interest group; others consider such groups the "professional" public and feel they have agendas that differ from those of the less organized "general" public. As OECD stated:[8]

> Public participation is a concept in search of a definition. Because it means different things to different people, agreement on what constitutes "the public" and what delineates "participation" is difficult to achieve. The public is not of course homogeneous; it is comprised of many heterogeneous elements, interests, and preoccupations. The emergence over the last several decades of new and sometime vocal special interest groups, each with its own set of competing claims and demands, attests to the inherent difficulty of achieving social and political consensus on policy goals and programmes purporting to serve the common interest.

[8]Nichols, op. cit., p. 7.

Because publics differ with each issue, no definition will be attempted here. It is assumed that "the public" is demanding a greater role in decisions about science and technology, and that it will continue to do so. The different publics that coalesce around different issues vary widely in their basic interests, their skills, and their ultimate objectives. They are the groups that will be heard in the widening debate about scientific and technological issues, and are part of what has been called the "social system of science."[9]

The public has already become involved in the decisionmaking process involving genetic research. As the science develops, new issues in which the public will demand involvement will arise. The question is therefore: What is the best way to involve the public in decisionmaking?

[9]J. M. Ziman, *Public Knowledge* (Cambridge: Cambridge University Press, 1968).

Issues and Options

Three issues are considered. The first is an issue of process, concerning public involvement in policymaking; the second is a technical issue; and the third reflects the complexity of some issues associated with genetics that may arise in the future.

ISSUE: How should the public be involved in determining policy related to new applications of genetics?

The question as to whether the public should be involved is no longer an issue. Groups demand to be involved when people feel that their interests are threatened in ways that cannot be resolved by representative democracy.

The more relevant questions are whether current mechanisms are adequate to meet public desires to participate and whether a deliberate effort should be made to increase public knowledge. The last can only be accomplished by educating the public and increasing its exposure both to the issues and to how people may be affected by different decisions.

OPTIONS:

A. *Congress could specify that the opinion of the public must be sought in formulating all major policies concerning new applications of genetics, including decisions on funding of specific research projects. A "public participation statement" could be mandated for all such decisions.*

B. *Congress could maintain the status quo, allowing the public to participate only when it decides to do so on its own initiative.*

If option A were followed, there would be no cause for claiming that public involvement was inadequate (as occurred after the first set of Guidelines for Recombinant DNA Research were promulgated). However, option A can be implemented in two ways. In the first, the opportunity for public involvement is always provided, but need not be taken if there is no public interest in the topic. In the second, public involvement is required. A requirement for public involvement would pose the problem that if the public does not wish to participate in a particular decision, then opinion will sometimes be sought from an uninterested (and therefore probably uninformed) public simply to meet the requirement. Option A poses additional problems: What is a "major" policy? At what stage would public involvement be required—only when technological development and application are imminent or at the stage of basic research? Finally, it should be noted that under option A, if the public's contribution significantly influences policy, the trend away from decisionmaking by elected representatives (representative democracy) and toward decisionmaking by the people directly ("participatory" democracy) may be accelerated.

Option B would be less cumbersome and would permit the establishment of ad hoc mechanisms when necessary. On the other hand, by the time some issues are raised, strong vested interests would already be in place. The growing role of single-issue advocates in U.S. politics, and their skill in influencing citizens and policymakers, might abort certain scientific developments in the future.

Regardless of which option is selected, it would be desirable to encourage different forms of structuring public participation and to evaluate the success of each method. Many different approaches to public participation have been tried in the United States and Western Europe in attempts to resolve conflicts over science and technology. Some have worked better than others, but most have had rather limited success.[10] Because public demands for involvement are not likely to diminish, the best

ways to accommodate them need to be identified.

ISSUE: How can the level of public knowledge concerning genetics and its potential be raised?

If public involvement is expected, an informed public is clearly desirable. Increasing the treatment of the subject, both within and outside the traditional educational system, is the only way to accomplish this.

Within the traditional educational system, at least some educators feel that too little time is spent on genetics. Some, such as members of the Biological Sciences Curriculum Study Program, are considering increasing the share of the curriculum devoted to genetics. Because science and technology cause broad changes in society, not only is a clearer perception of genetics in particular needed, but more understanding of science in general. For about half the U.S. population, high school biology is their last science course. Educators must focus on this course to increase public understanding of science. Because students generally find people more interesting than rats, and because human genetics is a very popular topic in the high school biology course, educators responsible for the Biological Sciences Curriculum Study Program are considering increasing time spent on its study in hopes of increasing public knowledge not only of genetics but of science in general.

At the university level, more funds could be provided to develop courses on the relationships between science, technology, and society, which could be designed both for students and for the general public.

Several sources outside the traditional school system already work to increase public understanding of science and the relationships between science and society. Among them are:

- Three programs developed by the National Science Foundation to improve public understanding of and involvement in science: Science for the Citizen; Public Understanding of Science; and Ethics and Values in Science and Technology.

[10]Dorothy Nelkin and Michael Pollack, "Problems and Procedures in the Regulation of Technological Risk," in *Societal Risk Assessment*, R. Schwing, and W. Albers (eds.) (New York: Plenum Press, 1980).

- Science Centers and similar projects specifically designed to present science information in an appealing fashion.
- New magazines that offer science information to the lay reader—another indication of increasing interest in science.
- Television programs dealing with science and technology. Examples are the two PBS series, *NOVA* and *Cosmos,* and the BBC series, *Connections.* CBS has also begun a new series called *The Universe.*
- Television programs dealing with social issues and value conflicts. Particularly interesting is the concept behind *The Baxters.* In this half-hour prime time show, the network provides the first half of the show, which is a dramatization of a family in conflict over a social or ethical issue. The second half of the show consists either of a discussion about what has been seen or of comments from people who call in.

One interesting possibility would be to combine a series of Baxter-type episodes on genetic issues with audience reaction using the QUBE system, a two-way cable television system in Columbus, Ohio (now expanding to other cities). In this system, television viewers are provided with a simple device that enables them to answer questions asked over the television. A computer tabulates the responses, which can either be used by the studio or immediately transmitted back to the audience. QUBE permits its viewers to do comparison shopping in discount stores, take college courses at home, and provide opinion to elected officials. It could be effectively combined with a program like *The Baxters,* to study social issues. If several such programs on genetics were shown to QUBE subscribers, audience learning and interest could be measured.

Any efforts to increase public understanding should, of course, be combined with carefully designed evaluation studies so that the effectiveness of the program can be assessed.

OPTIONS:

A. *Programs could be developed to increase public understanding of science and the rela-tionships between science, technology, and society.*

Public understanding of science in today's world is essential, and there is concern about the adequacy of the public's knowledge.

B. *Programs could be established to monitor the level of public understanding of genetics and of science in general and to determine whether public concern with decisionmaking in science and technology is increasing.*

Selecting this option would indicate that there is need for additional information, and that Congress is interested in involving the public in developing science policy.

C. *The copyright laws could be amended to permit schools to videotape television programs for educational purposes.*

Under current copyright law, videotaping television programs as they are being broadcast may infringe the rights of the program's owner, generally its producer. The legal status of such tapes is presently the subject of litigation. As a matter of policy, the Public Broadcasting Service negotiates, with the producers of the programs that it broadcasts, a limited right for schools to tape the program for educational uses. This permits a school to keep the tape for a given period of time, most often one week, after which it must be erased. Otherwise, a school must rent or purchase a copy of the videotape from the owner.

In favor of this option, it should be noted that many of the programs are made at least in part with public funds. Removing the copyright constraint on schools would make these programs more available for another public good, education. On the other hand, this option could have significant economic consequences to the copyright owner, whose market is often limited to educational institutions. An ad hoc committee of producers, educators, broadcasters, and talent unions is attempting to develop guidelines in this area.

ISSUE: **Should Congress begin prepar-**
ing now to resolve issues that
have not yet aroused much pub-
lic debate but that may in the
future?

As scientific understanding of genetics and the ability to manipulate inherited characteristics develop, society may face some difficult questions that could involve tradeoffs between individual freedom and societal need. This will be increasingly the case as genetic technologies are applied to humans. Developments are occurring rapidly. Recombinant DNA technology was developed in the 1970's. In the spring of 1980, the first application of gene replacement therapy in mammals succeeded. Resistance to the toxic effect of methotrexate, a drug used in cancer chemotherapy, was transferred to sensitive mice by substituting the gene for resistance for the sensitive gene in tissue-cultured bone marrow cells obtained from the sensitive mice. Transplanted back into the sensitive mice, the bone marrow cells now conferred resistance to the drug.[11] In the fall of 1980, the first gene substitution in humans was attempted.[12]

Although this study was restricted to non-human applications, many people assume from the above and other examples that what can be done with lower animals can be done with humans, and will be. Therefore, some action might be taken to better prepare society for decisions on the application of genetic technologies to humans.

OPTIONS:

A. *A commission could be established to identify central issues, the probable time-frame for application of various genetic technologies to humans, and the probable effects on society, and to suggest courses of action. The commission might also consider the related area of how participatory democracy might be combined with representative democracy in decisionmaking.*

B. *The life of the President's Commission for the Study of Ethical Problems in Medicine and Biomedical and Behavioral Research could be extended for the purpose of addressing these issues.*

The 11-member Commission was established by Public Law 95-622 in November 1978 and terminates on December 31, 1982. Its purpose is to consider ethical and legal issues associated with the protection of human subjects in research; the definition of death; and voluntary testing, counseling, information, and education programs for genetic diseases as well as any other appropriate topics related to medicine and to biomedical or behavioral research.

In July and September 1980, the Commission considered how to respond to a statement from the general secretaries of the National Council of Churches, the Synagogue Council of America, and the United States Catholic Conference that the Federal Government should consider ethical issues raised by genetic engineering. The request was prompted by the Supreme Court decision allowing patents on "new life forms." The general secretaries stated that "no government agency or committee is currently exercising adequate oversight or control, nor addressing the fundamental ethical questions (of genetic engineering) in a major way," and asked that the President "provide a way for representatives of a broad spectrum of our society to consider these matters and advise the government on its necessary role."[13]

After testimony from various experts, the Commission found that the Government is already exercising adequate oversight of the "biohazards" associated with rDNA research and industrial production. The Commission decided to prepare a report identifying what are and are not realistic problems. It will concentrate on the ethical and social aspects of genetic technology that are most relevant to medicine and biomedical research.

The Commission could be asked to study the areas it identifies and to broaden its coverage to

[11]Jean L. Marx, "Gene Transfer Given a New Twist," *Science* 208:25, April 1980, p. 386.
[12]Gina Bari Kolata and Nicholas Wade, "Human Gene Treatment Stirs New Debate," *Science* 210:24, October 1980, p. 407.

[13]Statement by the general secretaries, U.S. Catholic Conference, Origins, NC Documentary Service, vol. 10, No. 7, July 3, 1980.

additional areas. This would require that its term be extended and that additional funds be appropriated. The Commission operated on $1.2 million for 9 months of fiscal year 1980 and $1.5 million for fiscal year 1981. Given the complexity of the issues involved, the adequacy of this level of funding should be reviewed if additional tasks are undertaken.

A potential disadvantage of using the existing Commission to address societal issues associated with genetic engineering is that a number of issues already exist and more are likely to appear in the years ahead. Yet there are also other issues in medicine and biomedical and behavioral research not associated with genetic engineering that need review. Whether all these issues can be addressed by one Commission should be considered. There are obvious advantages and disadvantages to two Commissions, one for genetic engineering and one for other issues associated with medicine and biomedical and behavioral research. Comments from the existing Commission would assist in reaching a decision on the most appropriate course of action.

Bibliography: suggested further reading

Dobzhansky, Theodosium, *Genetic Diversity and Human Equality* (New York: Basic Tools, 1973).

A discussion of conflicts between the findings of science and democratic social goals. Detailed coverage of the scientific basis for present debates about intelligence and the misconceptions often involved in genetic v. environmental determinants of certain human traits.

Francoeur, Robert T., "We Can - We Must: Reflections on the Technological Imperative," *Theological Studies* 33 (#3): 428-439, 1972.

Argues that man is a creator by virtue of his special position in nature, and that humans must participate in deciding the course of their evolution.

Goodfield, June, *Playing God: Genetic Engineering and the Manipulation of Life* (New York: Harper Colophon Books, 1977).

Discusses the benefits, problems and potential of genetic engineering. Describes the moral dilemmas posed by the new technology. Suggests that the "social contract" between science and society is being "renegotiated.

Harmon, Willis, *An Incomplete Guide to the Future* (New York: Simon and Schuster, 1976).

Surveys how social attitudes and values have changed throughout history and how they may be changing today. Argues that mankind is in the midst of a transition to new values that will affect our world view as profoundly as did the industrial revolution in the 19th century.

Holton, Gerald, and William A. Blanpeid (eds.), *Science and Its Public: The Changing Relationship* (Boston: D. Reidel, 1976).

A collection of essays on the way science and the society of which it is a part interact, and how that interaction may be changing.

Hutton, Richard, *Bio-Revolution: DNA and the Ethics of Man-Made Life* (New York: New American Library (Mentor), 1978).

Reviews the history of the debate about recombinant DNA, discusses the scientific basis for the new technologies, and discusses the changing relationship between science and society. Suggests how the controversies might be resolved.

Monod, Jacques, *Chance and Necessity* (New York: Alfred Knopf, 1971).

A philosophical essay on biology. Two seemingly contradictory laws of science, the constancy of inheritance ("necessity") and spontaneous mutation ("chance") are compared with more vitalistic and deontological views of the universe. An affirmation of scientific knowledge as the only "truth" available to man.

Nichols, K. Guild, *Technology on Trial: Public Participation in Decision-Making Related to Science and Technology* (Paris: Organization for Economic Co-operation and Development, 1971).

Reviews mechanisms that have been used by countries in Europe and North America to settle disputes involving science and technology.

Sinsheimer, Robert, "Two Lectures on Recombinant DNA Research," in *The Recombinant DNA Debate,* D. Jackson and Stephen Stich (eds.) (Englewood Cliffs, N.J.: Prentice Hall, 1979), pp. 85-99.

Argues for proceeding slowly and thoughtfully with genetic engineering, for it potentially has far-reaching consequences.

Tribe, Laurence, "Technology Assessment and the Fourth Discontinuity: The Limits of Instrumental Rationality," *Southern California Law Review* 46 (#3): 617-660, June 1973.

An essay on the fundamental task facing mankind in the late 20th century: the problem of choice of tools. New knowledge, especially from biology, will increasingly offer options for technology, the use of which will cause changes in human values.

A Case Study of Acetaminophen Production

Summary

The objective of this case study is to demonstrate the economic feasibility of applying a genetically engineered strain to make a chemical product not now produced by fermentation.

BACKGROUND

Acetaminophen (APAP) was chosen for the case study. As an analgesic, it lacks some of the side effects of aspirin, and is the largest aspirin substitute on the market. Around 20 million pounds (lb) are manufactured annually. Mallinckrodt, Inc., produces 60 to 70 percent; the remainder is manufactured primarily by CPC International and Monsanto Co. APAP is sold to health care companies, which market it to retailers.

The McNeil Consumer Products division of Johnson & Johnson, which markets APAP under the trade name, Tylenol, has the largest share of the trade market. Over a dozen other companies in the United States sell it under other trade names.

One chemical manufacturer's bulk selling price for APAP is around $2.65/lb.[1] By the time the consumer purchases it at the drug store, the markup results in a selling price of around $25 to $50/lb, depending on dosage and package sizes. Thus, the total value of APAP to the manufactures is some $50 million annually, while the total retail value falls in the range of $500 million to $1 billion.

APPROACHES

- A conservative approach was taken, in that only a conventional batch fermentation process was considered.
- Variables were selected pertaining to the choice of the microbial pathway; the nature of the feedstock; conversion efficiencies of feedstock to APAP; and the final yield of APAP.
- Costs were based on proprietary processes involving startup, large-scale fermentation, and recovery of APAP.
- Costs were itemized for materials and supplies; labor distribution; utilities (broken down by specific energy requirements according to process and equipment); equipment (grouped according to process); and building requirements (space needs allocated according to process).

CONCLUSIONS

- The projected cost for manufacturing APAP by means of batch fermentation, using a genetically engineered strain, amounts to $1.05/lb. This cost is based on a plant producing 10 million lb of APAP annually.
- As a rule of thumb, the gross margin for manufacture of a chemical such as APAP should approximate 50 percent of sales. The gross margin represents the profit before general and administrative, marketing and selling, and research and development expenses. The gross margin for all of the products made by Mallinckrodt, the largest manufacturer of APAP, amounted to 39 and 37 percent of sales in 1977 and 1978, respectively.[2] The gross margin for Monsanto, a much larger company than Mallinckrodt but a smaller manufacturer of APAP, amounted to 27 and 26 percent of all sales in 1976 and 1977, respectively.[3] If the gross margin for APAP is as high as 50 percent of sales, its current cost of manufacture should amount to $1.325/lb, based on a bulk selling price of $2.65/lb. Therefore, its projected cost when produced by fermentation is around 20 percent lower than its estimated cost when produced by chemical synthesis.
- If the selling price of APAP produced by fermentation is marked up 100 percent, the bulk selling price becomes $2.10/lb. This decrease of $0.55/lb could be transformed into cost savings of around $5 to $10/lb to the consumer. These economies would result in an annual cost saving to the consumer of $100 million to $200 million.
- Current processes for synthesizing APAP from nitrobenzene do not appear to pose significant pollution problems, although a number of side products are formed and must be removed.[4][5][6] However, a fermentation process would be even

[1] *Chemical Marketing Reporter*, November and December 1979.

[2] Mallinckrodt, Inc., *Annual Report*, 1978.

[3] Monsanto Co., *Annual Report*, 1977.

[4] R. G. Benner, "Process for Preparing Aminophenol," U.S. Patent 3,383,416, 1968.

[5] F. A. Baron, R. G. Benner, and A. E. Weinberg, "Purification of p-Aminophenol," U.S. Patent 3,694,508, 1972.

[6] F. A. Baron and R. G. Benner, "Purification of p-Aminophenol," U.S. Patent 3,717,680, 1973.

cleaner. Only APAP would accumulate: all other metabolites are naturally occurring. Even micro-organisms could be collected after each batch and processed into a cake for use as a high protein animal feed.

Biological parameters

MICROBIAL PATHWAY

A proposed pathway for converting aniline to APAP via the acetylation of an intermediate, p-amino-phenol, is shown in figure I-A-1. Various fungi have been identified in which these reactions occur.[7][8][9]

[7]R. V. Smith and J. P. Rosazza, "Microbial Models of Mammalian Metabolism," *J. Pharmaceut. Sci.* 64:1737-1759, 1975.

[8]R. V. Smith and J. P. Rosazza, "Microbial Models of Mammalian Metabolism: Aromatic Hydroxylation," *Arch. Biochem. Biophys.* 161:551-558, 1974.

[9]V. R. Munzner, E. Mutschler, and M. Rummel, "Uber die mikro-biologische unwandlung N-haltiger substrate" (Concerning the Microbiological Transformation of N- containing Substrate), *Plant Medica* 15:97-103, 1967.

Figure I-A-1. Bioconversion of Aniline to APAP[a]

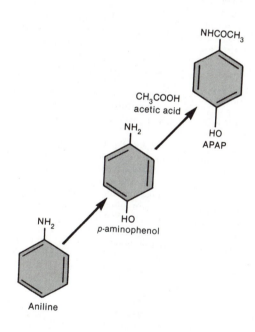

[a]APAP = N-acetyl-p-aminophenol = acetaminophen = p-acetamidophenol = p-hydroxyacetanilide = Tylenol (trade name of McNeil Laboratories).

SOURCE: Genex Corp.

Alternatively, aniline could be acetylated directly forming acetanilide, which in turn would be hydrox-ylated to APAP.[10][11][12] A number of *Streptomyces* species have been found to convert acetanilide to APAP.[13] The pathway involving p-aminophenol was chosen simply because the conversion efficiency of acetic acid to APAP would be slightly higher if acetic acid entered the overall reaction at the second step rather than at the first step.

HOST MICRO-ORGANISMS

The most suitable micro-organism for production of APAP in large-scale fermentation may not neces-sarily be one that normally metabolizes aniline or p-aminophenol. While a bacterium might serve as a suitable host for insertion and expression of the req-uisite genes, a yeast may represent a better choice. It will probably more closely resemble the organism from which the genes are isolated.

Fermentation efficiencies

CONVERSION EFFICIENCIES

The molar and weight conversion efficiencies for the bioconversion of feedstock to product are pro-jected in table I-A-1. The bioconversion of aniline to

[10]Smith, et al., op. cit.

[11]Munzner, et al., op. cit.

[12]R. J. Theriault and T. H. Longfield, "Microbial Conversion of Acetanilide to 2'-Hydroxyacetanilide and 4'- Hydroxyacetanilide," *Apl. Microbiol.* 15:1431-1436, 1967.

[13]Ibid.

Table I-A-1.—Fermentation Efficiencies to Meet the Requirements for the Production of Acetaminophen (APAP) From Aniline

Overall molar conversion efficiency of:	
(a) Aniline to APAP.	90.25%
(b) Acetic Acid to APAP.	95.0
Overall weight conversion efficiency of:	
(a) Aniline to APAP[a].	146.5
(b) Acetic Acid to APAP[a].	239.1
Utilization of:	
(a) Aniline in fermentation broth	2.28 lb/gal
(b) Acetic acid in fermentation broth	1.39/gal
Production of APAP in broth	3.34 lb/gal
Batch volume	33,500 gal
Recovery efficiency.	90.0 %
Yield of APAP/batch	100,701 lb
Number of batches/year	100
Annual yield of product.	10,070,100 lb

[a]Overall weight conversion of precursor to APAP = $\frac{\text{molecular weight of APAP}}{\text{molecular weight of precursor}} \times$ molar conversion efficiency of precursor to APAP

SOURCE: Genex Corp.

Appendixes

APAP involves two steps. The product of the individual reactions for each step represents the overall conversion efficiency. A molar conversion efficiency of 95 percent was assumed for each step. This value is based on a multitude of reports demonstrating similar molar conversion efficiencies for analogous biochemical reactions under actual fermentation conditions.[14]

PRODUCT YIELD

The yield of APAP projected in table I-A-1 is based on estimating a ratio of 40 percent weight to volume (i.e., 40 lb per 100 gallons (gal) of fermentation broth) prior to 90 percent recovery. Such a high yield is permitted because of the poor solubility of APAP under operating conditions. As a result, high levels of APAP would have no adverse effect on the host micro-organism. Use of insoluble systems in fermentation has in fact been reported in recent years—e.g., in certain microbial transformations of steroids, yields of 40 percent may result due to the insolubility of the product.

[14]B. J. Abbott, "Immobilized Cells," in *Annual Reports on Fermentation Processes*, vol. 1, D. Perlman (ed.) (New York: Academic Press, 1977), pp. 205-233.

Economics

PRODUCTION REQUIREMENTS

How the various production requirements would be met during the microbial transformation of aniline to APAP is summarized in tables I-A-1 and 2. Aniline and acetic acid would not be added to the fermentation broth all at once but rather step-wise according to their rates of conversion. The plant would contain two 50,000-gal fermenters, which in the course of a year would yield 10 million lb of APAP.

PRODUCTION COSTS

The costs for the annual production are summarized in table I-A-3. They are broken down into their major components and are expressed both as annual costs and as unit costs. Detailed budgets for the various cost centers are shown in tables I-A-4 through I-A-10. Materials and supplies are described in table I-A-4; labor distribution in table I-A-5; utility requirements in tables I-A-6 through I-A-8; equipment in table I-A-9; and space requirements in table I-A-10. This analysis reveals a unit cost of APAP equal to $1.05/lb.

Table I-A-2.—Summary of Production Conditions of APAP

Number of fermenters	2
Size of fermenters	50,000 gal
Operating volume	33,500 gal
Cycle	7[a]
Batches	100

[a]6-day fermentation, 1-day turn around.

SOURCE: Genex Corp.

Table I-A-3.—Summary of Costs of Production of APAP

	Annual cost	Cost/lb
Materials and supplies	$ 6,133,802	$0.6091
Labor	2,012,140	0.1998
Utilities	630,200	0.0626
Equipment	1,377,590	0.1368
Building	439,399	0.0436
Total	$10,593,131	$1.05/lb

Annual production = 10,070,100 lb

SOURCE: Genex Corp.

Table I-A-4.—Materials and Supplies for Production of APAP

Materials	Cost/batch	Cost/year
Fermentation		
Fishmeal (1.5% @ $0.155)	$ 648.68	$ 64,868
Glucose (1.5% @ $0.1535)	642.40	64,240
Lard oil (2.5% @ $0.325)	2,266.88	226,888
Mineral salts (4,215 lb @ $0.05074)	213.77	21,377
Aniline (76,250 lb @ $0.42)	32,027.52	3,202,752
Acetic acid (46,680 lb @ $0.245)	11,436.60	1,143,660
Miscellaneous (10% of basic materials)	377.17	37,717
Subtotal	$47,613.02	$4,761,302
Recovery		
Filter aid (0.2 lb/gal @ $13)	$ 871.00	$ 87,100
Other chemicals and supplies	1,600.00	$ 160,000
Subtotal	$ 2,471.00	$ 247,100
Finishing		
Packaging (1,255 bag units at $0.80)	$ 1,004.00	$ 100,400
Other (labels, stencils, etc.)	1,004.00	$ 100,400
Subtotal	$ 2,008.00	$ 200,800
General supplies		
Maintenance (4% of capital investment)		$ 425,900
Other (laboratory office, plant miscellaneous)		498,700
Total		$6,133,802

SOURCE: Genex Corp.

Table I-A-5.—Labor Distribution for Production of APAP

Category	Man-hours per week	Hourly rate	Salary and wage cost $/week	Salary and wage cost $/year
Supervision				
General manager	40	20	$ 800	$ 41,600
Superintendents	80	17	1,360	70,720
Managers	80	15	1,200	62,400
Supervisors	320	12	3,840	199,680
Hourly rated employees, services				
Laboratory				
Level I	80	10	800	41,600
Level II	80	8	640	33,280
Level III	120	6	720	37.440
Level IV	40	5	200	10,400
Maintenance and engineering				
Level I	240	10	2,400	124,800
Level II	240	8	1,920	99,840
Level III	240	6	1,440	74,880
Level IV	160	5	800	41,600
Hourly rated employees, production				
Fermentation department				
Level I	200	10	2,000	104,000
Level II	240	8	1,920	99,840
Level III	80	6	480	24,960
Level IV	80	5	400	20,800
Recovery department				
Level I	320	10	3,200	166,400
Level II	400	8	3,200	166,400
Level III	80	6	480	24,960
Level IV	120	5	600	31,200
Subtotal				$1,476,800
Add overtime @ 6% × 1.5				132,912
Subtotal				$1,609,712
Add fringe benefits @ 25%				402,428
Total salaries and wages				$2,012,140

SOURCE: Genex Corp.

Table I-A-6.—Steam Requirements for Production of APAP

Operation	Lb/batch
Sterilization, fermenters, and seed tanks:	
Heating	52,100
Holding	20,000
Sterilization, piping, and equipment (other)	20,000
Heating acetaminophen solution (recovery)	163,500
Drying, turbo dryer	200,300
General purpose usage	50,000
Total	505,900

Cost at $5.00/M lb:
Per fermenter batch = $ 2,530
Per year (100 batches) = $253,000

SOURCE: Genex Corp.

Table I-A-7.—Electricity Requirements for Production of APAP

Connected load	HP	kW	Units/batch (hours operation)	kWh
Fermenters 200		149	144	21,456
Seed tanks 47.5		35	24	840
Chillers 580		433	11	4,763
Air compressor 275		205	86	17,630
Harvest tank 100		75	11	825
Decanter centrifuge 120		90	52	4,680
Process tanks 300		224	19	4,256
Crystallizing tanks 300		224	11	2,464
Turbo dryer 30		22	23	506
Cooling tower 40		30	144	4,320
Pumps (est. = 6 @ 7.5) 45		34	144	4,896
Lighting, instruments and general load 25		19	144	2,736
Total kWh				69,372

@ 0.05/kWh   = $ 3,469 per batch
@ 100 batches/yr = $346,900 per year

SOURCE: Genex Corp.

Table I-A-8.—Water Requirements for Production of APAP

	Gal/batch
Fermentation	35,000
Tower makeup	63,000
Process loss	100,000
Chilled water makeup	30,000
Direct cooling	50,000
General use	25,000
Total	303,000 gal

Process water rate = $1.00/M gals
Cost   = $303/batch
100 batches/yr   = $30,300/year

SOURCE: Genex Corp.

Table I-A-9.—Equipment Costs for Production of APAP

Receiving and batching area

3 20,000 gal steel aniline storage tanks, insulated and cooled - @ $47,000	$ 341,000
2 20,000 gal aluminum acetic acid storage tanks, insulated and cooled - @ $71,300.	342,600
1 10,000 gal steel nitrogen storage tank with controls and instruments.	47,000
1 10,000 gal steel lard oil storage tank, insulated and heated	22,300
1 10,000 gal stainless steel Batch tank with programable controller and agitator.	59,500
2 1,700 ft³ stainless steel Hopper bins with conveyors - @ $58,100	116,200
1 Electric forklift truck	11,400

Fermentation and seed area

1 150 gal stainless steel seed vessel, fully instrumented.	125,000
1 2,500 gal stainless steel seed vessel, fully instrumented.	169,000
2 50,000 gal stainless steel fermenters, fully instrumented with central control room - @ $399,000	798,000

Recovery area

1 50,000 gal stainless steel process tank, cooled, agitated and insulated	195,000
1 3,000 gal steel filter aid slurry tank with agitator	11,300
1 Stainless steel continuous decanter centrifuge	167,000
2 100,000 gal stainless steel process tanks, insulated with external steam injection heater, pump and agitator - @ $333,000	666,000

1 20,000 gal stainless steel side-entering surge tank with agitator . . . . $	56,100
3 50,000 gal stainless steel crystallizing tanks, insulated with heavy duty cooling coils and top-mounted agitator - @ $195,000	585,000
1 Stainless steel turbo tray dryer.	653,000
2 3,500 ft³ stainless steel hopper bins - @ $66,000	132,000
1 Bagging unit	20,000
4 Stainless steel finished product conveyors - @ $12,000	48,000

Auxiliary equipment

3 1,500 c.f.m. reciprocating air compressors - @ $166,000	498,000
Laboratory and office equipment	650,000
Chillers, 500 ton total capacity	575,000
1 Cooling tower, 1,500 g.p.m..	210,000
35 Pumps and motors, various sizes.	104,700
2 Dump trucks - $12,000	24,000
Ventilation, general and spot - @ 7.5% of equipment	583,791
Piping, general, materials and installation - @ 7.5% of equipment.	583,790
Miscellaneous equipment (hand tools, etc.) - @ 5% of equipment	389,194
Total	$7,783,875

Annual charge for capital recovery over 10-year period, with 12% interest compounded annually ($7,783,875 × 0.17698) $1,377,590

SOURCE: Genex Corp.

Table I-A-10.—Building Requirements for Production of APAP

Area	Gross space ft²/ft³	Unit value[a]	Cost
Central office	940	41.00[b]	$ 38,540
Laboratories	4,500	70.00[b]	315,000
Warehouse	2,000/36,000	27.00[b]	54,000
Batching	1,000/30,000	1.75	52,500
Fermentation (including seed)	6,000/320,000	1.75	560,000
Harvest, filter	3,500/169,000	1.75	295,750
Processing, crystallization	8,700/470,000	1.75	822,500
Drying, finishing.	5,000/270,000	1.75	472,500
Warehouse, finished product	11,000/200,000	27.00[b]	297,000
Auxiliary equipment.	4,300/154,000	1.75	269,500
Maintenance, engineering.	11,500/207,000	1.75	362,250
Total			$3,539,540

Amortization over 30 years @ 12% compound interest $439,399[c]

[a]Unit values in cubic feet except where noted by "b."
[b]Unit value in square feet.
[c]Amortization = 0.12414 × total.

SOURCE: Genex Corp.

A Timetable for the Commercial Production of Compounds Using Genetically Engineered Micro-Organisms in Biotechnology

Objectives

- The estimation of the proportions of various groups of commercial products and processes for which recombinant DNA (rDNA) technology could be applicable.
- The construction of timetables to indicate plausible sequences of commercial developments that would result from the application of rDNA technology.

Approaches

The following five industries were evaluated:
1. pharmaceutical,
2. agricultural,
3. food,
4. chemical, and
5. energy.

The manufacturing processes that would result from the application of rDNA technology would be based on fermentation technology. Therefore, a set of parameters was developed to serve as a guide to assess the economics of applying fermentation technology to the manufacture of products currently manufactured by other means.

The chemical industry generates a large number of products that could be attributed to (and is in this study) the other four industries cited, this particular industry was focused on more closely than the others. The following factors were considered in constructing the timetables showing the applicability of rDNA technology:

- the current state of the art of genetic engineering;
- the current economic limitations of fermentation technology;
- the length of time to progress from a laboratory process to the pilot plant to large-scale production;
- the plant construction time; and
- the Government regulatory agency approval re-

quired (of the products and manufacturing processes, not of the rDNA technology per se).

Sources of information

While much of the information compiled for this report was obtained from published sources, a considerable amount came from prior proprietary studies performed by Genex Corp. In the latter case, information is used that is not proprietary, although the sources must remain confidential. In this connection Genex has had numerous discussions with the technical and corporate management of more than 100 large companies (generally multibillion dollar companies), concerning research interests, product lines, and market trends. Production costs are extrapolated for four fermentation plants of various sizes and capabilities. (See table I-B-1.)

A group of Genex scientists, consisting of a biochemical engineer, two organic chemists, a biochemist, and four molecular geneticists rated the feasibility of devising micro-organisms to produce various chemicals in accordance with the fermentation conditions specified in table I-B-1. For those chemicals that appeared to be capable of being produced microbiologically, dates were assigned for the times when the necessary technology would be achieved in the laboratory. By combining both technical and economic factors, it then became possible to project a timetable for commercial production. (See table I-B-2.)

It should be emphasized that an extremely conservative approach was taken in considering fermentation economics over the next 10 years. Only the relatively poor economics of conventional batch fermentation was considered. Immobilized cell processes were projected to be 15 years away, and even then, the incremental cost savings projected (see table I-B-1) are lower than the incremental cost savings currently obtained with immobilized cell processes. The assumptions made here, however, did include reasonably high product yields and highly effi-

Table I-B-1.—Unit Cost Assumptions for the Production of Chemicals by Fermentation After Various Intervals of Time

Earliest date (year)	Size of plant (lb)	Type of fermentation	Product yield (%)	Annual c[a] excluding precursor ($ millions)	Unit cost excluding precursor ($/lb)	Precursor	Complete unit cost ($/lb)
5	50	Ordinary batch	12	23.5	0.47	Petrochemical[b]	0.66
10	100	Ordinary batch	40	24.5	0.25	Petrochemical	0.44
15	200	Immobilized cells	40	25.5	0.13	Petrochemical	0.32[d]
20	200	Immobilized cells	40	25.5	0.13	Carbohydrate[c]	0.24[d]

[a]Annual costs for ordinary batch fermentation were estimated from proprietary data. Values obtained for the immobilized cell examples are computed at 31.2 percent below the comparable values for ordinary batch fermentation.

[b]Average cost of petrochemical equals $0.17/lb. At 90 percent conversion efficiency, cost contribution of petrochemical equals $0.19/lb of product.

[c]Average cost of carbohydrate assumed at $0.04/lb of molasses or $0.02/lb of cellulose-containing pellets from biomass residue. For 50 percent free sugar content of molasses, cost of sugar equals $0.08/lb. At 70 percent conversion efficiency from the sugar, cost contribution of molasses equals $0.11/lb of product. For 50 percent cellulose content in the biomass pellets, cost of cellulose equals $0.04/lb. For 50 percent conversion efficiency to free sugar, followed by 70 percent conversion efficiency from the sugar, cost contribution of the pellets also equals $0.11/lb of product.

[d]These unit costs may be further reduced to $0.26 and $0.17/lb., respectively, for products whose annual U.S. production currently exceeds 1 billion lb. Assumptions include reduction in precursor cost by 20 percent (presumably because manufacturer controls supply of precursor); reduction in unit cost of immobilized cell process by 13 percent (d) and 42 percent (e), respectively; maximum of 80 percent product yield (e); and a nearly 100 percent bioconversion efficiency from the petrochemical precursor.

SOURCE: Genex Corp.

Table I-B-2.—Basis for Estimating the Timetable for Manufacture of Chemicals by Means of Microbial Processes

Earliest date for commercial production[a] is:	If all the technology[b] is achieved by:	And if bulk selling prices[c] (in 1979 dollars) equal or exceed:	Assuming unit costs[d] (in 1979 dollars) equal or exceed:
5 years	2 years	$1.32/lb	$0.66/lb
10	7	0.88	0.44
15	12	0.64 (0.43)	0.32 (0.26)
20	17	0.48 (0.28)	0.24 (0.17)

[a]It is assumed that development of the appropriate manufacturing facilities begins at least 5 years prior to the onset of production.

[b]Technology refers to both genetic and biochemical engineering. Technology would be achieved on demonstrating that the chemical could be biologically produced in the laboratory at commercially desirable yields and reaction efficiencies.

[c]It is assumed that all bulk selling prices are marked up 100 percent from the corresponding unit costs, except for chemicals whose annual U.S. production currently exceeds 1 billion lb. In these cases the bulk selling prices (numbers in parentheses) are assumed to be marked up only 67 percent.

[d]Unit costs were obtained from table I-B-1. See footnote of table I-B-1 for explanation of numbers in parentheses.

SOURCE: Genex Corp.

cient transformations of precursor to product, but nothing exceptional with respect to current fermentation technology. Indeed, high product yields and highly efficient reactions would be expected with genetically engineered micro-organisms.

Two points should be stressed that place these projections on the low side. First, they exclude certain groups of products, the end products of which could not be microbially processed, although their basic constituents could be produced microbiologically (e.g., monomers of microbial origin could form chemically synthesized polymers). Second, the projections exclude naturally occurring products of

microbial origin, which could be effective or superior substitutes for chemically synthesized products that could not be manufactured microbiologically. As examples, dyes of microbial origin, such as prodigiosin, might advantageously replace those synthesized chemically, because their toxicity is lower than their chemical counterparts. In the case of plastics, a new generation of plastics of microbial origin, e.g., pullulans, would not have to be made from petrochemical feedstocks and would be biodegradable.

Explanation of tables

Tables I-B-3 through I-B-32 present the compounds from two points of view. Tables I-B-3 to I-B-10 group the compounds by industry subgrouped by product category. Tables I-B-11 to I-B-32 group the compounds by product category irrespective of industry.

The tables based on industry present end use data for each compound; e.g., in the pharmaceutical industry aspirin is listed as an aromatic used as an analgesic, whereas in the chemical industry aniline is listed as an aromatic used as a cyclic intermediate. Thus, the similarities and differences between compounds of similar origin, i.e., product category, are revealed.

The tables based solely on product category are divided into two types; one type pertaining to market data (tables I-B-10, 11, and the subsequent odd numbered ones through table I-B-33), and the other pertaining to technical data (the even numbered tables from I-B-12 through I-B-32.)

The market data were obtained both from published sources and from prior proprietary studies

Table I-B-3.—Pharmaceuticals: Small Molecules

Product category	End use
Amino acids	
Phenylalanine	Intravenous solutions
Tryptophan	Intravenous solutions
Arginine	Therapeutic: liver disease and hyperammonemia
Cysteine	Therapeutic: bronchitis and nasal catarrh
Vitamins	
Vitamin E	Intravenous solutions, prophylactic
Vitamin B_{12}	Intravenous solutions
Aromatics	
Aspirin	Analgesic
p-acetaminophenol	Analgesic
Steroid hormones	
Corticoids	
Cortisone	Therapeutic: anti-inflammatory agent
Prednisone	Therapeutic: anti-inflammatory agent
Prednisolone	Therapeutic: anti-inflammatory agent
Aldosterone	Therapeutic: control of electrolyte imbalance
Androgens	
Testosterone	Therapeutic: infertility, hypogonadism, and hypopituitarism
Estrogens	
Estradiol	Prophylactic, therapeutic: vaginitis
Antibiotics	
Penicillins	Control of infectious diseases
Tetracyclines	Control of infectious diseases
Cephalosporins	Control of infectious diseases
Short peptides	
Glycine-Histidine-Lysine	Manufacturing processes: tissue culture

SOURCE: Genex Corp.

Table I-B-4.—Pharmaceuticals: Large Molecules

Product category	End use
Peptide hormones	
Insulin	Control of diabetes
Endorphins	Analgesics, narcotics, prophylactics
Enkephalins	Analgesics, narcotics, prophylactics
ACTH[a]	Diagnostic: adrenal instability
Glucagon	Therapeutic: diabetes-induced hypoglycemia
Vasopressin	Therapeutic: antidiuretic
Human growth hormone	Therapeutic: dwarfism
Enzymes	
Glucose oxidase	Diagnostic: measurement of blood sugar
Urokinase	Therapeutic: antithrombotic
Asparaginase	Therapeutic: antineoplastic
Tyrosine hydroxylase	Therapeutic: Parkinson's disease
Viral antigens	
Hepatitis viruses	Vaccine
Influenza viruses	Vaccine
Herpes viruses	Vaccine
Varicella virus	Vaccine
Rubella virus	Vaccine
Reoviruses	Vaccine: common cold
Epstein-Barr virus	Vaccine: infectious mononucleosis, nasopharyngeal carcinoma, Burkitt lymphoma
Miscellaneous proteins	
Interferon	Control of infectious diseases
Human serum albumin	Therapeutic: shock and burns
Monoclonal antibodies	Diagnostics: hepatitis, cancer, etc; therapeutics
Gene preparations	
Sickle-cell anemia	Control of hereditary disorder
Hemophilias	Control of hereditary disorder
Thallasemias	Control of hereditary disorder

[a]Adrenocorticotropic hormone.
SOURCE: Genex Corp.

performed by Genex. In the latter case, data are used that are not proprietary although the sources must remain confidential. Market values were estimated by multiplying the market volume (total amount of product sold in 1978) by the bulk cost (unit bulk selling price in 1980). Except for aromatics and aliphatics, all market data represent worldwide estimates. Market data for aromatics and aliphatics are restricted to the United States. Data that could not be found were marked not available (N/A). Compounds with a high market value were identified, and those that could be produced biologically were selected for this report.

High market values were relative to the industry and end use listing of each compound. For example, with respect to chemicals, normally only cyclic in-

termediates with production volumes (which differ from market volumes) exceeding 50 million lb were selected, but in the case of flavor and perfume materials, compounds with production values generally exceeding 1 million lb were selected. In the case of many pharmaceuticals, clinical importance was weighed heavily in their selection process.

The technical data were also obtained both from published and proprietary sources. With respect to the timetable for commercial production, the stated length of time is the time required to develop existing technology (including both genetic and biochemical engineering) to the point where it may be applied to appropriate manufacturing facilities for the large-scale production of the desired compounds. These time intervals should be sufficient for undertaking

Table I-B-5.—Food Products

Product category	End use
Amino acids	
Glutamate	Food enrichment agent, flavoring agent
Cysteine	Food enrichment agent, manufacturing processes
Aspartate	Flavoring agent
Vitamins	
Vitamin C	Food additive, food enrichment agent
Vitamin D	Food enrichment agent
Aromatics	
Benzoic acid	Food preservative
Aliphatics	
Propionic acid	Food preservative
Short peptides	
Aspartame	Artificial sweetener
Enzymes	
Rennin	Manufacturing processes
Amyloglucosidase	Manufacturing processes
α-amylase	Food enrichment agent, manufacturing processes
Glucose isomerase	Manufacturing processes: sweetener
Nucleotides	
5'-IMP[a]	Flavoring agent
5'-GMP[b]	Flavoring agent

[a]5'-inosinic acid.
[b]5'-guanylic acid.
SOURCE: Genex Corp.

Table I-B-6.—Agricultural Products

Product category	End use
Amino acids	
Lysine	Feed additive
Methionine	Feed additive
Threonine	Feed additive
Tryptophan	Feed additive
Vitamins	
Nicotinic acid	Feed additive
Riboflavin (B$_2$)	Feed additive
Vitamin C	Feed additive
Aliphatics	
Sorbic acid	Feed preservative
Antibiotics	
Penicillins	Feed additive, prophylactic
Erythromycins	Feed additive, prophylactic
Peptide hormones	
Bovine growth hormone	Growth promoter
Porcine growth hormone	Growth promoter
Ovine growth hormone	Growth promoter
Viral antigens	
Foot-and-mouth disease virus	Vaccine
Rous sarcoma virus	Vaccine
Avian leukemia virus	Vaccine
Avian myeloblastosis virus	Vaccine
Enzymes	
Papain	Feed additive
Glucose oxidase	Feed preservative
Pesticides	
Microbial	Insecticide
Aromatic	Insecticide
Inorganics	
Ammonia	Fertilizer

SOURCE: Genex Corp.

all the R&D starting from the current knowledge base necessary to demonstrate that the desired compounds can be biologically produced first in the laboratory and then in the pilot plant at commercially desirable yields and reaction efficiencies. The timetable does not consider delays caused by construction of new facilities nor delays required to obtain Government regulatory approval of new products.

It should be noted that in the technical data charts, when glucose is listed as an alternate precursor by fermentation, other carbohydrates, e.g., cellulose and cornstarch, could be used. Moreover, if glucose were the precursor of choice, the actual feedstock would probably be a commodity like molasses as opposed to pure glucose.

Summary

Over 100 compounds representing 17 different product categories that span the five industries under evaluation are represented in table I-B-10. The current market value of all these products exceeds $27 billion. One particular compound, methane, accounts for over $12 billion. The even-numbered tables from I-B-12 to I-B-32 project that within 20 years all these products could be manufactured using genetically engineered microbial strains on a more economical basis than using today's conventional technologies. In many cases, the time required to apply genetically engineered strains in commercial fermentations could be reduced to as little as 5 years.

The impact of genetic engineering on selected markets is shown in table I-B-33. Only five product categories are considered here, and in one, amino acids, only a few of the compounds comprising it are evaluated. The products represented in the five categories currently have a total market value exceeding $800 million. However, within 20 years this market value could rise to over $5 billion (in 1980 dollars) due largely to the application of genetic engineering. In a number of cases, the desired products would most likely not be available in significant quantities if not for the application of genetic engineering technology.

Table I-B-7.—Chemicals: Aliphatics

Compound	End use
Acetic acid[a]	Miscellaneous acyclic
Acrylic acid[a]	Miscellaneous acyclic
Adipic acid[a]	Miscellaneous acyclic
Bis (2-ethylhexyl) adipate	Plasticizer
Citronellal	Flavor/perfume material
Citronellol	Flavor/perfume material
Ethanol[a]	Miscellaneous acylic
Ethanolamine	Miscellaneous acyclic
Ethylene glycol[a]	Miscellaneous acyclic
Ethylene oxide[a]	Miscellaneous acyclic
Geraniol	Flavor/perfume material
Glycerol[a]	Miscellaneous acyclic
Isobutylene	Miscellaneous acyclic, flavor/perfume material
Itaconic acid	Plastics/resin
Linalool	Flavor/perfume material
Linalyl acetate	Flavor/perfume material
Methane	Primary petroleum product
Nerol	Flavor/perfume material
Pentaerythritol	Miscellaneous acyclic
Propylene glycol[a]	Miscellaneous acylic
Sorbitol	Miscellaneous acyclic
α-terpineol	Flavor/perfume material
α-terpinyl acetate	Flavor/perfume material

[a]Indicates compounds also identified by the Massachusetts Institute of Technology. The following additional chemicals were identified by MIT as amenable to biotechnological production methods: acetaldehyde, acetoin, acetone, acetylene, acrylic acid, butadiene, butanol, butyl acetate, butyraldehyde, dihydroxyacetone, ethyl acetate, ethyl acrylate, ethylene, formaldehyde, isoprene, isopropanol, methanol, methyl ethyl ketone, methyl acrylate, propylene, propylene oxide, styrene, vinyl acetate.

SOURCE: Genex Corp. and the Massachusetts Institute of Technology.

Table I-B-8.—Chemicals: Aromatics and Miscellaneous

Product category	End use
Aromatics	
Aniline	Cyclic intermediate
Benzoic acid	Cyclic intermediate
Cresols	Cyclic intermediate
Phenol	Cyclic intermediate
Phthalic anhydride	Cyclic intermediate
Cinnamaldehyde	Flavor/perfume material
Diisodecyl phthalate	Plasticizer
Dioctyle phthalate	Plasticizer
Inorganics	
Ammonia	Manufacturing processes
Hydrogen	Manufacturing processes
Enzymes	
Pepsin	Manufacturing processes
Bacillus protease	Manufacturing processes
Mineral leaching	
Transition metals (cobalt, nickel, manganese, iron)	Inorganic intermediates; catalysts
Biodegradation	Removal of organic phosphates, aryl sulfonates, and haloaromatics

SOURCE: Genex Corp.

Table I-B-9.—Energy Products

Product category	End use
Enzymes	
Ethanol dehydrogenase	Manufacturing processes
Hydrogenase	Manufacturing processes
Biodegradation	Petroleum byproducts removal
Aliphatics	
Methane	Fuel
Ethanol	Fuel
Inorganics	
Hydrogen	Fuel
Mineral leaching	
Uranium	Fuel

SOURCE: Genx Corp.

Table I-B-10.—Total Market Values for the Various Product Categories

Product category	Number of compounds	Current value ($ millions)
Amino acids	9	$ 1,703.0
Vitamins	6	667.7
Enzymes	11	217.7
Steroid hormones	6	376.8
Peptide hormones	9	263.7
Viral antigens	9	N/A
Short peptides	2	4.4
Nucleotides	2	72.0
Miscellaneous proteins	2[a]	300.0
Antibiotics	4[b]	4,240.0
Gene preparations	3	N/A
Pesticides	2[b]	100.0
Aliphatics:		
Methane	1	12,572.0
Other	24[c]	2,737.5
Aromatics	10[c]	1,250.9
Inorganics	2	2,681.0
Mineral leaching	5	N/A
Biodegradation	N/A	N/A
Totals	107	$27,186.7[d]

[a]Only two of a number of compounds are considered here.
[b]These numbers refer to major classes of compounds; not actual numbers of compounds.
[c]These numbers refer only to those compounds representing the largest market volume in classes specified in the text.
[d]Current value excluding methane = $14,614,700,000.

SOURCE: Genex Corp.

Table I-B-11.—Amino Acids: Market Information

| | Current market data | | |
Compound	Market volume 1,000 lb	Bulk cost $/lb	Market value ($ millions)
Arginine	900	12.73	11.46
Aspartate	3,000	2.86	8.6
Cysteine	600	22.75	13.6
Glutamate.	600,000	1.80	1,080.0
Lysine	129,000	2.10	258.0
Methionine	210,000	1.40	294.0
Phenylalanine. . . .	300	38.18	11.46
Threonine	300	58.18	16.2
Tryptophan	225	43.18	9.71

SOURCE: Compiled by Genex Corp. from data in references 1, 2, and 3.

Table I-B-12.—Amino Acids: Technical Information

Compound	Typical synthetic process	Typical precursor	Is precursor renewable/non-renewable limited	Alternate precursor by fermentation	Time to implement commercial fermentation by genetically engineered strain
Arginine	fermentation	glucose and NH_4[a]	renewable	—	5 yrs.
Aspartate	fermentation	fumaric acid and ammonia	limited	—	5 yrs.
Cysteine	extraction	protein hydrolysis	renewable	—	5 yrs.
Glutamate.	fermentation	glucose and NH_4^+	renewable	—	5 yrs.
Lysine	fermentation	glucose and NH_4^+	renewable	—	5 yrs.
Methionine	chemical	β-methylmercapto propionaldehyde	nonrenewable	glucose and NH_4^+	10 yrs.
Phenylalanine. . . .	chemical	α-acetamino-cinnamic acid	limited	glucose and NH_4^+	5 yrs. 5 yrs.
Threonine	fermentation	glucose and NH_4^+	renewable	—	5 yrs.
Tryptophan	chemical	α-ketoglutaric phenylhydrazone	nonrenewable	glucose and NH_4^+	5 yrs.

[a]Ammonium ion.

SOURCE: Compiled by Genex Corp. from data in references 2, 3, 4, and 5.

Table I-B-13.—Vitamins: Market Information

| | Current market data | | |
Compound	Market volume 1,000 lb	Bulk cost $/lb	Market value ($ millions)
Nicotinic acid. . . .	1,400	1.82	2.5
Riboflavin (B_2). . . .	22	15.40	0.34
Vitamin B_{12}	22	6,991.60	153.8
Vitamin C	90,000	4.50	405.0
Vitamin D	12	42.50	0.51
Vitamin E	3,641	29.00	105.6

SOURCE: Compiled by Genex Corp. from data in references 1, 6, 7, 8, and 9.

Table I-B-14.—Vitamins: Technical Information

Compound	Typical synthetic process	Typical precursor	Is precursor renewable/non-renewable limited	Alternate precursor by fermentation	Time to implement commercial fermentation by genetically engineered strain
Nicotinic Acid . . .	chemical	alkyl α-subst.	nonrenewable	glucose and NH_4^+ [a]	10 yrs.
Riboflavin (B₂). . . .	fermentation	pyridines glucose	renewable	—	10 yrs.
Vitamin B₁₂	fermentation	carbohydrates	renewable	—	10 yrs.
Vitamin C	semisynthetic	glucose or sorbitol	renewable	—	10 yrs.
Vitamin D	fermentation	glucose	renewable	glucose	10 yrs.
Vitamin E	extraction	wheat germ oil	limited	glucose	15 yrs.

[a]Ammonium ion.

SOURCE: Compiled by Genex Corp. from data in references 4, 7, 8, 10, 11, and 12.

Table I-B-15.—Enzymes: Market Information

	Current market data		
Compound	Market volume 1,000 lb	Bulk cost $/lb	Market value ($ millions)
α-amylase	600	19.33	11.6
Amyloglucosidase	600	.00	12.0
Asparaginase. . . .	(Information not available)		
Bacillus protease.	1,000	8.28	8.2
Ethanol dehydrogenase .	(Information not available)		
Glucose isomerase	100	400 00	40 0
Glucose oxidase .	5	160.00	0.80
Hydrogenase	(Information not available)		
Papain	200	59.00	11.8
Pepsin	10	380.00	3.8
Rennin.	24	696.00	40.0
Tyrosine hydroxylase	(Information not available)		
Urokinase	60,000 IU[a]		89.5

[a]IU = international units.

SOURCE: Compiled by Genex Corp. from data in references 9, 13, 14, 15, and 16.

Table I-B-16.—Enzymes: Technical Information

Compound	Typical synthetic process	Typical precursor	Is precursor renewable/non-renewable limited	Alternate precursor by fermentation	Time to implement commercial fermentation by genetically engineered strain
α-amylase	fermentation	molasses	renewable	—	5 yrs.
Amyloglucosidase	fermentation	molasses	renewable	—	5 yrs.
Asparaginase	extraction	tissue culture	renewable	glucose and NH_4^+	5 yrs.
Bacillus protease.	fermentation	molasses	renewable	—	5 yrs.
Ethanol dehydrogenase .		(Information not available)		glucose and NH_4^+	10 yrs.
Glucose isomerase.....	fermentation	glucose and NH_4^+ [a]	renewable	—	5 yrs.
Glucose oxidase .	fermentation	molasses	renewable	—	5 yrs.
Hydrogenase		(Information not available)		glucose and NH_4^+	10 yrs.
Papain..........	extraction	papaya	renewable	glucose and NH_4^+	5 yrs.
Pepsin..........	fermentation	molasses	renewable	—	5 yrs.
Rennin..........	fermentation	molasses	renewable	—	5 yrs.
Tyrosine	extraction	tissue culture	renewable	glucose and NH_4^+	5 yrs.
Urokinase	extraction	tissue culture	renewable	glucose	5 yrs.

[a] Ammonium ion.

SOURCE: Compiled by Genex Corp. from data in references 4, 5, 13, 14, 16, 17, and 18.

Table I-B-17.—Steroid Hormones: Market Information

Compound	Current market data		
	Market volume 1,000 lb	Bulk cost $/lb	Market value ($ millions)
Corticoids.......			305.8
Cortisone	N/A	208.84	N/A
Prednisone......	N/A	467.62	N/A
Prenisolone	N/A	463.08	N/A
Aldosterone	N/A	N/A	N/A
Androgens			10.8
Testosterone	(Information not available)		
Estrogens			60.2
Estradiol	(Information not available)		

SOURCE: Compiled by Genex Corp. from data in references 1 and 4.

Table I-B-18.—Steroid Hormones: Technical Information

Compound	Typical synthetic process	Typical precursor	Is precursor renewable/non-renewable limited	Alternate precursor by fermentation	Time to implement commercial fermentation by genetically engineered strain
Corticoids					
Cortisone					
Prednisone......	semisynthetic	diosgenin or stigmasterol	renewable	glucose	10 yrs.
Predisolone					
Aldosterone					
Androgens					
Testosterone	semisynthetic	chemical modification of cholesterol	renewable	glucose	10 yrs.
Estrogens					
Estradiol........	semisynthetic	chemical modification of cholesterol	renewable	glucose	10 yrs.

SOURCE: Compiled by Genex Corp. from data in references 4, 19, 20, 21, and 22.

Table I-B-19.—Peptide Hormones: Market Information

	Current market data		
Compound	Market volume 1,000 lb	Bulk cost $/lb	Market value ($ millions)
ACTH[a].........	N/A	N/A	5.6
Bovine growth hormone......	0.0	0.0	0.0
Endorphins......	(Information not available)		
Enkephalins.....	(Information not available)		
Glucagon	(Information not available)		
Human growth hormone......	N/A	N/A	75.0
Insulin..........	N/A	N/A	183.1
Ovine growth hormone......	0.0	0.0	0.0
Porcine growth hormone......	0.0	0.0	0.0
Vasopressin.....	(Information not available)		

[a]Adrenocorticotropic hormone.

SOURCE: Compiled by Genex Corp. from data in reference 4.

Table I-B-20.—Peptide Hormones: Technical Information

Compound	Typical synthetic process	Typical precursor	Is precursor renewable/non-renewable limited	Alternate precursor by fermentation	Time to implement commercial fermentation by genetically engineered strain
ACTH[a]	extraction	adrenal cortex	limited	glucose and NH_4^+ [b]	5 yrs.
Bovine growth hormone	extraction	anterior pituitary	limited	glucose and NH_4^+	5 yrs.
Endorphins	extraction	brain	limited	glucose and NH_4^+	5 yrs.
Enkephalins	extraction	brain	limited	glucose and NH_4^+	5 yrs.
Glucagon	extraction	pancreas	limited	glucose and NH_4^+	5 yrs.
Human growth hormone	extraction	anterior pituitary	limited	glucose and NH_4^+	5 yrs.
Insulin	extraction	pancreas	limited	glucose and NH_4^+	5 yrs.
Ovine growth hormone	extraction	anterior pituitary	limited	glucose and NH_4^+	10 yrs.
Porcine growth hormone	extraction	anterior pituitary	limited	glucose and NH_4^+	10 yrs.
Vasopressin	extraction	posterior pituitary	limited	glucose and NH_4^+	5 yrs.

[a]Adrenocorticotropic hormone.
[b]Ammonium ion.
SOURCE: Compiled by Genex Corp. from data in references 4, 23, and 24.

Table I-B-21.—Viral Antigens: Market Information

Compound	Current market data		
	Market volume 1,000 lb	Bulk cost $/lb	Market value ($ millions)
Avian leukemia virus	(Information not available)		
Avian myeloblastosis virus	(Information not available)		
Epstein-Barr virus	0.0	0.0	0.0
Hepatitis virus	0.0	0.0	0.0
Herpes virus	0.0	0.0	0.0
Hoof and mouth disease virus	0.0	0.0	0.0
Influenza virus	(Information not available)		
Reoviruses	0.0	0.0	0.0
Rous sarcoma virus	(Information not available)		
Rubella virus	(Information not available)		
Varicella virus	(Information not available)		

SOURCE: Compiled by Genex Corp. from data in reference 4.

Table I-B-22.—Viral Antigens: Technical Information

Compound	Typical synthetic process	Typical precursor	Is precursor renewable/non-renewable limited	Alternate precursor by fermentation	Time to implement commercial fermentation by genetically engineered strain
Avian leukemia... virus		(Information not available)		glucose and NH_4^+ [a]	5 yrs.
Avian myeloblastosis. virus		(Information not available)		glucose and NH_4^+	5 yrs.
Epstein-Barr virus	tissue culture	lymphoblasts	renewable	glucose and NH_4^+	5 yrs.
Hepatitis........ viruses		(Information not available)		glucose and NH_4^+	5 yrs.
Herpes .... viruses		(Information not available)		glucose and NH_4^+	5 yrs.
Hoof and mouth.. disease virus		(Information not available)		glucose and NH_4^+	5 yrs.
Influenza........ viruses		(Information not available)		glucose and NH_4^+	10 yrs.
Reoviruses		(Information not available)		glucose and NH_4^+	15 yrs.
Rous sarcoma ... virus		(Information not available)		glucose and NH_4^+	5 yrs.
Rubella .... virus	tissue culture	duck embryonic cells	renewable	glucose and NH_4^+	5 yrs.
Varicella .... virus		(Information not available)		glucose NH_4^+	5 yrs.

[a]Ammonium ion.

SOURCE: Compiled by Genex Corp. from data in references 4 and 25.

Table I-B-23.—Short Peptides, Nucleotides, and Miscellaneous Proteins: Market Information

Product category	Current market data		
	Market volume 1,000 lb	Bulk cost $/lb	Market value ($ millions)
Short peptides[a]			
Aspartame	40	110.00	4.4
Glycine-histidine-lysine..........	(Information not available)		
Nucleotides[b]			
5'-IMP[c]	4,000	12.00	48.0
5'-GMP[d]	2,000	12.00	24.0
Miscellaneous proteins[e]			
Interferon	N/A	N/A	50.0
Human serum albumin........	250	1,000.00	250.0
Monoclonal antibodies......	(Information not available)		

[a]Data from references 4 and 26.
[b]Data from references 4 and 27.
[c]5'-inosinic acid.
[d]5'-guanylic acid.
[e]Data from reference 4.

SOURCE: Compiled by Genex Corp.

Table I-B-24.—Short Peptides, Nucleotides, and Miscellaneous Proteins: Technical Information

Product category	Typical synthetic process	Typical precursor	Is precursor renewable/non-renewable limited	Alternate precursor by fermentation	Time to implement commercial fermentation by genetically engineered strain
Short peptides[a]					
Aspartame	chemical	phenylalanine & aspartic acid	renewable	glucose and NH_4^+[b]	5 yrs.
Glycine-histidine-lysine.	extraction	human serum	renewable	glucose and NH_4^+	5 yrs.
Nucleotides[c]					
5'-IMP[d]	extraction	yeast	renewable	glucose and NH_4^+	10 yrs.
5'-GMP[e].	extraction	yeast	renewable	glucose and NH_4^+	10 yrs.
Miscellaneous proteins[f]					
Interferon	extraction or tissue culture	leukocytes, lymphoblasts, or fibroblasts	renewable	glucose and NH_4^+	5 yrs.
Human serum albumin.	extraction	human serum	renewable	glucose and NH_4^+	5 yrs.
Monoclonal antibodies.	somatic cell hybridization	various cells	renewable	glucose and NH_4^+	10 yrs.

[a]Data from references 4 and 27.
[b]Ammonium ion.
[c]Data from references 4 and 28.
[d]5'-inosinic acid.
[e]5'-guanylic acid.
[f]Data from reference 4.

SOURCE: Compiled by Genex Corp.

Table I-B-25.—Antibiotics, Gene Preparations, and Pesticides: Market Information

Product category	Current market data		
	Market volume 1,000 lb	Bulk cost $/lb	Market value ($ millions)
Antibiotics[a]			
Penicillins.	49,300	22.11	1,080.0
Tetracyclines	29,300	34.13	1,000.0
Cephalosporins . .	4,210	114.00	480.0
Erythromycins . . .	(Information not available)		
Gene preparations[b]			
Sickle cell anemia	0.0	0.0	0.0
Hemophilias.	0.0	0.0	0.0
Thallasemias	0.0	0.0	0.0
Pesticides[c]			
Microbial.	N/A	N/A	25.0
Aromatics.	N/A	N/A	75.0

[a]Data from references 4, 28, and 9.
[b]Data from references 4 and 29.
[c]Data from references 4 and 30.

SOURCE: Compiled by Genex Corp.

Table I-B-26.—Antibiotics, Gene Preparations, and Pesticides: Technical Information

Product category	Typical synthetic process	Typical precursor	Is precursor renewable/non-renewable limited	Alternate precursor by fermentation	Time to implement commercial fermentation by genetically engineered strain
Antibiotics[a]					
Penicillins.......	fermentation semisynthetic	lactose & nitrogenous oils	limited	—	10 yrs.
Tetracyclines	fermentation	lactose & nitrogenous oils	limited	—	10 yrs.
Cephalosporins ..	fermentation	lactose & nitrogenous oils	limited	—	10 yrs.
Erythromycins ...	fermentation	lactose & nitrogenous oils	limited	—	10 yrs.
Gene preprations[b]					
Sickle cell anemia		(No process exists currently)		glucose and NH_4^{+} [d]	15 yrs.
Hemophilias.....		(No process exists currently)		glucose and NH_4^{+}	20 yrs.
Thallasemias		(No process exists currently)		glucose and NH_4^{+}	20 yrs.
Pesticides[c]					
Microbial........	fermentation	molasses & fishmeal	renewable	—	5 yrs.
Aromatics.......	semisynthetic	naphthalene	nonrenewable	—	10 yrs.

[a]Data from references 4, 5, 28, 31, and 32.　[c]Data from references 4 and 30.
[b]Data from reference 4.　[d]Ammonium ion.

SOURCE: Compiled by Genex Corp.

Table I-B-27.—Aliphatics: Market Information

Compound	Current market data		
	Market volume 1,000 lb	Bulk cost $/lb	Market value ($ millions)
Acetic acid	823,274	0.23	189.4
Acrylic acid......	46,503	0.43	20.0
Adipic acid	181,097	0.50	90.5
Bis (2-ethylhexyl) adipate	43,015	0.49	21.1
Citronellal.......	394	3.90	1.5
Citronellol.......	1,443	4.50	6.5
Ethanol	1,048,000	0.24	251.5
Ethanolamine....	320,236	0.46	147.3
Ethylene glycol ..	3,137,000	0.31	972.5
Ethylene oxide...	525,113	0.36	189.0
Geraniol	2,307	3.25	7.5
Glycerol	116,612	0.54	63.0
Isobutylene......	597,712	0.95	567.2
Itaconic acid.....	200	0.83	0.2
Linalool.........	3,341	2.60	8.7
Linalyl acetate ...	1,535	3.50	5.4
Methane	878,000,000	0.013	11,573.0
Nerol	462	4.20	1.9
Pentaerythritol...	117,085	0.62	72.6
Propionic acid ...	62,848	0.21	13.2
Propylene glycol .	525,527	0.73	173.4
Sorbic acid	20,000	2.15	43.0
Sorbitol.........	160,267	0.36	57.7
α-terpineol	2,416	1.28	3.0
α-terpinyl acetate.	1,066	1.30	1.4

SOURCE: Compiled by Genex Corp. from data in references 1, 4, 9, and 33.

Table I-B-28.—Aliphatics: Technical Information

Compound	Typical synthetic process	Typical precursor	Is precursor renewable/non-renewable limited	Alternate precursor by fermentation[a]	Time to implement commercial fermentation by genetically engineered strain[b]
Acetic acid	chemical	methanol or ethanol	nonrenewable	glucose	10 yrs.
Acrylic acid	chemical	ethylene	nonrenewable	glucose	10 yrs.
Adipic acid	chemical	phenol	nonrenewable	glucose	10 yrs.
Bis (2-ethylhexyl) adipate	chemical	phenol	nonrenewable	glucose	20 yrs.
Citronellal	chemical	isobutylene	nonrenewable	glucose	20 yrs.
Citronellol	chemical	isobutylene	nonrenewable	glucose	20 yrs.
Ethanol	chemical	ethylene	nonrenewable	glucose	5 yrs.
Ethanolamine	chemical	ethylene	nonrenewable	glucose	10 yrs.
Ethylene glycol	chemical	ethylene	nonrenewable	glucose	5 yrs.
Ethylene oxide	chemical	ethylene	nonrenewable	glucose	5 yrs.
Geraniol	chemical	isobutylene	nonrenewable	glucose	20 yrs.
Glycerol	chemical	soap manuf.	nonrenewable	glucose	5 yrs.
Isobutylene	chemical	petroleum	nonrenewable	glucose	10 yrs.
Itaconic acid	fermentation	molasses	renewable	----	5 yrs.
Linalool	chemical	isobutylene	nonrenewable	glucose	20 yrs.
Linalyl acetate	chemical	isobutylene	nonrenewable	glucose	20 yrs.
Methane	chemical	natural gas	nonrenewable	sewage	10 yrs.
Nerol	chemical	isobutylene	nonrenewable	glucose	20 yrs.
Pentaerythritol	chemical	acetaldehyde & formaldehyde	nonrenewable	glucose	10 yrs.
Propionic acid	chemical	ethanol & carbon monoxide	limited	glucose	10 yrs.
Propylene glycol	chemical	propylene	nonrenewable	glucose	10 yrs.
Sorbic acid	chemical	crotonaldehyde & malonic acid	nonrenewable	glucose	15 yrs.
Sorbitol	chemical	glucose	renewable	----	10 yrs.
α-terpineol	chemical	isobutylene	nonrenewable	glucose	20 yrs.
α-terpinyl acetate	chemical	isobutylene	nonrenewable	glucose	20 yrs.

[a]Wherever glucose is mentioned, other carbohydrates may be substituted, including starch and cellulose.
[b]In many cases these times are based on more readily developed fermentations using nonrenewable or limited hydrocarbons as precursors.

SOURCE: Compiled by Genex Corp. from data in references 4, 33, 34, and 35.

Table I-B-29.—Aromatics: Market Information

Compound	Current market data		
	Market volume 1,000 lb	Bulk cost $/lb	Market value ($ millions)
Aniline	187,767	0.42	78.9
Aspirin	32,247	1.41	45.5
Benzoic acid	36,822	0.47	17.3
Cinnamaldehyde	1,098	2.10	3.4
Cresols	94,932	0.54	51.2
Diisodecyl phthalate	151,319	0.42	63.6
Dioctyl phthalate	391,131	0.42	164.3
p-acetaminophenol	20,000	2.65	53.0
Phenol	1,431,000	0.36	515.2
Phthalic anhydride	646,289	0.40	258.5

SOURCE: Compiled by Genex Corp. from data in references 1 and 9.

Table I-B-30.—Aromatics: Technical Information

Compound	Typical synthetic process	Typical precursor	Is precursor renewable/non-renewable limited	Alternate precursor by fermentation	Time to implement commercial fermentation by genetically engineered strain
Aniline	chemical	benzene	nonrenewable	aromatic[a]	10 yrs.
Aspirin	chemical	phenol	nonrenewable	aromatic	5 yrs.
Benzoic acid	chemical	tar oil	nonrenewable	aromatic	10 yrs.
Cinnamaldehyde	chemical	benzaldehyde acetaldehyde	nonrenewable	aromatic	20 yrs.
Cresols.	chemical	phthalic anhydride	nonrenewable	aromatic	10 yrs.
Diisodecyl . . . . phthalate	chemical	coal tar	nonrenewable	aromatic	20 yrs.
Dioctyl . . . . phthalate	chemical	coal tar	nonrenewable	aromatic	20 yrs.
p-acetaminophenol . .	chemical	nitrobenzene	nonrenewable	aromatic	5 yrs.
Phenol	chemical	coal tar	nonrenewable	aromatic	10 yrs.
Phthalic . . . . anhydride	chemical	coal tar	nonrenewable	aromatic	15 yrs.

[a]Aromatic refers to benzene or benzene derivative. Eventually it is anticipated that lignin, a renewable resource, would serve as a precursor.

SOURCE: Compiled by Genex Corp. from data in references 4 and 35.

Table I-B-31.—Inorganics and Mineral Leaching: Market Information

	Current market data		
Product category	Market volume 1,000 lb	Bulk cost $/lb	Market value ($ millions)
Inorganics			
Ammonia	33,400,000	0.06	2,004.0
Hydrogen	451,000	0.15	677.0
Mineral leaching			
Uranium	(Information not available)		
Transition metals.	(Information not available)		
(cobalt, nickel, manganese, iron)			

SOURCE: Compiled by Genex Corp. from data in reference 4.

Table I-B-32.—Inorganics and Mineral Leaching: Technical Information

Product category	Typical synthetic process	Typical precursor	Is precursor renewable/non-renewable limited	Alternate precursor by fermentation	Time to implement commercial fermentation by genetically engineered strain
Inorganics					
Ammonia	chemical	water and coke	nonrenewable	nitrogen(air)	15 yrs.
Hydrogen	catalytic reforming	petroleum	nonrenewable	water and air	15 yrs.
Mineral leaching					
Uranium		(Information not available)			
Transition metals. (cobalt, nickel, manganese, iron)		(Information not available)			

SOURCE: Compiled by Genex Corp. from data in references 4 and 35.

Table I-B-33.—Projected Growth of Selected Markets Involving Applications of Genetic Engineering

Product category	Current market $ millions	Projected market in 20 yrs. $ millions
Amino acids[a]	300	900
Miscellaneous proteins	300	1,000
Gene preparations	0	100
Short peptides	5	2,100
Peptide hormones	260	1,000
Totals	865	5,100

[a]Only four amino acids are considered here.

SOURCE: Genex Corp.

References

1. *Chemical Marketing Reporter*, March and April 1980.
2. Hirose, Y., and Okada, H., "Microbial Production of Amino Acids," in *Microbial Technology: Microbial Processes*, vol. 1, H. J. Peppler and D. Perlman (eds.) (New York: Academic Press, 1979), pp. 211-240.
3. Hirose, Y., Sano, K., and Shibai, H., "Amino Acids," in *Annual Reports on Fermentation Processes*, vol. 2, G. T. Tsao and D. Perlman (eds.) (New York: Academic Press, 1978), pp. 155-190.
4. Genex Corp. proprietary information.
5. Weinstein, L., "Chemotherapy of Microbial Diseases," in *The Pharmacological Basis of Therapeutics*, 5th ed., L. S. Goodman and A. Gilman (eds.) (New York: MacMillan Publishing Co., Inc., 1975), pp. 1,090-1,247.
6. Cohn, V. H., "Fat-Soluble Vitamins: Vitamin K and Vitamin E," in *The Pharmacological Basis of Therapeutics*, 5th ed., L. S. Goodman and A. Gilman (eds.) (New York: MacMillan Publishing Co., Inc., 1975), pp. 1,591-1,600.
7. Florent, J., and Nitiet, L., "Vitamin B_{12}," in *Microbial Technology: Microbial Processes*, vol. 1, H. J. Peppler and D. Perlman (eds.) (New York: Academic Press, 1979), pp. 497-520.
8. Perlman, D., "Microbial Process for Riboflavin Production," in *Microbial Technology: Microbial Processes*, vol. 1, H. J. Peppler (eds.) (New York: Academic Press, 1979), pp. 521-528.
9. *Synthetic Organic Chemicals*, March and April 1980.
10. Burns, J. J., "Water Soluble Vitamins, The Vitamin B Complex," in *The Pharmacological Basis of Therapeutics*, 5th ed., L. S. Goodman and A. Gilman (eds.) (New York: MacMillan Publishing Co., Inc., 1975), pp. 1,549-1,563.
11. Greengard, P., "Water Soluble Vitamins, The Vitamin B Complex," in *The Pharmacological Basis of Therapeutics*, 5th ed., L. S. Goodman and A. Gilman (eds.) (New York: MacMillan Publishing Co., Inc., 1975), pp. 1,549-1,563.
12. Straw, J. A., "Fat-Soluble Vitamins: Vitamin D," in *The Pharmacological Basis of Therapeutics*, 5th ed., L. S. Goodman and A. Gilman (eds.) (New York: MacMillan Publishing Co., Inc., 1975), pp. 1,579-1,590.
13. Aunstrup, K., "Industrial Approach to Enzyme Production," in *Biotechnological Applications of Proteins and Enzymes*, Z. Bohak and N. Sharon (eds.) (New York: Academic Press, 1977), pp. 39-50.
14. Aunstrup, K., Andresen, O., Falch, E. Z., and Mielsen, T. K., "Production of Microbial Enzymes," in *Microbial Technology: Microbial Processes*, vol. 1, H. J. Peppler and D. Perlman (eds.) (New York: Academic Press, 1979), pp. 282-311.
15. Bernard Wolnak & Associates, in *Food Prod.*, July 1978, p. 41.
16. Solomons, G. L., "The Microbial Production of Enzymes," in *Biotechnological Applications of Proteins and Enzymes*, Z. Bohak and N. Sharon (eds.) (New York: Academic Press, 1977), pp. 51-61.
17. *Chemical and Engineering News*, Apr. 14, 1980, p. 6.
18. Levine, W. G., "Anticoagulants, Antithrombotic, and Thrombolytic Drugs," in *The Pharmacological Basis of Therapeutics*, 5th ed., L. S. Goodman and A. Gilman (:eds.) (New York: MacMillan Publishing Co., Inc., 1975), p. 1,365.
19. Gilman, A. G., and Muriad, F., "Androgens and Anabolic Steroids," in *The Pharmacological Basis of Therapeutics*, 5th ed., L. S. Goodman and A. Gilman (eds.) (New York: MacMillan Publishing Co., Inc., 1975), pp. 1,451-1,471.
20. Gilman A. G., and Muriad, F., "Estrogens and Progestins," in *The Pharmacological Basis of Therapeutics*, 5th ed., L. S. Goodman and A. Gilman (eds.) (New York: MacMillan Publishing Co., Inc., 1975), pp. 1,423-1,450.
21. Haynes, R. C., and Larner, J., "Adrenocorticotropic Hormone: Adrenocorticol Steroids and Their Synthetic Analogs: Inhibitors of Adrenocorticol Steroid Biosynthesis," in *The Pharmacological Basis of Therapeutics*, 5th ed., L. S. Goodman and A. Gilman (eds.) (New York: MacMillan Publishing Co., Inc., 1975), pp. 1,472-1,506.
22. Sebek, O. K., and Perlman, D., "Microbial Transformation of Steroids and Sterols," in *Microbial Technology: Microbial Processes*, vol. 1, H. J. Peppler and D. Perlman (eds.) (New York: Academic Press, 1979), pp. 483-496.
23. Brazeau P., "Agents Affecting the Renal Conser-

vation of Water," in *The Pharmacological Basis of Therapeutics*, 5th ed., L. S. Goodman and A. Gilman (eds.) (New York: MacMillan Publishing Co., Inc., 1975), pp. 858-859.

24. Haynes, R. C., and Larner, J., "Insulin and Oral Hypoglycemic Drugs: Glucagon," in *The Pharmacological Basis of Therapeutics*, 5th ed., L. S. Goodman and A. Gilman (eds.) (New York: MacMillan Publishing Co., Inc., 1975), pp. 1,507-1,533.

25. Jawetz, E., Melnick, J. L., and Adelberg, E. Z., *Reviews of Medical Microbiology*, 11th ed., (Los Altos, Calif.: Lange Medical Publications, 1974).

26. Thaler, M. M., *Biochem. Biophys. Res. Comm.* 54:562, 1973.

27. Nakao, Y., "Microbial Production of Nucleosides and Nucleotides," in *Microbial Technology: Microbial Processes*, vol. 1, H. J. Peppler and D. Perlman (eds.) (New York: Academic Press, 1979), pp. 312-355.

28. Perlman, D., "Microbial Production of Antibiotics," in *Microbial Technology: Microbial Processes*, vol. 1, H. J. Peppler and D. Perlman (eds.) (New York: Academic Press, 1979), pp. 241-281.

29. Weatherall, D. J., and Clegg, J. B., "Recent Developments in the Molecular Genetics of Human Hemoglobin," *Cell* 16:467-480, 1979.

30. Ignoffo, C. M., and Anderson, R. F., "Bioinsecticides," in *Microbial Technology: Microbial Processes*, vol. 1, H. J. Peppler and D. Perlman (eds.) (New York: Academic Press, 1979), pp. 211-240.

31. Gorman, M., and Huber, F. M., "β-Lactam Antibiotics," in *Annual Reports on Fermentation Processes*, vol. 2, G. T. Tsao and D. Perlman (eds.) (New York: Academic Press, 1978), pp. 203-222.

32. Shibata, M., and Uyeda, M., "Microbial Transformation of Antibiotics," in *Annual Reports on Fermentation Processes*, vol. 2, G. T. Tsao and D. Perlman (eds.) (New York: Academic Press, 1978), pp. 267-304.

33. Lockwood, L. B., "Production of Organic Acids by Fermentation," in *Microbial Technology: Microbial Processes*, vol. 1, H. J. Peppler and D. Perlman (eds.) (New York: Academic Press, 1979), pp. 367-372.

34. Roberts, J. D., and Caserio, M. C., *Basic Principles of Organic Chemistry* (New York: Benjamin, Inc., 1964), p. 1,096.

35. Windholz, M. (ed.), *The Merck Index*, 9th ed. (Rahway, N.J.: Merck & Co., 1976).

Chemical and Biological Processes

A comparison was made of waste stream pollution for chemical and biological processes. Ideally, the comparison should be between the two processes used in the production of the same end product. Since such data do not currently exist at the industrial level, the comparison was made between the chemical production of a mixture of chemicals and the biological production of alcohol and antibiotics. One noteworthy parameter is the 5-day biochemical oxygen demand (BOD5) —the oxygen required over a 5-day period by organisms that consume degradable organics in the waste stream. If the oxygen demand is too high, the discharge of such a stream into a body of water will deplete the dissolved oxygen to the point that it threatens aquatic life. An important variable that must be considered along with the BOD is the COD (the chemical oxygen demand). Large differences between the COD and BOD of a waste system can indicate the presence of nonbiodegradable substances. Although the conventional process stream shown in table I-C-1 has less BOD5 than the biological process stream, its COD content probably means that nonbiodegradables are present, and specialized waste treatment is necessary.

BOD is one area where traditional fermentation based processes have posed pollution problems. Batch fermentation processes typically generate large quantities of dead cells and residual nutrients that cause a large BOD if they are dumped directly into a dynamic aquatic environment. (See table I-C-1.) This difficulty can be circumvented by the use of spent cell material as an animal feed supplement or

Table I-C-1.—Waste Stream Pollution Parameters: Current Processes v. Biological Processes

Compounds: Mixed chemicals, including ethylene oxide, propylene oxide, glycols, amines, and ethers

Pollution parameters	Current processes	Biological processes
Alkalinity (mg/l)	4,060	0
BOD5[a] (mg/l)	1,950	4,000-12,000
Chlorides (mg/l)	430-800	0
COD[b] (mg/l)	7,970-8540	5,000-13,000
Oils (mg/l)	547	0
pH	9.4-9.8	4-7
Sulfates (mg/l)	655	0
Total nitrogen (mg/l).	1,160-1,253	50-200
Phosphates (mg/l)	0	50-200

[a]5-day biological oxygen demand.
[b]Chemical oxygen demand.

SOURCE: Office of Technology Assessment.

as a fertilizer. These applications have been intensively investigated and have met with success.

Because of the renewed interest generated by the potentials of genetic engineering, some traditional fermentation systems are being redesigned. Immobilization allows the reuse of cells that would otherwise be discarded. These systems can be used continuously for several months as compared with the usual fermentation time in a batch process of about one week or less. Immobilized operations create waste cells much less often than batch systems, and therefore generate less BOD.

The Impact of Genetics on Ethanol—
A Case Study

Objective

This study examines how genetics can and will affect the utilization of biomass for liquid fuels production. There are two major areas where genetics are applicable. One is in plant breeding to improve availability (both quantity and quality) of biomass resources (with existing and previously unused land); the second is in the application of both classical mutation and selection procedures and the new genetic engineering techniques to develop more efficient microbial strains for biomass conversion. Examples of goals in a plant breeding program would include improvements in photosynthetic efficiencies, increased carbohydrate content, decreased or modified lignin content, adaptation of high productivity plants to poor quality land, improved disease resistance, and so forth. However, the focus here is entirely on the second area, the use of genetics to improve microbial-based conversion to produce ethanol.

In order to assess the type and extent of improvements in micro-organisms that might benefit ethanol production, its process technology and economics must first be examined. An overview of the biomass conversion technology is presented in figure I-D-1; processes are defined mainly on the basis of the primary raw material and the type of pretreatment required to produce mono- or disaccharides prior to fermentation. In addition, there are several alternative fermentation routes to produce ethanol; these are characterized by the type of micro-organisms and will be examined with the in-

Figure I-D-1.—An Overview of Alternative Routes for Conversion of Biomass to Ethanol

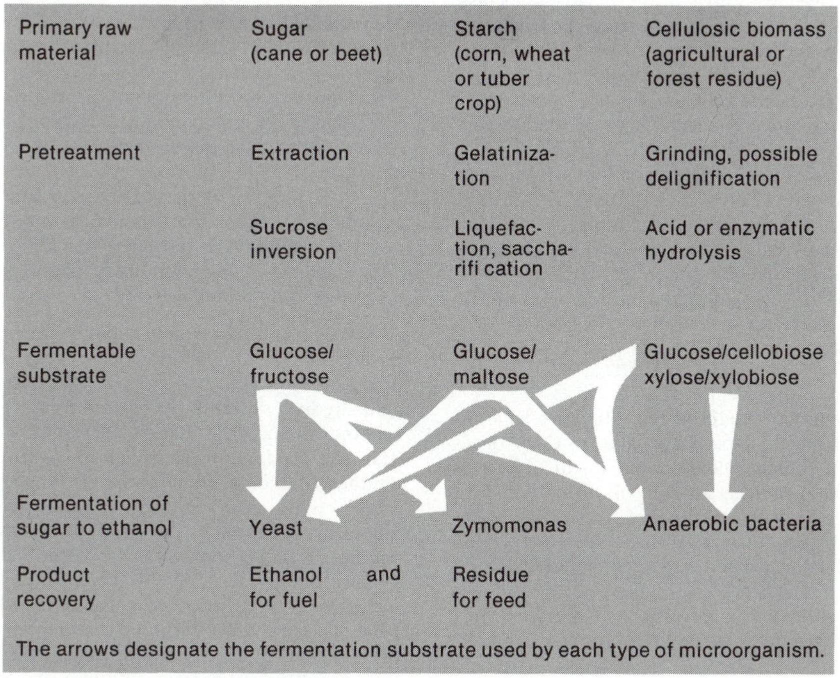

Primary raw material	Sugar (cane or beet)	Starch (corn, wheat or tuber crop)	Cellulosic biomass (agricultural or forest residue)
Pretreatment	Extraction	Gelatinization	Grinding, possible delignification
	Sucrose inversion	Liquefaction, saccharification	Acid or enzymatic hydrolysis
Fermentable substrate	Glucose/fructose	Glucose/maltose	Glucose/cellobiose xylose/xylobiose
Fermentation of sugar to ethanol	Yeast	Zymomonas	Anaerobic bacteria
Product recovery	Ethanol and for fuel	Residue for feed	

The arrows designate the fermentation substrate used by each type of microorganism.

SOURCE: Massachusetts Institute of Technology.

tent of quantifying the potential impact of genetic improvement on each one. It is interesting to note that each type of organism has its substrate restrictions, and only the anaerobic bacteria such as *Clostridium thermosaccharolyticum* and *C. thermohydrosulforicum* can utilize all of the available substrate.

Substrate pretreatment

Pretreatment refers to the processing that is required to convert a raw material such as sugarcane, starch, or cellulosic biomass to a product that is fermentable to ethanol. In most cases, the pretreatment is either extraction of a sugar or hydrolysis of a polysaccharide to yield a mono- or disaccharide.

EXTRACTION OF SUGAR

Sugar crops such as sugarcane, sugar beets, or sweet sorghum are highly desirable raw materials for producing ethanol. These crops contain high amounts of sugars as sucrose. In addition, the yield of fermentable material per acre is high; sugarcane and sugar beets yield 7.5 and 4.1 dry tons of biomass per acre, respectively.[1]

Sugar is extracted from cane or beets with hot water and then recrystallized. The resulting sugars are utilized directly by organisms having invertase activity (to split sucrose to glucose plus fructose). Molasses, a sugary byproduct of the crystallization of sucrose, may also contain sucrose although in most cases it is inverted with acid.

The primary use for sugar crops is food sugar. Sugar sells for over 20 cents/lb. Molasses, which currently sells for about $100/ton (about 10 cents/lb sugar) is used extensively as an animal feed. Substantial amounts of both sugar and molasses are imported into the United States for food uses and are therefore unavailable for ethanol production. There are proposals to increase sugar production for use as an energy crop; however, this will require the development of new land for sugar production.

STARCH

The primary raw material for ethanol fermentation in the United States is cornstarch. Corn processed by wet milling, yields about 36 lb of starch from each 56 lb bu; this amount of starch will produce 2.5 gal of absolute ethanol. Corn yields are typically 80 to 120 bu/acre so that 200 to 300 gal of ethanol can be derived per acre of corn per year.

Pretreatment of starch is initiated by a gelatinization step whereby a starch slurry is heated for 5 min at 105° C. After cooling to 98° C, α-amylase is added

---

[1] Paul B. Weisz and John F. Marshall, *Science* 206:24, 1979.

to break down the starch to about 15DE (dextrose equivalents). This process of liquefaction reduces the viscosity such that the solution can be easily mixed. After further cooling to 30° C, glucoamylase is added along with a starting culture of yeast so that saccharification and fermentation proceed simultaneously. The resulting fermentation, to produce typically 8 to 10 percent ethanol (volume per volume), requires 42 to 48 hr for completion. This compares with a 16- to 20-hr fermentation if sugar as molasses or cane juice is used as the substrate. Thus, the use of starch requires the addition of enzymes prior to and during fermentation, as well as large fermenter capacity as a consequence of the slower fermentation time compared with sugar substrates.

Improvement in the economy of ethanol fermentation based on starch is possible by developing a micro-organism that can produce α-amylase and glucoamylase and thus eliminate the need to add these enzymes. Since the rate of fermentation depends on the rate of starch hydrolysis, increased levels of glucoamylase may enhance the rate of starch hydrolysis and thus increase the rate of ethanol production. This would lower the capital requirements as well as the cost of enzyme addition.

CELLULOSIC BIOMASS

Processes for the utilization of cellulosic biomass to produce liquid fuels all have three features in common:

1. They employ some means of pretreatment to at least effect some initial size reduction and, more often, cause a disassociation of lignin and cellulose;
2. they involve either acid or enzymatic hydrolysis of the cellulose and hemicellulose to produce mono- and disaccharides; and
3. they employ fermentation to produce ethanol or some other chemical.

A wide variety of process schemes have been proposed for the conversion of cellulosic biomass to liquid fuels; a summary of the major steps in two acid hydrolysis and three enzymatic hydrolysis schemes in shown in figures I-D-2 and I-D-3. The initial size reduction is required to increase the amount of biomass surface area that can be contacted with acid, solvent, steam, enzymes, or chemicals that might be used to disassociate the cellulose and hemicellulose from the lignin. Pretreatments that have been investigated to facilitate the process are summarized in table I-D-1. The problems with pretreatment are that they require energy, equipment, and often chemicals; they result in an irretrievable loss of sugar, and in undesirable side-reactions and byproduct forma-

Figure I-D-2.—Alternative Schemes for Acid Hydrolysis of Cellulosic Biomass for Ethanol Production

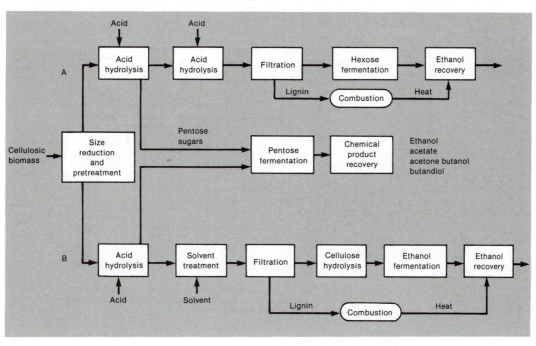

SOURCE: Massachusetts Institute of Technology.

tion. Furthermore, if acids, alkali, or organic chemicals are used, they must be recycled to minimize cost or disposed of in order to prevent pollution.

In starch processing, prior to ethanol fermentation, mechanical grinding, steam, and enzymes are employed. The energy requirements are small and contribute relatively little to the final ethanol cost. The objective in the development of cellulose-based processes should be to minimize both energy and chemical requirements. The development and scale-up of effective pretreatment technology are under active investigation[2] and require continued financial support to better develop several alternative routes. The most promising routes are: steam treatment, solvent delignification, dilute acid, cellulose dissolution, and direct fermentation.

Several different acid hydrolysis schemes have been proposed. However, most appear as in flow scheme A or B in figure I-D-2. Dilute acid is used to hydrolyze the hemicellulose to pentose sugars primarily and then stronger acid at higher tempera-

[2]*Proceedings of 3rd Annual Biomass Energy System Conference,* National Technical Information Service, SERI/TP-33-285, 1979.

tures is used to cause cellulose hydrolysis (scheme A). A major problem with this approach is the irreversible loss of sugars to undesirable side-product formation. After separation of residual solids (mostly lignin), which can be burned to provide energy for distillation, the sugar solution is fermented by yeast to ethanol. The pentose sugars also can be fermented, but by organisms other than the ethanol producing yeast, to other chemicals, some of which could be used as fuels (e.g., ethanol, acetic acid, acetone, butanol, 2,3-butanediol, etc.).

An alternative (scheme B, figure I-D-2) to the above is to use a solvent, after pentose sugar removal, to dissolve the cellulose, allowing its separation from lignin. This cellulose solution is easily and efficiently hydrolyzed to sugars. The advantage of this approach over the direct acid hydrolysis is that the yield of sugar is much higher. In the harsh acid hydrolysis, considerable sugar is destroyed. However, the major disadvantage of both these schemes is that they require recycling or disposal of acids and solvents. A second problem is that almost nothing is known about how to scale-up some of the newly de-

Figure I-D-3.—Alternative Schemes for Enzymatic Hydrolysis of Cellulosic Biomass for Ethanol Production

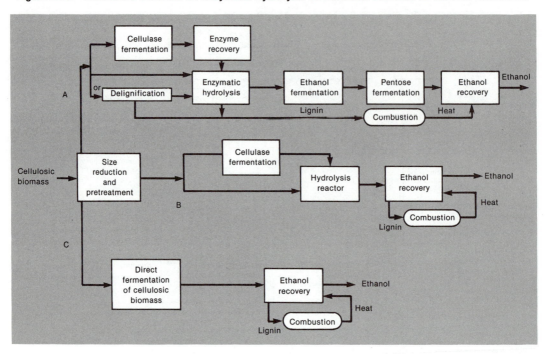

SOURCE: Massachusetts Institute of Technology.

Table I-D-1.—Alternative Pretreatment Methods for Lignocellulose Materials

Chemical methods	Physical methods
Sodium hydroxide (alkali)	Steam
Ammonia	Grinding and milling
Chemical pulping	Irradiation
Ammonium bisulfite	Freezing
Sulfite	
Sodium chlorite	
Organic solvents	
Acids	

SOURCE: Office of Technology Assessment.

veloped technology, such as that developed by groups working at Purdue University, New York University, and Dartmouth College. There are several engineering problems involving both heat and mass transfer and acid/solvent recycle that need to be evaluated at larger scale. At least some of this work will be done at the process development unit now being built at the Georgia Institute of Technology. The

most promising directions that need development are:

- the scale-up of high rates and high yield laboratory hydrolysis systems, and
- the development of methods for acid and chemical recycle schemes.

There are three types of approaches that have been employed for enzymatic hydrolysis of cellulosic biomass. These are summarized in figure I-D-3. They all involve some initial size reduction to increase the surface area available for enzymatic attack. In schemes A and B, the incoming cellulosic biomass is split into two streams; one is used to grow organisms that produce cellulolytic enzymes called cellulases, and the other is used to produce sugar.

In scheme A, the cellulases are recovered and then added to a separate enzyme hydrolysis reaction. They hydrolyze both the cellulose and hemicellulose, and the resulting sugar solution is then passed to an ethanol fermentation stage where hexoses are converted by a yeast fermentation to ethanol. Utilization of the pentose requires a separate fermentation. Residual lignin, which is removed before (by solvents extraction) or after hydrolysis, is used to provide energy for ethanol recovery. Extensive work on this approach has been done at the University of California, Berkeley, and the U.S. Army Natick Laboratories.

In scheme B, the cellulase is not recovered but rather, the whole fermentation broth from cellulase production is added to the cellulosic biomass along with ethanol-producing yeast. The result is a simultaneous cellulose hydrolysis (saccharification) and fermentation. (In the production of ethanol from starch, the starch hydrolyzing enzymes are added at the same time as the yeast for simultaneous saccharification and fermentation.) This technology has been demonstrated by the Gulf Oil Co. After fermentation, the ethanol is recovered and the residual lignin can again be used for energy for distillation. The problem of unused pentose sugar still remains and will require a separate fermentation step.

A third alternative (scheme C, figure I-D-3) shows a simpler approach, namely a direct fermentation on cellulose. This approach has been developed at the Massachusetts Institute of Technology. It utilizes bacteria that will produce cellulase to hydrolyze the cellulose and hemicellulose and ferment both the hexose and pentose sugars to ethanol in a single-stage reactor. The advantage of this approach is a minimal requirement for pretreatment, a combined enzyme production, cellulose hydrolysis and ethanol fermentation, and simultaneous conversion of both pentose and hexose sugars to ethanol. This concept is new and work still needs to be done to increase the ethanol concentration, minimize side product formation, and increase the rate of ethanol production. Again, residual lignin will be used to provide the energy for ethanol distillation.

FERMENTATION OF ETHANOL

An examination of the economics for ethanol production shows that the dominant cost is the process raw material. As seen in table I-D-2 the feedstock represents 60 to 70 percent of the manufacturing cost. Thus, it is clear that any improvement in substrate utilization efficiency is of substantial benefit. The theoretical yields of ethanol from glucose, sucrose, and starch or cellulose are 0.51, 0.54 and 0.57 gram (g) ethanol/g material, respectively; the differences result from the addition of a molecule of water on hydrolysis. There are several approaches to improve the yield above the typical value of 90 to 95 percent currently achieved. These are:

- increase the ratio of ethanol produced per unit weight of cells, e.g., through cell recycle, vacuum fermentation, immobilized cells, or improvement in specific productivity (g ethanol/g

Table I-D-2.—A Comparison of the Distribution of Manufacturing Costs for Several Ethanol Production Processes

Substrate	Molasses	Corn	Grain Sorghum
Cost component (%)			
Capital.	9	12	10
Operating	20	26	30
Feedstock.	71	62	60
Total	100	100	100
Cost on energy basis ($MMBtu)	12.5	14.9	12.7
Cost/gal ethanol ($/gal)	1.05	1.25	1.07
Capital investment ($/annual gal)	1.02	1.05	1.75

SOURCE: "Comparative Economic Assessment of Ethanol From Biomass," Mitre Corp., report HCP/ET-2854).

cell hr), by increasing the content and/or activity of those enzymes in the pathway to ethanol;
- increase the utilization of other materials in the substrate, e.g., the use of oligosaccharides, especially branched, in starch, and the use of contaminating sugars such as galactose or mannose for hemicellulose; and
- develop a route for the utilization of pentose sugars, especially xylose, present in hemicellulose.

The potential effect of oligosaccharides or contaminating sugar utilization is relatively small, since they represent typically 1 to 3 percent of the total sugar content. However, if cellulosic biomass containing 15 to 25 percent hemicellulose is used, then the impact of pentose conversion to ethanol is great

Cellulosic biomass is made up primarily of cellulose, hemicellulose (mostly xylan) and lignin. Other components such as protein, ash, fats, etc., typically comprise about 10 percent. The composition of biomass can be expressed in terms of the following equation:

$$\frac{F_c + F_H + F_L}{1 - F_A} = 1.0 \qquad (1)$$

where F_c, F_H, F_L, and F_A are the weight fractions of cellulose, hemicellulose, lignin, and ash, respectively. Assuming that the ash is 10 percent ($F_A = 0.1$) and that F_c and F_H are the only fermentable components in the biomass, then:

$$F_c = F_H = 0.9 - F_L \qquad (2)$$

The maximum amount of ethanol from one unit of biomass ($Y_{E/B}$) is:

$$Y_{E/c}F_c = Y_{E/H}F_H = Y_{E/B} \qquad (3)$$

Where $Y_{E/c}$ and $Y_{E/H}$ are the yield of ethanol for cellulose and hemicellulose, respectively. Equation 2 can be rearranged to relate the fractions of cellulose:

$$F_c = 0.9 - F_L - F_H \qquad (4)$$

Substituting this into equation 3 gives:

$$Y_{E/B} + Y_{E/c}(0.9 - F_L - F_H) + Y_{E/H}F_H \qquad (5)$$

From equation 5, the effect can be calculated of hemicellulose content and conversion yield on the overall conversion of biomass to ethanol. Assuming a lignin content of 15 percent ($F_L = 0.15$) and using $Y_{E/c} = 0.57$ g/g the following equation is obtained:

$$Y_{E/B} = 0.43 + F_H(Y_{E/H} - 0.57) \qquad (6)$$

The theoretical yield value on hemicellulose, $Y_{E/H}$, is not well-defined because so little is known about the biochemistry of anaerobic pentose metabolism. If one mole of ethanol is produced per mole of xylose, the yield is 0.3 g ethanol/g xylose. It two moles of ethanol could be obained, $Y_{E/H}$ would be 0.61; however, neither the mechanism nor the thermodynamics of the conversion is sufficiently well-defined to allow one to expect this value. The maximum observed values are about 0.41 g ethanol/g xylose.[3] The sensitivity of the overall yield to this value is shown in figure I-D-4. The impact of pentose utilization depends on the amount

[3]S. D. Wang and C. Cooney, Massachusetts Institute of Technology, unpublished results.

Figure I-D-4.—Effect of Pentose Yield ($Y_{E/H}$) on Overall Yield of Ethanol from Cellulosic Biomass ($Y_{E/B}$) with Varying Fractions of Hemicellulose (F_H).

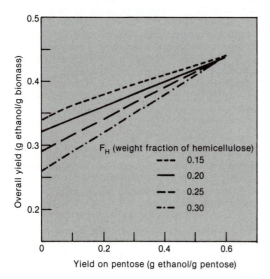

SOURCE: Massachusetts Institute of Technology.

of hemicellulose present. From the value in figure I-D-4 and the observation that 70 percent of the manufacturing cost is the raw material cost, it is possible to estimate the economic benefit of pentose utilization. Equation 7 relates the overall ethanol yield to the manufacturing cost:

$$C_E = \frac{C_B}{Y_{E/B}} \times \frac{6.6}{0.7} \qquad (7)$$

where C_E is the manufacturing cost per gallon of ethanol, C_B is biomass cost (cents/lb), 6.6 is the conversion from pound to gallon of ethanol, and 0.7 is the 70-percent factor for relative biomass cost to ethanol cost. For a biomass costing 2 cents/lb and containing 20 percent hemicellulose, the manufacturing cost is reduced from 59 to 43 cents/gal, when the yield on pentose goes from zero to 0.6.

At the present time, there are few organisms that produce more than one mole of ethanol per mole of pentose and none of the usual alcohol producing yeasts will ferment pentoses to ethanol. Addition or improvement of the ability to use pentose will have a major impact on the economics of ethanol production.

The second major cost in ethanol production relates to the cost of operation. Typically, 20 to 30 percent of the final manufacturing cost is accounted for by the sum of labor, plant overhead, administration, chemical supplies, and fuel costs. The chemical supplies represent less than 1 cent/gal ethanol and may be neglected. The labor, overhead, and marketing costs vary with plant size, but represent 11 to 7 cents/gal for a 20 to 100 million gal/yr plant, respectively. Any improvement in the reduction of plant size or complexity will reduce this cost; however, the economic impact is small. The major component of the operating cost is the fuel charge for plant operation and for distillation. Plant operations, e.g., mixing, pumping, sterilization, starch gelatinization, biomass grinding, etc., represent about 20 to 30 percent of the energy cost. The remainder is for ethanol distillation and residual solids drying. Considerable effort has been focused on methods to improve the energy efficiency of distillation to reduce it from the 160,000 Btu/gal required for beverage alcohol. While considerable differences in opinion exist as to the minimum, a reasonable expectation is about 40,000 Btu/gal although current technology requires 69,000 Btu/gal.[4] Forty thousand Btu is about half of the energy content of ethanol per gallon.

A discussion of process improvements relating to ethanol recovery has two components. The first is

[4]Report of the Gasohol Study Group of the Energy Research Advisory Board, Department of Energy, Washington, D.C., 1980. [5]M. Gibbs, and R. D. DeMoss, "Ethanol Formation in *Pseudomonas lindneri*," *Arch. Biochem. Biophys.* 34:478-479, 1951.

related to operating costs and the second is related to energy efficiency. If coal is used to provide energy for distillation, and it is valued at $30/ton, with 10,500 Btu/lb or $1.50/million Btu, then the energy cost for distillation (optimistically assuming 40,000 Btu/gal) is $6/gal. If lignin from cellulosic biomass is used as a fuel, the cost is reduced further. On the other hand, if oil at $40/bbl (130,000 Btu/gal and 42 gal/bbl) or $7/million Btu is used, then the energy cost is 28 cents/gal of ethanol.

From a common sense, economic, and political point of view, it does not seem reasonable to utilize liquid fuel to produce liquid fuel from biomass. Therefore, it will be assumed that petroleum will not be used for distillation and that either coal or biomass will be employed.

In order to assess the impact of process improvements on the energy demand, it is necessary to look at an overall material balance. This is summarized in figure I-D-5. Only a portion of the entering biomass feedstock is fermented to ethanol and there are two product streams, one containing ethanol and the other solids, both must be separated from water. It is important to note that as the ethanol concentration is increased, the energy requirement for both ethanol recovery from the water and for drying will decrease. Therefore, the impact of developing ethanol tolerant micro-organisms is seen as a reduction in energy cost.

Figure I-D-5.—Process Schematic for Material and Energy Balance

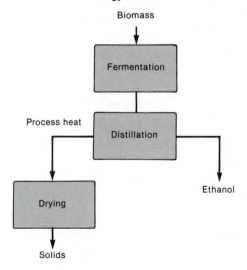

SOURCE: Massachusetts Institute of Technology.

The third major cost for ethanol manufacturing is the capital investment, which represents about 4 to 12 percent of the manufacturing cost. The capital investment is determined by the complexity of the processes and the volumetric productivity of ethanol production. Thus, the development of a micro-organism that will require a minimum amount of feedstock pretreatment and will produce ethanol at a higher rate will reduce the net capital investment.

The volumetric productivity (Q_E) for ethanol production is given by:

$$Q_E = q_p X$$

where q_p is the specific productivity expressed in g ethanol per g cell hr, and X is the culture density. Therefore, there are two approaches to obtain high productivity: first, to choose or create an organism with a high specific rate of ethanol production and second, to design a process with high cell density.

The application of genetics can be used to enhance the intracellular enzyme activity of the enzymes used for ethanol production. The resulting increase in q_p will result in reduced capital investment requirements.

There are four types of ethanol processes based on different organisms; they are:

1. *Saccharomyces cerevisiae* and related yeast,
2. *Saccharomyces cerevisiae/Trichoderma reesei,*
3. *Zymomonas mobilis,* and
4. *Clostridium thermocellum/thermosaccharolyti-cum,* or *thermohydrosulfuricum.*

The first is the traditional yeast based process using *S. cerevisiae* to ferment soluble hexose sugar to ethanol. In the second, the substrate range is extended to cellulose by the use of cellulase produced by *T. reesei.* The third approach utilizes *Z. mobilis;* this organism is a particularly fast and high ethanol yielding one. Its range of fermentable substrates, however, is limited to soluble hexose sugars.

In many tropical areas of the Americas, Africa, and Asia, alcoholic beverages prepared from a mixed fermentation of plant steeps are popular. Bacteria from the genus *Zymomonas* are commonly employed. In the early 1950's, the genus *Zymononas* acquired a certain fame among biochemists by the discovery that the anaerobic catabolism of glucose follows the Enter-Doudoroff mechanism.[5] This was very surprising, since *Zymomonas* was the first example of an anaerobic organism using a pathway mainly in strictly aerobic bateria.[6]

In spite of its extensive use in many parts of the world, its great social implications as an ethanol pro-

[5]M. Gibbs and R. D. de Moss, "Ethanol Formation, in *Psuedomonas lindneri,"* Arch. Biochem. Biophys., 34:478-479, 1951.

[6]J. Swings and J. DeLey, "The Biology of *Zymomonas,"* *Bacteriological Reviews* 41:1-46, 1977.

ducer, and its unique biochemical position, *Zymomonas* has not been studied extensively.[7]

The organism most often studied is *Zymomonas mobilis*, which can produce up to 1.9 moles of ethanol per mole of glucose. Recent studies reported from Australia, have established the *Z. mobilis* can ferment high concentrations of glucose rapidly to ethanol in both batch and continuous culture with higher specific glucose uptakes rates for glucose and ethanol production rates than for yeasts currently used in alcohol fermentations in Australia.[8][9]

For example, several kinetic parameters for a *Z. mobilis* fermentation were compared with *Saccharomyces carlsbergensis*[10] specially selected for its sugar and alcohol tolerance.[11] Both specific ethanol productivity and specific glucose uptake rate are several times greater for *Z. mobilis.* This result is mainly due to lower levels of biomass formation and glucose consumption. The lower biomass produced would seem to be a consequence of the lower energy available for growth with *Zymomonas* than with yeasts— the Enter-Doudoroff pathway producing only 1 mole of adenosine triphosphate (ATP) per mole of glucose, compared to glycolysis with 2 moles ATP per mole glucose. In none of the first three examples can ethanol be produced from pentose sugar.

The fourth approach utilizes a mixed culture of *Clostridia*, which will utilize cellulose and hemicellulose, hexoses, and pentoses for ethanol production.

The application of genetics for improving microbial strains

In the previous sections, the process steps have been identified that are particularly sensitive to the quality of the microbial strains. The following are improvements of microbial characteristics that are either now possible or might be so in the future and that will have an impact on the overall economics of the process. The effect of new genetic techniques requiring future research is similar for all micro-organisms in two ways.

1. Manipulations could be attempted today with less effort and greater chance of success if tools like cell fusion and recombinant DNA (rDNA) techniques were available for all of the microbes of interest.

2. Manipulations require further knowledge in a specific area or the development of an entirely new genetic system in ethanol producing microbes—e.g., there is no genetic system for the thermophilic anaerobic bacteria. Knowledge on how to genetically alter ethanol tolerance of both bacteria and yeast is lacking.

The economics of the fermentation of a substrate into alcohol is primarily controlled by three factors:

1. *Ethanol yield.*—The amount of product produced per unit of substrate determines the major raw materials cost of the fermentation.
2. *Final ethanol concentration.*—The cost of separating the ethanol from the fermentation broth is a function of the ethanol concentration in that broth.
3. *Productivity.*—The amount of ethanol produced per liter of fermenter volume per hour determines the capital cost of the fermentation step, once the type of fermenter and the annual output have been chosen. Productivity is not independent of the final ethanol concentration, and so an optimum compromise between these variables must be chosen.

The impact of genetics on ethanol yield

Most microbes that are chosen for making ethanol already produce nearly the theoretical maximum yield. In these cases little improvement can be made.

The yield may be lower when the microbe has been chosen for its other technical advantages such as ability to degrade cellulose. Lower yield of a microbial end product, like ethanol, can result from the diversion of substrate to cell mass or to an alternative product. Both of these faults can be readily attacked. A number of cell changes (e.g., leaky membranes) can cause the microbe to waste energy, requiring it to metabolize more substrate into alcohol to make the same cell mass. Where the thermodynamics and redox balance of the fermentation allow, unwanted waste products can be eliminated by mutation of the relevant pathways. Only limited work has been done on this type of research with industrially sigificant bacteria.

The impact of genetics on final ethanol concentration

This is amenable to genetic manipulation, both empirical and planned. An improvement in ethanol tolerance decreased both separation costs and fermenter capital cost (through increased productivity).

When traditional distillation is used, the effect on

[7]Gibbs, et al., op. cit.

[8]K. J. Lee, D. E. Tribe, and P. L. Rogers, "Ethanol Production by *Zymomonas mobilis* in Continuous Culture at High Glucose Concentrations," *Biotechnology Lett.* 421-426, 1979.

[9]P. L. Rogers, K. J. Lee, and D. E. Tribe, *Biotechnol. Lett.* 1:165-170, 1979.

[10]Ibid.

[11]D. Rose, *Proc. Bichem.* 11(2), 1976, pp. 10-12.

the separation cost of increased ethanol tolerance is smaller once ethanol concentrations have reached approximately 6 percent. However, the importance of increased ethanol concentration to fermenter productivity remains.

It is likely that the most important inhibitory action of ethanol takes place at the cell membrane. Strategies for manipulating the cell membrane composition and properties, and understanding in this area, are increasing rapidly.

Genetics and ethanol tolerance

The study of ethanol tolerance by micro-organisms has been approached using strains with altered genetic makeup. Several kinds of *Escherichia coli* mutants have been isolated having different tolerances to ethyl alcohol.[12] Solvent resistant strains either had larger amounts of total phospholipid (type III) or had an altered phospholipid and membrane-bound protein composition (type II). On the other hand, mutants with a lesion mapping close to *pss* gene (which codes for phosphotidylserine synthetase) were either solvent sensitive or resistant.[13]

The physiology of an *E. coli* ethanol resistant mutant has been characterized similarly.[14] This strain had pleiotropic growth defects including abnormal cell division and morphology. It also had an altered *lac* permease that was not due to a mutation in the Y gene. It was concluded that altered membrane composition was responsible for this abnormal behavior.

More recently, ethanol tolerant mutants have been isolated from *C. thermocellum*.[15] Indirect evidence lead to the conclusion that strain S-4 was defective in hydrogenase, since this strain produced lower amounts of acetic acid.[16] A different ethanol resistant isolate of the same bacterium, strain C9, proved to have a lower activation energy for growth than the wild type, a property that has been related to membrane composition.

There are three categories of changes that could influence the fermentation process:

1. Manipulate the existing controls on metabolism. Consider an example. In many organisms the energy level of the cell, expressed through adenosine monophosphate (AMP), adenosine diphosphate (ADP), and adenosine triphosphate (ATP) levels, partially controls the rate of glycolysis. A defective cell membrane would provide an energy sink, to keep glycolysis at its maximum rate. Strategies such as this could be attempted now.

2. Increase the amount of each transport and catabolic enzyme in the fermentation pathway. This requires the ability to isolate the genes of interest and to amplify them with in vivo or in vitro recombinant techniques in the microbe of interest. This is not an immediate prospect.

3. Accomplish complete deregulation of the fermentation pathway in the microbe of interest. Essential catabolic enzymes are difficult to manipulate, and this is also not an immediate prospect.

Genetic manipulation of the microbe can influence fermentation processes in other ways as well. These are less important than improvements in yield, final ethanol concentration, and productivity, but they also affect the cost. Examples are:

- type of fermenter used;
- nonsubstrate nutrients;
- strain stability;
- cell separations for byproducts, recycle, or ethanol recovery (i.e., increased size for recovery);
- operating conditions, i.e., higher growth temperatures for yeast and mesophilic bacteria; and
- range and efficiency of substrate utilization (i.e., complete utilization of all sugars).

More detailed examples are:

- *Type of fermenter.*—If the organism, whether it be a yeast or a bacterium, can be made to grow under conditions of pH, ethanol concentration, temperature, etc., that preclude contamination, inexpensive lined basins can be used instead of tanks, since steam sterilization of the fermenter is not required. In this case, some operating and capital costs associated with sterilization are avoided as well.

A type of continuous beer fermenter requires growth in the form of fast-settling pellets. In other fermenters, fast-settling particles (such as mycelia) present problems that are best avoided by agglomeration of the cell mass. This type of control over the growth form of micro-organisms is amenable to genetic manipulations.

- *Nonsubstrate medium costs.*—In addition to the carbon-energy substrate and water, growing cells must be supplied with other nutrients. Some organisms can make all of their biochemicals from quite simple sources of nitrogen, phosphorus, sulfur, magnesium and trace metals. Others require more

[12]D. P. Clark and J. P. Beard, "Altered Phospholipid Composition in Mutants of *Escherichia Coli* Sensitive or Resistant to Organic Solvents," *J. Gen. Microbiol.* 113:267-274, 1979.

[13]A. Ohta and I. Shibuya, "Membrane Phospholipid Synthesis and Phenotypic Correlation of an *E. Coli pss* Mutant," *J. Bacteriol.* 132:434-443, 1977.

[14]V. A. Fried and A. Novick, "Organic Solvents as Probes for the Structure and Function of the Bacterial Membrane: Effects of Ethanol on the Wild Type and as Ethanol Resistant Mutant of *Escherichia Coli*," *J. Bacteriol.* 114:239-248, 1973.

[15]S. D. Wang, "Production of Ethanol From Cellulose by *Clostridium Thermocellum*," M.S. Thesis, Department of Nutrition and Food Science, Massachusetts Institute of Technology, 1979.

[16]Ibid.

complex molecules, ready-made, such as amino acids and vitamins.

The more cheaply these nutrient needs can be provided, the better. Whenever an organism can be given genes from another source by applied biotechnology techniques, there is a possibility that complex nutrient requirements can be obviated. However, this requires that all the genes in a given pathway be located in the source and be made to function in the new microbes. The feasibility of this is uncertain, but solutions would decrease the cost of producing ethanol with yeast as well as *clostridia*.

• *Stain stability.*—Many of the suggested ethanol processes propose to employ continuous culture. Although this offers several advantages over batch culture, it is somewhat vulnerable to deleterious mutations of the microbe used, particularly if the microbe has been extensively altered in ways that make it less competitive.

These deleterious genetic changes are almost entirely catalysed by biological systems in the microbe. Alteration of these systems, so that the frequency of unwanted genetic changes is decreased, could greatly extend the period of operation that is possible before having to shut down and restart the fermentation. So far, this is a possibility only in microbes that have a highly developed genetics. It may be that strain stabilization of this sort would not be possible in other microbes until their genetics are highly developed.

It is also possible to design strategies using current strain development techniques that might lead to genetically stable strains, but these are unproven.

• *Cell separations.*—Many fermentation schemes incorporate cell recycle to boost productivity. This requires that cells be separated from effluent broth. Others need to separate cells from other residue as a byproduct. In addition, some of the low-energy alternatives to distillation, such as adsorption, could require separation of the cells from the broth prior to ethanol recovery.

In these cases, microbes that can be made to flocculate and redisperse, or that can be made to reversibly change their morphology would allow cheap gravity separations (settling or flotation).

• *Operating conditions.*—An increase in the temperature an organism will tolerate is advantageous for heat removal and in situ ethanol removal schemes. The feasibility of accomplishing this is uncertain.

The extreme of productivity improvement via cell recycle is an immobilized cell reactor. It is conceivable that cells could be made less prone to degradation under the conditions of immobilization, by modifying sensitive components and degradation

systems, and by adding protective systems. This is not at all a near-term possibility.

• *Range and efficiency of substrate utilization.*—A single-step conversion of a substrate to ethanol is highly desirable. This often requires that the ethanol fermenting organism possess a degradation capability it does not have.

As an example, consider ligno-cellulose. It consists of hexosans, pentosans, and lignin. All of these components should be used. Assume that one cellulase-producing candidate does not use pentoses, while a related noncellulase producing organism does, this is exactly the situation with clostridia. If the second organism can be given the cellulase genes of the first, a microbe better-suited to direct conversion could be created. The pace at which such a manipulation could be developed cannot be predicted with confidence, although this is not necessarily a long-term prospect.

Another obvious area that merits attention is the enhancement of cellulase activity. Classical genetic manipulations, employing mutation and selection or screening, should result in micro-organisms better equipped to degrade cellulose. E.g, it should be possible to isolate strains that are deregulated in cellulase production (hyperproducers) as well as those in which the cellulase is not subject to product inhibition. In addition, it is tempting to think about the possibilities of amplifying cellulase genes by means of DNA technology and cloning. However, this latter approach must await further understanding of the biochemistry and genetics of the cellulase system as well as the development of the appropriate genetic systems in cellulolytic micro-organisms.

Utilization of fermentation byproducts

Presently for each gallon of ethanol produced, approximately 14 liters of stillage is formed.[17] If ethanol is mixed with gasoline to make gasohol (10 percent ethanol), the total stillage produced annually in the United States would be in the billions of liters. Surely a problem of this magnitude deserves serious attention. The utilization of stillage or fermentation byproducts could be greatly improved by genetic means in several ways. In actuality, only a rational long-range genetic approach can increase the value of such a fermentation byproduct. Value can be increased in two main ways. The first is to increase the nutritive value of the fermentation byproduct followed by developing economical processing technol-

[17]W. E. Tyner, "The Potential of Obtaining Energy From Agriculture, *Symposium on Biotechnology: The Energy Production and Conservation*, Gatlinberg, Tenn., 1979.

ogies that stabilize and preserve nutritive value. The second approach is to increase the functionality of the byproducts so that more useful products can be developed.

For this one can envisage clever and novel ways to utilize mutants to increase the value in a manner similar to those described.[18][19][20] Ethanol production is not compatible with producing a valuable byproduct. E.g., a filamentous yeast may be useful for direct texturization or fortification of an animal food but production of ethanol may not be suitable with such an organism. A possible solution to this type of conflict involves the development and engineering of two-stage fermentation processes. In the first stage, ethanol producing organisms are propagated under optimal economic conditions for ethanol production. After the production phase is over, the organisms are then transferred to a second-stage reactor, where desirable phenotypic properties are then expressed. Signals for expression of phenotypic properties can be extrinsic environmental parameters, such as temperature, or levels of oxygen or carbon dioxide, or intrinsic parameters, such as specific nutrient requirements.

Thus the large-scale utilization of fermentation byproducts as feed or other materials will then become more valuable when genetic engineering can decrease processing costs and increase product quality. Most of these types of studies remain to be done. However, the potential for innovative applications is great, but such applications may not result because of the current lack of any Government agency that has a sound program for funding biotechnology research.

Recommendations and areas in which applied genetics should have an impact

There has been little published research done in the United States on the genetic improvement of ethanol production processes with bacteria such as *Zymomonas* and *clostridia*, and only limited studies with yeast. In light of previous discussion, the following points have been identified as being the most important and relevant in the application of genetics for improving ethanol-producing processes:

- improvements on ethanol yield;
- increased ethanol tolerance to achieve higher final ethanol concentrations in the fermentation broth;
- increased rates of ethanol production;
- elimination of unwanted products of anaerobic catabolism, that is, direction of catabolism towards ethanol;
- enhanced cellulolytic and/or saccharolytic capabilities to improve rates of conversion of cellulose and/or starch to fermentable sugars;
- incorporation of pentose catabolic capabilities into ethanol producers;
- development of strains capable of hydrolyzing cellulose and starch as well as of producing ethanol from pentoses and hexoses;
- improved temperature stability of micro-organisms and/or their enzymes; and
- improved harvesting properties of cellular biomass produced during fermentation.

---

[18]A. J. Sinskey, J. Boudrant, C. Lee, J. DeAngelo, Y. Miyasaka, C. Rha, and S. R. Tannenbaum, "Applications of Temperature-Sensitive Mutants for Single-Cell Protein Production," in *Proceedings of U.S./U.S.S.R. Conference on Mechanisms and Kinetics of Uptake and Utilization of Substrates in Processes for the Production of Substances by Microbiological Means*, Moscow-Pushchino, p. 362, June 4-11, 1977. PB. 283-330-T.

[19]J. Boudrant, J. DeAngelo, A. J. Sinskey, and S. R. Tannenbaum, "Process Characteristics of Cell Lysis Mutants of *Saccharomyces cerviciae*," *Biotech. Bioeng.* 21:659, 1979.

[20]Y. Miyasaka, A. J. Sinskey, J. DeAngelo, and C. Rha, "Characterization of a Morphological Mutant of *Saccharomyces cervisiae* for Single-Cell Protein Production," *J. Food Science* 45:558;563, 1980.

A Case Study of Wheat

Wheat is a major food staple in the diet of a large percentage of the world's population. Wheat grain in the United States is used almost exclusively for human consumption, although temporary localized oversupply may result in some wheat feeding to livestock.

Attempts to improve wheat plant populations by selection began several thousand years ago. The desirable attributes selected included the ability to withstand severe environmental stresses such as heat, cold, and drought and the stability of the seed head (which tends to disarticulate in wild forms).

Wheat seeds moved from country to country along with explorers and colonists. New varieties played major roles in the establishment of many productive wheat cultures—e.g., the Mennonite settlers introduced hard red winter (Turkey Red) wheat into the Kansas area from Russia in the late 19th century. And two private breeders—E. G. Clark of Sedgenick, Kans., and Danne of Elreno, Okla.—developed varieties that set new levels of productivity and straw strength in hard winter wheats which were sought by millers for their excellent flour recovery.

Breeding programs expanded during the first half of the 20th century. At first, the U.S. Department of Agriculture (USDA) played a lead role; but the emergence of the Land Grant System and the establishment of the State experiment station concept prompted individual States to launch breeding programs designed to address the particular production problems faced by farmers within their respective boundaries.

As the State experiment stations began to assume more responsibility, USDA programs and personnel began to concentrate in central locations to assemble the optimal number of personnel for the greatest interaction and productive output. If the present trend continues, there will be virtually no USDA scientists engaged in actual wheat breeding. Instead they will have assumed the roles of basic researchers and regional coordinators supplying information to the public and private breeders.

Disease and insect resistance have been the primary breeding goals of many programs. The dramatic losses associated with severe pest problems have focused the attention of producers, researchers, and legislators on these areas of need. Other traditional breeding objectives have included improved use properties, tolerance to environmental stresses such

as cold, wheat, wind dessication, and excessive moisture, and inherent yield capacity in the absence of significant production limitations.

The quality of wheat's end products has been improved significantly through breeding. Varieties have been tailored to meet the demands of various industries. The bread bakeries needed a higher protein and more gluten strength to make a lighter, larger loaf, while the cookie producer needed a low-protein flour with desirable dough-spreading properties.

Wheat productivity and management

The pattern of wheat productivity (yield per acre) in developed countries is remarkably similar. When yields are plotted over the centuries, there is a long period of barely perceptible increases in yield, from the time of first records of production to the end of the first third of this century (the period of 1925-35). Since around 1935, yield has increased sharply. Recent data suggest that yield increases may be leveling off. Why increases have been so substantial after generations of little success, is a complex question involving genetic resources, economic development, social interaction, and adoption of mechanical and biological innovations.

Until recently, the U.S. commercial seed companies, with one or two exceptions, have not been interested in wheat breeding programs as a profitmaking venture. Since wheat has a perfect flower and can fertilize itself, the farmer can purchase seed of a new variety and reproduce it from generation to generation. However, the discovery of cytoplasmic male sterility and nuclear restorer genes has stimulated industry interest in the possibility of developing hybrid wheat. The farmer would purchase the hybrid seed each year; the inbred lines used to make the hybrid would be the exclusive property of the originating company. Although progress has been good, problems exist with the sterility and restorer systems, the ability to produce adequate amounts of hybrid seed, and the identification of economic levels of hybrid vigor. The next 5 years should reveal the potential for success in hybrid wheat.

Several milestones of progress have been set in wheat. Yield has risen dramatically. Genetic protection against pests and other hazards has been a major contributor to increased yields. In addition, recent advances using semidwarf genes have been associated with significant yield improvement. The shorter, stiffer stems of the semidwarf plants allow maximization of resources without yield reductions. Improvement in the inherent yield components of stems per unit area, kernels per stem, and kernel weight has also contributed extensively to yield improvement.

The use of applied genetics in wheat improvement occurs in close harmony with total wheat management systems. The farmer must integrate a huge assortment of alternatives in each decision—e.g., an individual producer may be deciding on a nitrogen program. If the farm is irrigated, the producer selects nitrogen amounts and application timing based on soil tests, intended crop and variety, the end use of that crop, and watering schedules. If the farm is rainfed, the producer takes into account soil tests, crop considerations, and rainfall probabilities.

In both cases product prices at the time of sale must be predicted since they govern potential gross return, which in turn affects the costs of maintaining a profit margin. Genetic interaction in this system is intricate. The farmer must first select the variety most likely to produce at the maximum economic level. For irrigated land, it may be a short high-yielding semidwarf either for the cookie trade or the export market. The farmer knows that part of the value of his product is dependent on low protein. However, inappropriately high levels of nitrogen, which greatly improve yield, will also raise the protein of the crop beyond acceptable levels. If the export market is strong and the total U.S. supply reduced, the higher protein may be of little economic consequence.

In the case of the dryland farmer, the variety selected may be taller with lower yield potential but with much better levels of adaptation and tolerance to adverse environments. It may be designed for the bread industry or the export market. Part of the value is related to high-protein content. Since moisture conservation and use is critical, nitrogen applications and amounts must be selected so that the plants do not waste their moisture reserve. However, nitrogen applied too late may not receive enough rain to penetrate the soil and become available to the plants. If the plants "burn up" because of unwise water use early in the season, the seeds will be high in protein but low in yield. If inadequate nitrogen is available, the crop will generally be low in protein.

The abbreviated protein story is but one of many examples of farm management interaction with applied genetics in wheat production. Recent changes in energy price and availability, environmental restraints, marketing structures, and technology development are producing a new array of complex problems.

Genetic vulnerability in wheat

Genetic vulnerability is defined as a high degree of genetic uniformity in a crop grown over a wide acreage. Wheat, which is produced on about 62 million acres annually in the United States, has a relatively high level of uniformity and genetic vulnuerability. In 1974, 102 hard red winter wheat varieties were grown on 36.6 million acres, with four varieties occupying 40 percent of the acreage. Hard red spring wheat varieties totaled 80 percent on 14.7 million acres, with three varieties occupying 52 percent of the acreage. Similar situations occurred with other classes of wheat. Plant pests, including diseases and insects, have periodically caused moderate to severe wheat crop losses in years favorable to the development of strains capable of attacking current forms of resistance.

Incorporating genetic resistance to pests has traditionally been the responsibility of public breeders. Wheat is a self-fertilized plant that can be faithfully reproduced from generation to generation. Private industry has been reluctant to invest R&D money in improvements since the farmer, following the initial seed purchase, can reproduce the crop without returning to the seed company. Thus, public breeders have been the main source of new varieties and have had the responsibility of delivering genetic improvements to the producer. Wheat breeding programs are generally designed to respond to State production needs. Goals and objectives are established by technical advisory groups that include breeders and scientists, growers, use industry representatives, and extension workers.

Genetic variability is available to the breeder from naturally occurring sources and artificially induced mutations. Naturally occurring variability has been collected from native plant populations throughout the world and is maintained in the World Wheat Collection by the Science and Education Administration of USDA located in Beltsville, Md. Currently, about 37,000 accessions are contained in the collection. Breeders use the collection as a reservoir from which to draw exotic genes needed to improve the value of their breeding programs. In addition to variability within wheat varieties, the breeders can use special genetic techniques to draw valuable genes from related species such as rye and various forage grasses. This approach, while time-consuming and costly, has been used in a number of variety development programs. Mutations induced by artificial means have not been used extensively by the breeders, since desired mutations without detrimental effects are very difficult to obtain. Enough natural genetic variability seems to exist to satisfy needs in the foreseeable future.

The National Wheat Improvement Committee has stated that the World Wheat Collection is inadequately evaluated, characterized, and documented, forcing breeders to spend time and resources carrying out their own evaluation work. The committee has proposed a standard set of descriptors for all accessions in the collection, as well as an information management system to efficiently bring the information to the breeders.

Genetics and the Forest Products Industry Case Study

The Weyerhaeuser Co.

The Weyerhaeuser Co., which has its main headquarters in Centralia, Wash., is the largest forest products company in the United States. In 1970, Weyerhaeuser initiated a program to research the mass propagation of Douglas fir trees by tissue culture. Douglas firs are the main species in many of the Nation's forests, over $3.1 billion (or about 8.5 billion board feet) worth were harvested in 1979. While they are normally propagated by seed in the field, the classical development of improved seed does not adequately satisfy the criteria of the rapid availability of trees of superior quality.

Specially selected clones have the potential to double the productivity of forestlands; each year that unimproved trees are planted is another year of "suboptimum" harvests 40 years from now. With the steadily increasing demand for forest products, planting substantially improved trees as soon as possible is of great economic importance.

Weyerhaeuser's tissue culture research began in 1974 with a project at the Institute of Paper Chemistry to produce Douglas firs. The project was expanded with a contract for additional research at the Oregon Graduate Center. Although the intention was to propagate select strains of mature trees, the main focus of the program, in 1974 to 1978, was to develop a basic, consistent system for propagation. From 1978 to the present, Weyerhaeuser has been conducting most of its applied research into Douglas firs at its own research facilities in Centralia, Wash. Basic research is still being funded at the Institute of Paper Chemistry, which services the entire forest industry. While specific figures for the tissue culture systems research have not been made available, the annual research and development budget at Weyerhaeuser specifically for biological work with forest species is on the order of $7 million to $8 million.[1]

The project in mass propagation of Douglas fir by tissue culture was initiated to establish a reliable, economic means for mass production of superior trees. The cloning of these trees could bring higher yields and shorter harvest cycles, as well as rapid production of tree stands for seed production.

The immediate results of 10 years of research are not overly impressive at first glance. To date, 3,000 tissue-cultured Douglas firs have been planted for comparison analysis and research of handling techniques, transfer procedures, etc.

The cost effectiveness of a tissue culture program is determined by several factors, of which labor intensity varies the most. The more streamlined the system can be made, the fewer labor-requiring steps that are needed—the less direct costs will be incurred. Ideally, cells would be cultured in sterile conditions and then planted for the direct embryogenesis of plantlets that are ready for the field. Steps that involve cutting shoots and rooting them on another media or repeated subculturing procedures are costly and cumbersome. The major problem affecting cost so far is the difficulty of achieving high volume plant regeneration from the tissue cultures. Efficient systems with more successful regeneration will reduce the labor and materials involved in culturing and result ultimately in a lower cost per plant.

In addition to problems of cost, Weyerhaeuser has run into the classic difficulty with woody species—the inability to obtain required results from plants more than 1 year old. In addition, the risk of induced genetic variability increases with every subculture of the tissues. The triggering techniques for effective manipulation of mature versus embryonic and immature tree tissues are not well understood, and unlocking the Douglas fir system may well provide insight into some basic physiological questions.

Some commercial companies do not want to get deeply involved in basic research because it is extremely expensive and time-consuming. However, it has been up to the major forestry companies, such as Weyerhaeuser, to independently fund essentially basic research into the biological triggers for organogenesis and embryogenesis of Douglas fir.

By comparison, no other plant has been as intensely researched for mass propagation purposes and proved so unyielding. Among other things, this indicates that questions of basic plant cell physiology will have to be addressed before major breakthroughs can be expected. The goals of the Weyerhaeuser program are exacting and demand the refinement of present techniques into a precise in-

[1]Rex McCullough, The Weyerhaeuser Co., personal communication (May 1980) with the Plant Resources Institute in the working report, *Commercial Uses of Plant Tissue Culture and Potential Impact of Genetic Engineering on Forestry*, prepared under contract to OTA, 1980.

dustrial science. While it may seem that the investment has been disproportional to the returns at this point, it must be remembered that they are the forerunners of a new technology, both in terms of working with mature tree tissues of an especially intricate species and in terms of imposing stringent industrial standards on a mass biological production system.

Simpson Timber Co.

The Simpson Timber Co., whose central headquarters are in Seattle, Wash., is a large producer of redwood and other forest products, and has been involved over the past 5 years in a program to develop a mass production system for the coast redwoods through tissue culture. Approximately $250,000 has been invested in research performed at the University of California, Irvine, by Dr. Ernest Ball, a recognized authority in the field of tissue-cultured redwoods.[2]

Coastal redwoods are normally a field-seeded crop and have a production cycle of around 50 years. The major reason for consideration of tissue culture over seed is the greater speed with which superior trees might be developed through tissue culture as compared to using seed stock. Simpson Timber Co., which has been involved in a controlled breeding program along conventional lines as well, and is approaching the creation of homozygous strains. Since a sequoia seedling does not reach sexual maturity before it is 15 to 20 years old, and since about 10 generations are normally required to produce a true homozygous strain,[3] the classical process is time-consuming and contains no guarantees that the end products will be better than the clones selected through tissue culture.

Elite trees are selected from wild stands for straightness of trunks, height, specific gravity of wood, and proper branch drop (branches that drop without tearing the stem). There are no major pests in redwoods, so pest and disease resistance have not been a concern. Two methods of selection are used. Clones of special trees are produced by rooting the uppermost branches of the tree, a process that takes up to 1 year. Although the rooting percentage may be as low as 10 percent, this method has the advantage of producing mature cloned plants that can continue to produce flowers and seed. Simpson is using roughly 200 elite trees for these clones.

Elite trees can also provide clones through tissue cultures of their needles, a process that is less time-consuming but which produces seed very slowly because of the time involved in maturation. Simpson Timber Co. has planted out 2,500 tissue cultured redwoods for field comparisons with seedling material. The results so far have been encouraging, but it may take another 10 to 15 years before definite conclusions can be drawn. The major factors being analyzed are field growth rates and outplanting survival percentages. Clones of elite varieties will also have to be compared to the parent trees for the traits originally selected for, such as wood quality. Since the operational cost of tissue-cultured plantlets is about twice that of seedlings, the quality of tissue cultured plants must be markedly superior if the program is to be cost effective.

Dr. Ball is confident that the tissue culture system which has been developed for the rapid multiplication of elite redwood trees is ready for implementation at a commercial production facility.[4] Simpson Timber Co. is planning the construction of a tissue culture lab at their California headquarters within the next 2 years. The pilot plant is expected to cost $250,000 and produce upwards of 200,000 plantlets in its first year of production. As mass production techniques are perfected, the company plans to expand the facility to a production capacity of over 1 million plantlets per year.[5]

[2]Ernest Ball, University of California, Irvine, personal communication (May 1980) with the Plant Resources Institute in the working report, *Commercial Uses of Plant Tissue Culture and Potential Impact of Genetic Engineering on Forestry*, prepared under contract to OTA, 1980.

[3]James Radelius, Simpson Timber Co., personal communication (May 1980) with the Plant Resources Institute in the working report, *Commercial Uses of Plant Tissue Culture and Potential Impact of Genetic Engineering on Forestry*, prepared under contract to OTA, 1980.

[4]Ball, op. cit.

[5]Radelius, op. cit.

Animal Fertilization Technologies

Sperm storage

DEFINITION

The freezing of semen to $-196°$ C, storage for an indefinite time, followed by thawing and successful insemination.

STATE OF THE ART

Conception rates at first insemination with frozen sperm average between 30 to 65 percent for most species. This technology is not a key to the success of artificial insemination (AI), but because of the convenience it is now an essential ingredient. Current operational procedures are adequate for the dairy industry.

ADVANTAGES

1. Greater use of selected bulls as AI studs.
2. Elimination of the need to maintain expensive and dangerous bulls on dairy farms.
3. Sperm can be tested for disease and treated for venereally transmitted diseases.
4. Ease of transport and therefore of increasing potential offspring.

FUTURE

Little change is anticipated in semen processing. Freeze-dried semen is unlikely to be successful enough to use. Sperm banking can be expected to increase, especially on AI studs. Banking provides cheap storage while bulls (slaughtered) are being progeny tested, and insurance against loss of bulls through natural causes. For preservation of semen from bulls of less populous breeds, banking can be completed in about a year, after which the bull can be slaughtered.

Artificial insemination

DEFINITION

Manual placing of sperm into the uterus.

STATE OF THE ART

Highly developed for most species. Representative use rates in the United States are: dairy cattle, 60 percent; beef cattle, 5 percent; turkeys, 100 percent. The major limitation to the use of AI is the low national average conception rate at first service, around 50 percent. The success or failure of AI is determined by a multiplicity of factors including estrus detection, quality of semen, timing of insemination, and semen handling.

DISADVANTAGES

1. With increased herd size, estrus detection has become a major problem.
2. Inexperienced dairymen are buying semen and inseminating their own cows, resulting in lowered fertility and no feedback on semen fertility.

ADVANTAGES

1. Widespread use of genetically superior sires.
2. Services of proven sires at a lower cost.
3. Elimination of cost and danger of keeping bulls on the farm.
4. Control of certain diseases.
5. Use of other breeding techniques including cross-breeding.
6. Continued use of valuable sire after his death.

FUTURE

Greater use of AI in beef cattle will depend on the availability of successful and inexpensive estrus synchronization technology, on relaxed restrictions of the various breed associations, and on accurate progeny records.

Estrus synchronization

DEFINITION

Estrus ("heat"), is the period during which the female will allow the male to mate her. This sexual behavior is subtle and varies widely among individuals. Thus the synchronization of estrus in a herd, using various drug treatmnts, greatly enhances AI and other reproduction programs.

STATE OF THE ART

Effective methods for synchronization of estrus periods for large numbers of animals have been available for more than two decades, and several approaches are now available which result in normal fertility. Several schemes involve use of prostaglandin F_2 (PGF_2) for the cow and ewe. However, FDA approves usage only for controlled breeding in beef cows and heifers, nonlactating dairy heifers, and in mares.

ADVANTAGES

1. Time a heifer's entry into a milking stream.
2. Increase productivity by breeding heifers earlier in life.

3. Ability to breed large numbers of cattle over a shorter calving interval.
4. Increase use of AI, especially in beef cattle, sheep, and swine.

FUTURE

Estrus cycle regulation should allow selected sires to be more widely used to improve important traits in beef cattle. It should also gain widespread and rapid acceptance among dairymen as well.

Superovulation

DEFINITION

Superovulation is the hormonal stimulation of multiple ovarian follicles resulting in release from the ovary of a larger number of oocytes (ova) than normal.

STATE OF THE ART

Superovulation with implantation into surrogate mothers increases the number of offspring, usually from highly selected dams. Adequate procedures are presently available for superovulation of laboratory and domestic animal species, except the horse. The drugs used to induce superovulation are the gonadotropins, pregnant mare's serum gonadotropin (PMSG) and follicle stimulating hormone (FSH), in some instances followed by other treatments to stimulate ovum maturation and ovulation. Superovulated ova result in normal offspring with the same success rates as achieved with normally ovulated ova.

DISADVANTAGES

1. Greatest drawback is that degree of success cannot be predicted for an individual animal.
2. Batches of hormones for ovulation treatment vary widely in quality.
3. PMSG is scarce, and has been declared a drug by the Food and Drug Administration (FDA). Thus, most use of PMSG is now illegal.
4. There is insufficient data to judge the effect of repeated superovulation.

FUTURE

Methods for superovulation will improve consistency of results. Additional understanding of basic physiological mechanisms will facilitate such efforts. New work in superovulatory technology involves active immunization against adrostenedione (a hormone involved in regulation of follicular development). This treatment prevents atresia and reliably increases the frequency of multiple ovulations. The technology has definite commercial potential for cattle husbandry and limited potential for sheep hus-

bandry, and much current effort is directed towards developing and testing a commercial procedure.

Embryo recovery

DEFINITION

The collection of the fertilized ova from the oviducts or uteri. Collection of embryos is a necessary step for embryo transfer or storage, and for many experiments in reproductive biology. Both surgical and nonsurgical methods are used.

STATE OF THE ART

Surgical.—Methods are available for recovering 40 to 80 percent of ovulations from cattle, sheep, goats, swine, and horses. The development of adhesions and scar tissue following surgery limits these techniques. Surgical recovery is the only method for sheep, goats, and pigs. It is presently practiced almost exclusively when a suspected pathology of the oviducts renders an individual subfertile, or when embryos must be recovered before the individual reaches puberty.

Nonsurgical.—Non-surgical embryo recovery techniques are preferred for the cow and horse. Fifty to eighty percent of cow ovulations can be recovered, and 40 to 90 percent of the operations on horses to recover the single ovulation are successful.

ADVANTAGES

1. Nonsurgical embryo transfer can be performed an unlimited number of times.
2. Requirements for equipment, personnel, and time are low in nonsurgical recovery. This is especially important in milk cattle: since the nonsurgical procedure is performed on the farm, milk production is not interrupted.
3. A single embryo can be obtained between superovulation treatments.
4. Embryos can be obtained from a young heifer before it reaches puberty.
5. The technology is especially important for research, e.g., in efforts to produce identical twins, embryo biopsies for sex determination, etc.

FUTURE

Methods of collecting embryos have not changed appreciably since about 1976, nor are significant advances predicted for the future.

Embryo transfer

DEFINITION

Implantation of an embryo into the oviduct or uterus.

STATE OF THE ART

Surgical.—Pregnancy rates of 50 to 75 percent are achievable in cows, sheep, goats, pigs, and horses. Surgical transfer is the only practical method in sheep, goats, and pigs, and is the predominant method for cows and horses. A number of factors determine the success of surgical transfer: age and quality of embryos, site of transfer, degree of synchrony between estrous cycles of the donor and recipients, number of embryos transferred, in vitro culture conditions, skill of personnel, and management techniques. The 50- to 60-percent success rate in cattle compares with AI success rates at first service. (Pregnancy rates should not be confused with survival rates, which may be much lower.)

Nonsurgical.—This method is an adaptation of AI. Reported success rates are much lower than those with surgical transfer. Nonsurgical transfer is not used in sheep, goats, or pigs.

ADVANTAGES

1. Obtaining offspring from females unable to support pregnancy.
2. Obtaining more offspring from valuable females.
3. With a homozygous donor, undesirable recessive traits among animals used for AI can be rapidly detected.
4. Introducing new genes into specific pathogen-free swine herds.
5. Coupled with short- or long-term embryo storage, transportation of animals as embryos.
6. Increasing the population base of rare or endangered breeds of animals by use of closely related breeds for recipients.
7. Separation of embryonic and maternal influences in research.

DISADVANTAGES

1. Personnel requirements in surgical transfer account for a large share of high costs and thus limit applicability in animal agriculture.
2. Provision of suitable recipients is the greatest single cost in embryo transfer.

FUTURE

Surgical transfers will remain the method of choice for sheep, goats, and pigs in the foreseeable future. For cows and horses, however, nonsurgical methods will be increasingly used rather than surgical techniques (and this will be apparent) within the next year or two. It is likely that half of the commercial transfer pregnancies in cattle in North America in 1980 will be done nonsurgically, even if success rates are only 60 to 80 percent of those obtainable with surgical transfer. Among future applications, a role for embryo transfer can be predicted

in progeny testing of females, obtaining twins in beef cows, obtaining progeny from prepubertal females, and in combination with in vitro fertilization and a variety of manipulative treatments (e.g., production of identical twins, selfing, genetic engineering, etc.)

Embryo storage

DEFINITION

Maintenance of embryos for several hours or days (short-term) or for an indefinite length of time (freezing).

STATE OF THE ART

Short-term.—The requirement for embryos from farm animal species has not been defined, although adequate culture systems for the short interval between recovery and transfer have been developed by trial and error. Whereas the important parameters of culture systems have been identified (e.g., temperature, pH, etc.), optimal conditions have not been determined. Cow embryos may be stored for three days in the ligated oviduct of the rabbit.

Long-term (freezing).—No completely adequate protocol exists for freezing embryos of farm species. One-third to two-thirds of embryos are killed using present methods. Pregnancy rates of 32 to 50 percent for cattle, sheep, and goats have been reported after freezing. No successful freezing of swine or horse embryos followed by development to term has been reported. Despite disadvantages (one-half of embryos are often killed) advantages are such that in some situations embryo freezing, and embryo selling, are already profitable.

ADVANTAGES

1. Amplification of advantages of embryo transfer.
2. Elimination of requirements for large recipient herds when embryo transfer is being used.
3. Reduction of costs in animal transport.
4. Control of genetic drift in animals over prolonged time intervals.

FUTURE

Anticipated development of embryo culture technology would be of significance in efforts toward in vitro maturation of gametes, in vitro fertilization, sex determination, cloning, and genetic engineering, all of which involve prolonged manipulation of gametes and embryos outside of the reproductive tract.

As freezing rates improve, nearly all embryos recovered from cattle in North America will be frozen. Probably as many as half of the embryos will be deep-frozen for 2 to 3 years. It is unlikely that success rates will ever approach 90 percent of those without freezing. However, 70- to 80-percent success rates may be attainable within several years. It appears that embryos can be stored indefinitely with little deterioration.

Sex selection

DEFINITION

Tests to determine the sex of the unborn or determination of sex at fertilization by separating x- bearing from y-bearing sperm.

STATE OF THE ART

Sexing of embryos.—Through karyotyping nearly two-thirds of embryos can be sexed. Techniques using identification of the condensed X chromosomes are unreliable. A third method, identification of sex-specific gene products, is under development.

Sexing of sperm.—A 100-percent method has not been achieved in any mammalian species; and no standard protocol for farm species exists.

FUTURE

Before this technology can be applied commercially, it must be simple, fast, inexpensive, reliable, and nonharmful for embryos. Such techniques could undoubtedly be developed. There would be numerous medical and experimental applications.

There is much interest in research in this area because of its use in understanding male fertility with AI in humans, and in enhancing sperm survival after frozen storage.

Twinning

DEFINITION

Artificial production of twins, either using embryo transfer or hormone treatments.

STATE OF THE ART

Currently, embryo transfer is the most effective method for inducing twin pregnancies in cattle, resulting in pregnancy rates of between 67 to 91 percent, of which 27 to 75 percent deliver twins. Other methods include transferring one embryo into a cow which has been artificially inseminated, and hormonal induction of twinning, which is a modification of superovulation. This latter method is not reliable.

ADVANTAGE

The advantage of twinning in nonlitter-bearing species is the improved feed conversion ratio of producing the extra offspring.

DISADVANTAGE

The major disadvantage of twinning is intensive management necessary for periparturient complications, unpredictable gestation periods, depressed lactation, etc.

FUTURE

Technical feasibility for twinning, in conjunction with embryo transfer, management adjustments, and selection for good recipients, can be predicted. A reliable procedure for twinning in sheep can also be expected. The technology would most likely be first used in Europe and Japan, where there are shortages of calves to fatten for beef.

In vitro fertilization

DEFINITION

The union of egg and sperm outside the reproductive tract. For some species, the technology includes successful development of the embryo to gestation and birth.

STATE OF THE ART

In vitro fertilization has been accomplished in several laboratory animal species, including the rabbit, mouse, rat, hamster, and guinea pig and nine other mammalian nonlaboratory species, including man, cat, dog, pig, sheep, and cow. However, normal development following in vitro fertilization and embryo transfer has only been accomplished in the rabbit, mouse, rat, and human. Consistent and repeatable success with in vitro fertilization in farm species has not yet been accomplished.

None of the cases of reported success of in vitro fertilization, embryo transfer, and normal development in man is well documented.

Most of the in vitro fertilization work to date has concentrated on the development of a research tool so that the physiological and biochemical events in fertilization and early development could be better understood. More practical application of in vitro fertilization techniques would include:

1. a means for assessing the fertility of ovum and/or sperm;
2. a means to overcome female infertility with embryo transfer into a recipient animal; and
3. when coupled with ovum and/or embryo storage and transfer, a means to facilitate combination of selected ova with selected sperm for production of individuals with predicted characteristics at an appropriate time.

FUTURE

Rapid progress in research is anticipated and many of the potential applications of in vitro fertilization to animal breeding should become practical within the next 10 to 20 years. With further development of in vitro fertilization methodology, along with storage of unfertilized oocytes (gamete banking), fertilization of desired crosses should become possible. In the more distant future, genetic engineering and sperm sexing along with in vitro fertilization may become possible.

Parthenogenesis

DEFINITION

The initiation of development in the absense of sperm.

STATE OF THE ART

Parthenogenesis has not been satisfactorily demonstrated or described for mammalian species. The best available information leads to the conclusion that maintenance of parthenogenetic development to produce normal offspring in mammals approximates impossibility.

Cloning: production of identical twins

DEFINITION

The production, using a variety of methods, of genetically identical individuals.

STATE OF THE ART

There are several ways to obtain genetically identical livestock. The natural way is identical twins, although these are rare in species other than cattle and primates. Both natural and laboratory methods depend on the fact that the blastomeres of early embryos are totipotent (i.e., each cell can develop into a complete individual if separated from the others.) For practical purposes, highly inbred lines of some mammals are already considered genetically identical; F_1 crosses of these lines are also considered genetically identical and do not suffer from the depressive effect of inbreeding.

ADVANTAGE

An advantage of identical twins is the experimental control provided by one animal through which two sets of environmental conditions can be compared for effects on certain end points, e.g., native v. surrogate uterine environments for gestational development, nutrition on milk production, etc.

Cloning: nuclear transplantation

DEFINITION

The production of genetically identical mammals by inserting the nucleus of one cell into another, before or after destroying the original genetic complement. These occur by separation of embryos or parts of embryos early in development but well after fertilization has occurred.

STATE OF THE ART

Experimentalists have found in certain amphibia that transplantation of a nucleus from a body cell of an embryonic (tadpole) stage into a zygote following destruction or removal of the normal nucleus can lead to development of a sexually mature frog.

FUTURE

The ideal technique for making genetic copies of any given outstanding adult mammal would involve inserting somatic (body) cell nuclei into ova, which may take years of work to perfect if indeed it is possible. There is some evidence that adult body cells are irreversibly differentiated.

How identical will clones be? They can be expected to be fairly similar in appearance. They would be less similar than identical twins, however, which share ooplasm and uterine and neonatal environments. Furthermore, certain components are inherited exclusively from the mother, e.g., the mitochondrial genome and perhaps the genome of centrioles. The random inactivation of one or the other of the X chromosomes may also limit similarities. Other differences among clones would result from the prenatal environment: in litter-bearing species even uterine position can affect offspring. In single-bearing species the maternal effect may be pronounced. Environmental differences in later life may greatly affect certain traits, even if those traits have a strong genetic component.

Serious technical barriers must be overcome before realistic speculation of possible advantages in animal production can be foreseen.

Cell fusion

DEFINITION

The fusion of two mature sex cells or the fertilization of one ovum with another. An analogous scheme for the male would be accomplished by microsurgical removal of the female pronucleus and substitution of nuclei from two sperm. Combining sex cells from the same animal is called "selfing."

STATE OF THE ART

Combination of ova has led to early development to the blastocyst stage in the mouse but no further development following transfer has been reported. Initial success in experimentation with manipulation of pronuclei has been reported.

FUTURE

Cell fusion technology may someday prove useful for getting genetic material from a somatic cell into a fertilized 1-cell embryo for the purpose of cloning. In conjunction with tissue culture technology the technology would have a role in gene mapping of chromosomes for the cow and perhaps other species.

Combining ova of the same animal, "selfing," would rapidly result in pure genetic (inbred) lines for use as breeding stocks. The technique would also lead to rapid identification of undesirable recessive traits which could be eliminated from the species.

Chimeras

DEFINITION

A chimera is an animal comprised of cell lines from a variety of sources. They can be formed by fusing two or more early embryos or by adding extra cells to blastocysts.

STATE OF THE ART

Live chimeras between two species of mouse have been produced. Such young have four parents instead of two; hexaparental chimeras have also been produced.

FUTURE

Practical applications of chimera technology to livestock are not obvious at this stage of development. The main objective of this research is to provide a genetic tool for better understanding of development, and maternal-fetal interactions.

Recombinant DNA

DEFINITION

The introduction of foreign DNA into the germplasm.

STATE OF THE ART

The mechanics of changing the DNA molecules of farm animals directly have not yet been worked out. The plasmid methods used in bacteria may not be applicable.

FUTURE

None of these techniques, no matter how great the potential, will be of any use in animal breeding until knowledge of genetics is greatly advanced. Before one can alter genes, they must be identified.

Prior to exploitation of recombinant DNA technology in animal breeding, it is necessary to identify gene loci on chromosomes, i.e., genetic mapping. Work toward this goal has only recently been initiated and rapid progress cannot be anticipated. Multivariate genetic determinants of characteristics of economic importance are anticipated to be the rule.

History of the
Recombinant DNA Debate

The history of the debate over the risks from rDNA techniques and the Government's response may be divided into four phases.* Phase I covered the period from the first awareness of risks to human health from experiments involving recombinant DNA (rDNA) in the summer of 1971 to the end of the Conference at the Asilomar Center in February 1975, which resulted in prototype guidelines covering the research. Phase II covered the period from Asilomar through the development by the National Institutes of Health (NIH) of the Guidelines of June 1976. In this period, the public first became significantly involved in the debate and most, if not all, of the policy issues were clearly framed. Phase III, from mid-1976 through mid-1978, involved congressional consideration of the issues in an atmosphere that went from almost imminent passage of legislation to the cessation of such efforts. Phase IV covers the postlegislative period, when NIH and its organizational parent, the Department of Health, Education, and Welfare (HEW) (now the Department of Health and Human Services) undertook to develop satisfactory voluntary standards in areas over which they had no legal authority and to accommodate growing pressure for public involvement, while avoiding a full regulatory role.

Phase I began in the summer of 1971, when several scientists became concerned about the safety of a proposed experiment to insert DNA from SV40 virus, a monkey tumor virus that also transforms human cells into tumor-like cells, into a type of bacteria naturally found in the human intestine. After months of discussion, the scientist who had proposed the experiment decided to defer it. Meanwhile, as rDNA techniques became more refined, debates about safety increased; at the June 1973 Gordon Research Conference, safety issues were discussed. The participants voted: to send a letter to the National Academy of Sciences (NAS) and the National In-

stitute of Medicine requesting the appointment of committees to study potential hazards to laboratory workers and the public; and by a narrow majority[4] to arrange for the letter to be published in the widely read journal, *Science,* to alert the broader scientific community.[5]

NAS appointed a committee of prominent scientists involved in rDNA research. In July 1974, the panel asked for a temporary worldwide moratorium on certain types of experiments, and called for an international conference on potential biohazards of the research through a letter published in *Science* and its British counterpart, *Nature.*[6] This letter also requested the Director of NIH to consider establishing an advisory committee to develop an experimental program to evaluate potential hazards and establish guidelines for experimenters.

In response, the Director of NIH, after authorization by the Secretary of HEW, established the Recombinant DNA Molecule Program Advisory Committee (later renamed the Recombinant DNA Advisory Committee, RAC) on October 7, 1974, along the lines suggested by the NAS Committee. The Committee's charter described its purpose as:[7]

The goal of the Committee is to investigate the current state of knowledge and technology regarding DNA recombinants, their survival in nature, and transferability to other organisms; to recommend programs of research to assess the possibility of spread of specific DNA recombinants and the possible hazards to public health and to the environment; and to recommend guidelines on the basis of the research results. *This Committee is a technical committee, established to look at a specific problem.* (Emphasis added.)

The international conference called for by the NAS Committee letter was held at the Asilomar Conference Center, Pacific Grove, Calif., in February 1975. The organizing committee made it clear that its purpose was to focus on scientific issues rather than to become involved in considering ethical and moral questions. However, in one session the few lawyers

---

*For a detailed history through 1977, see footnote 1. For a history and a discussion of the broader issues, see footnotes 2 and 3.

[1]J. Swazey, J. Sorenson, and C. Wong, "Risks and Benefits, Rights and Responsibilities: A History of the Recombinant DNA Research Controversy," *Southern California Law Review* 51:1019, September 1978.

[2]C. Grobstein, *A Double Image of the Double Helix* (San Francisco: W. H. Freeman Co., 1979).

[3]D. Jackson, and S. Stich (eds.), *The Recombinant DNA Debate* (Englewood Cliffs, N.J.: Prentice-Hall, Inc., 1979).

[4]Swazey, et al., op. cit., p. 1,023.

[5]Letter from Maxine Singer and Dieter Soll to the National Academy of Sciences (NAS) and the National Institute of Medicine, reprinted in *Science,* vol. 181, 1973, p. 1114.

[6]Letter from Paul Berg, et al. to the editor, reprinted in *Science,* vol. 185, 1974, p. 303.

[7]The charter of the Recombinant DNA Molecule Program Advisory Committee, Oct. 7, 1974.

invited confronted the scientists with some of these questions.[8] The conference report concluded that although a moratorium should continue on some experiments, most work involving rDNA could continue with appropriate safeguards in the form of physical and biological containment.

In *Phase II,* the debate widened to encompass broader social and ethical issues, such as the relationship between scientific freedom of inquiry and the protection of society's interests, in whatever manner those were defined. Such issues led naturally to questions about who makes the decisions and the role of the public in that process. Finally, decisionmaking mechanisms were developed. Issues raised and actions taken during this phase in many respects controlled the subsequent development of the Federal response to the debate, and created problems that continue to the present. At this stage, participation in the debate went beyond the scientific community.

Questions of ethics and public policy had been raised earlier, but they now received much wider attention. On April 22, 1975, Sen. Edward M. Kennedy, Chairman of the Subcommittee on Health of the Senate Committee on Labor and Public Welfare, held a half-day hearing on science policy issues arising from rDNA research. In May 1975, a 2-day conference on "Ethical and Scientific Issues Posed by Human Uses of Molecular Genetics" was held under the joint sponsorship of the New York Academy of Sciences and the Institute of Society, Ethics, and the Life Sciences. In addition to molecular biologists, participants included lawyers, sociologists, psychiatrists, and philosophers.

The issue of public participation arose as decisionmaking mechanisms were developed. RAC was originally composed of 12 members from "the fields of molecular biology, virology, genetics and microbiology."[9] Critics first noted the need for more expertise in the fields of epidemiology and infectious diseases, since most molecular biologists were trained as chemists.[*] RAC's membership was increased to 16 and the range of expertise was widened to include the fields of epidemiology, infectious diseases, and the biology of enteric organisms, by amendment to the charter on April 25, 1975.

Since some members were conducting the research in question, critics claimed that a conflict of interests existed. They also noted that the Committee

advised the Director of NIH, an agency whose mission was to foster biomedical research, not to stop or otherwise regulate it. These issues were brought out in a petition to NIH signed by 48 biologists in August 1975. Criticizing a proposed draft of the guidelines as setting substantially lower safety standards than those accepted at Asilomar, the petition argued for broader representation on RAC from other fields of scientific expertise and from the public-at-large. RAC itself had been sensitive to these limitations; in the summer of 1975, an attempt was made to recruit nonscientists.[10] One nonscientist was added in January 1976, and another was added in August 1976.

In December 1975, RAC submitted revised draft guidelines to the Director of NIH, Dr. Donald Fredrickson. Although they were stricter than those drafted at Asilomar, some criticized them as being "tailored to fit particular experiments that are already on the drawing boards."[11] The consensus of RAC, on the other hand, was that the guidelines were excessively strict, but that it was necessary to be overly cautious because of its limited expertise in public health.[12] In any event, Dr. Frederickson arranged for public hearings on the proposed guidelines at a 2-day meeting in February 1976 of the Advisory Committee to the Director, a diverse group of scientists, physicians, lawyers, philsophers, and others. A similarly diverse group of scientists and public interest advocates were invited to attend. Some modifications to the Guidelines proposed by Dr. Fredrickson as a result of that meeting were adopted and others were rejected by RAC in April 1976.[13]

The final major issue arising during this period concerned NIH's lack of authority to set conditions on research funded by other Federal agencies or by the private sector. In a June 2, 1976, meeting between Dr. Fredrickson and some 30 representatives of industry, including pharmaceutical and chemical companies, it became clear that some rDNA research was being done; however, the representatives appeared hesitant to commit themselves to voluntary compliance with the proposed guidelines.[14] The pri-

---

[8]Swazey, et al., op. cit., p. 1,034.

[9]The charter of the Recombinant DNA Molecule Program Advisory Committee, Oct. 7, 1974, op. cit.

[*]One of the members of the original RAC (Stanley Falkow) did have substantial expertise with enteric organisms and *E. coli* in particular.

---

[10]Dr. Elizabeth Kutter, a former RAC member, personal communication, Sept. 11, 1980.

[11]N. Wade, "Recombinant DNA: NIH Sets Strict Rules to Launch New Technology," *Science,* vol. 190, 1975, pp. 1175, 1179.

[12]Kutter, op. cit.

[13]Ibid.

[14]Subcommittee on Science, Research and Technology of the House Committee on Science and Technology, *Genetic Engineering, Human Genetics, and Cell Biology: DNA Recombinant Molecule Research* (Supp. Report II) 94th Cong., 2d sess., 1976, p. 51.

mary reason was their concern over protection of proprietary information.[15]

Phase II culminated with the promulgation on June 23, 1976, of the Guidelines for Research Involving Recombinant DNA Molecules ("1976 Guidelines") covering institutions and individuals receiving NIH funds for this research.

Phase III was characterized by attempts to remedy the limited applicability of the Guidelines. Soon after their publication, Senators Kennedy and Javits sent a letter to President Ford, calling his attention to the Guidelines. They noted that any risk was not limited to federally funded research, and urged him to take necessary steps to implement the Guidelines throughout the research community. In October 1976, the Secretary of HEW, with the approval of the President, formed the Federal Interagency Advisory Committee under the chairmanship of the Director of NIH to determine the extent to which the Guidelines could be applied to all research and to recommend necessary executive or legislative actions to ensure compliance.[16] In March 1977, the Committee concluded that existing Federal law would not permit the regulation of all rDNA research in the United States to the extent deemed necessary;[17] it further recommended new legislation, specifying the elements of that legislation.[18]

During 1977, several bills to deal with this and other problems were introduced in Congress. They addressed in different ways the issues of the extent of regulatory coverage, the mechanisms for regulation, and Federal preemption of State and local regulation. The major bills were those of Rep. Paul Rogers, H.R. 7897 (and its substitute, H.R. 11192) and of Sen. Edward Kennedy, S. 1217.*

While hearings were being held, three developments occurred which, by the end of 1977, had dissipated much of the impetus for legislation. The first was the expanded role of RAC. On September 24, 1976, its charter had been amended once more to provide for additional expertise in the areas of botany, plant pathology, and tissue culture. Moreover, its membership was increased from 16 to 20 so that four members would be "from other disciplines or representatives of the general public." This was the first official provision for public representation

although two nonscientists were already members. The number of nonscientists remained the same until December 1978.[20] Also, RAC's responsibilities were defined in greater detail, including the responsibility for reviewing large-scale experiments. Nevertheless, RAC continued formally at least to be "a technical committee, established to look at a specific problem."

The second development was a growing belief among scientists that the risks of the research were less than originally feared. This was based on the following: 1) a letter from Roy Curtiss at the University of Alabama to the Director of NIH, explaining risk assessment experiments using *Escherichia coli,* from which he concluded that the use of *E. coli* K-12 host-vectors posed no danger to humans; 2) the conclusions of a committee of experts in infectious diseases assembled by NIH in June 1977 in Falmouth, Mass., that the alleged hazards of the research were unsubstantiated; and 3) a prepublication report on experiments showing that genetic recombination occurs naturally between lower and higher life forms, and suggesting that the rDNA technique was not as novel as presumed.

The third development affecting the legislation was a concerted lobbying effort by scientists against what they considered to be some of the overly restrictive provisions of the bills, especially S. 1217.[21][22][23] The efforts included wide circulation of reports (including some in draft form) as soon as available, which supported the conclusion that the research was less hazardous than originally supposed.

By the end of 1977, the legislation was in limbo. This situation continued in early 1978, although some hearings were held. On June 1, 1978, Senators Kennedy, Javits, Nelson, Stevenson, Williams, and Schweiker addressed a letter to HEW Secretary Joseph Califano, which acknowledged the likelihood that legislation would not pass and urged that deficiencies in the regulatory system be addressed through executive action based on existing authority, if that were to be the case.

During *Phase IV*, NIH and its parent organization, HEW (now DHHS), have attempted to operate in the regulatory vacuum left by the lack of legislation. In response to the consensus that developed in 1977 on

---

[15]Ibid., pp. 52.

[16]*Interim Report of the Federal Interagency Committee on Recombinant DNA Research: Suggested Elements for Legislation,* Mar. 15, 1977, pp. 3-4.

[17]Ibid., pp. 9-10.

[18]Ibid., pp. 11-15.

*For a more complete discussion of the legislation, see footnote 19.

[19]."Recombinant DNA Molecule Research," Congressional Research Service, issue brief No. IB 77024, update of Jan. 2, 1979.

[20]William Gartland, Director of the Office of Recombinant DNA Activities, NIH, personal communication, June 19, 1980.

[21]B. Culliton, "Recombinant DNA Bills Derailed: Congress Still Trying to Pass Law," *Science,* vol. 199, Jan. 20, 1978. pp. 274-277.

[22]D. Dickson, "Friends of DNA Fight Back," *Nature,* vol. 272, April 1978, pp. 664-665.

[23]R. Lewin, "Recombinant DNA as a Political Pawn," *New Scientist,* vol. 79, Sept. 7, 1978, pp. 672-674.

the question of risk, RAC proposed revisions to the Guidelines, which placed most experiments at a lower containment level. They were published for public comment in September 1977.* As with the original Guidelines, public hearings were held in the course of a 2-day meeting of the Advisory Committee to the Director in December 1977, in which a diverse group of individuals and organizations were permitted to comment. However, at this point, HEW took a much more active role in a situation that had been handled almost entirely by NIH.[24]

When RAC's charter was renewed on June 30, 1978, Secretary Califano reserved the power to appoint its members instead of delegating it to the Director of NIH as in the past.** And the new proposed Guidelines, published in the Federal Register on July 28, 1978, were accompanied by an introductory statement by Secretary Califano announcing a 60 day public comment period to be followed by a public hearing before a departmental panel chaired by HEW General Counsel Peter Libassi.*** The Secretary was particularly interested in comments on: new mechanisms to provide for future discretionary revision of the Guidelines; and the composition of the various advisory bodies, especially the RAC and the local Institutional Biosafety Committees (IBCs).[25]

The public hearing called for by Secretary Califano and held on September 15, 1978, was a significant event in the history of Federal actions on the rDNA issue. Testimony was heard from representatives of industry, labor, the research community, and public interest groups; more than 170 letters of comment were received and subsequently reviewed. As a result, the revised final Guidelines of December 22, 1978, were significantly rewritten to increase public participation in the decisionmaking process:[26]

- Twenty percent of the members of the IBCs had to represent the general public and could have no connection with the institution.
- Most of the records of the IBCs had to be publicly available.

- Major actions, such as decisions to except otherwise prohibited experiments on a case-by-case basis or to change the Guidelines, could be made only on the advice of RAC and after public and Federal agency comment.

The increased public responsiveness of the IBC's was crucial, since the revised Guidelines placed major responsibility for compliance on them. This had been proposed in the July version and had not been changed by the hearings.* Califano also announced he would appoint 14 new members to the RAC, including people knowledgeable in fields such as law, public policy, ethics, the environment, and public health.[27]** All of these changes were envisioned to "provide the opportunity for those concerned to raise any ethical issues posed by recombinant DNA research" and to change the role of the RAC to "serve as the principal advisory body to the Director of NIH and the Secretary of HEW on recombinant DNA policy."[28]***

In addition to broadening public participation, Califano attempted to deal with a major limitation of the Federal response—the Guidelines did not cover private research. He directed the Food and Drug Administration (FDA) to take steps to require that any firm seeking approval of a product requiring the use of rDNA techniques in its development or manufacture, demonstrate compliance with the Guidelines for the work done on that product; an FDA notice of its intention to propose such regulations accompanied the revised Guidelines in the Federal Register. In addition, he requested the Environmental Protection Agency (EPA) to review its regulatory authority in that area. He believed if both agencies could regulate research on products within their jurisdiction, "virtually all recombinant DNA research in this country would be brought under the requirements of the revised guidelines."[29]* In the meantime, the

---

*Shortly thereafter, in October 1977, the Final Environmental Impact Statement for the 1976 Guidelines was published.

[24]D. Fredrickson, "A History of the Recombinant DNA Guidelines in the United States," *Recombinant DNA Technical Bulletin*, vol. 2, July 1979. pp. 87, 90.

**The statement providing for delegation of authority that accompanied the updated Charter was not signed by Califano. See also, footnote 24.

***The other members of the HEW panel were Dr. Fredrickson, Julius Richmond, who was the Assistant Secretary for Health, and Henry Aaron, who was the Assistant Secretary for Planning and Evaluation.

[25]43 F.R. 33042, July 28, 1978.

[26]Statement of Secretary Califano accompanying the revised Guidelines, 43 F.R. 60080, Dec. 22, 1978.

---

*As part of the revision process, HEW held a meeting in October 1978 for IBC chairpersons in order to exchange information and experiences gained under the 1976 Guidelines.

[27]Ibid.

**This was implemented by an amendment to the RAC Charter on Dec. 28, 1978, which increased the membership to 25 and changed the composition to the following categories: 1) at least eight specialists in molecular biology or rDNA research; 2) at least six specialists in other scientific fields; and 3) at least six persons knowledgeable in law, public policy, the environment, and public or occupational health. In addition, the Charter was amended to grant nonvoting representation to representatives of various Federal agencies.

[28]Ibid.

***The Charter was never amended to change or delete the final sentence of the "Purpose" section, which states, "This Committee is a technical committee, established to look at a specific problem."

[29]Ibid., p. 60081.

revised Guidelines provided, for the first time, for voluntary registration of projects with NIH, in which the registrant would agree to abide only by the containment standards of the Guidelines.[31]

Other major changes were embodied in the new Guidelines. Because of the consensus that the experiments posed lower risks than originally thought, some types of experiments were exempted, while containment levels were lowered for almost all others. In order to provide greater flexibility, these Guidelines permitted exceptions on a case-by-case basis, and included procedures for their change on a piecemeal basis without going through the whole internal process at HEW. For major changes, the procedure was: 1) publication of the proposed changes in the Federal Register at least 30 days prior to a RAC meeting; 2) RAC consideration of the proposed changes; and 3) publication in the Federal Register of the final decision of the Director, NIH. The standard for all actions of the Director under the Guidelines was "no significant risk to health or to the environment."[32] Lastly, the new Guidelines delegated project approval to the IBCs.

The problems posed by voluntary compliance and commercialization have continued to be addressed by NIH. In a second major revision to the Guidelines on January 29, 1980, a section (Part VI) was added to specify procedures for voluntary compliance.** On

April 11, 1980, NIH published Physical Containment Recommendations for Large Scale Uses of Organisms Containing Recombinant DNA Molecules in the form of Draft Part VII to the Guidelines.[33] Besides setting large scale containment levels, this document recommends that the institution: appoint a biological safety officer with specified duties; and establish a worker health surveillance program for work requiring a high (P_3) containment level. Finally, a more ad hoc requirement has been used since October 1979 for approvals of industrial requests for cultures up to 750 liters (l); the approvals were conditioned on NIH designated observers being permitted by the companies to inspect their facilities.[34] At least one inspection has taken place.

On November 21, 1980, NIH adopted the third major revision to the Guidelines.[35] It contained these significant changes: institutions sponsoring the research are no longer required to register their projects with NIH pursuant to an informational document called a Memorandum of Understanding (MUA) whenever the containment levels are specified in the Guidelines; and NIH will no longer review IBC decisions on experiments for which containment levels are specified in the Guidelines.

On November 21, 1980, NIH also promulgated revised application procedures for large-scale proposals. The application must include the following information: 1) the registration document submitted to the local IBC; 2) the reason for wanting to exceed the 10-l limit; 3) evidence that the rDNA to be used was rigorously characterized and free of harmful sequences; and 4) specification of the large-scale containment level proposed to be used as defined in the NIH Physical Containment Recommendations of April 11, 1980.

(continued from p. 318)

*Subsequently, Califano sent similar letters to the Secretaries of Agriculture (February 1979) and Labor (July 1979) requesting them to consider how their agencies' authorities could be used to require private sector rDNA research to comply with the Guidelines.[30]

[30]Minutes of the Interagency Committee on Recombinant DNA Research, p. 3, July 17, 1979, reprinted in *Recombinant DNA Research*, vol. 5, p. 132, et. seq.

[31]Sec. IV-F-3, 1978 Guidelines.

[32]Sec. IV-E-1-b.

**Several responses to the FDA notice had questioned the agency's legal authority to regulate private rDNA research. Consequently, Dr. Fredrickson and Dr. Donald Kennedy, then Commissioner of Food and Drugs, developed a draft supplement to the Guidelines, specifying procedures for voluntary compliance by industry. It was published for comment on Aug. 3, 1979 (44 F.R. 45868) and incorporated as part of the proposed revised Guidelines of November 30, 1979. (44 F.R. 69210, 69247).

In addition to adding part VI to the Guidelines, the most significant change in the January 1980 Guidelines was the addition of sec. III-0, which permitted most experiments using *E. coli* K-12 host-vector systems to be done at the lowest containment levels.

[33]45 F.R. 24968, Apr. 11, 1980.
[34]44 F.R. 69251, Nov. 30, 1979.
[35]45 F.R. 77372, Nov. 21, 1980.

Constitutional Constraints on Regulation

Under the checks and balances of our system of government, the Constitution, as ultimately interpreted by the Supreme Court, requires certain procedural and substantive standards to be met by statutory or other regulation imposed upon an activity. These requirements depend on the nature of the activity involved. In the present case, it will be useful to consider first the regulation of basic research and then the regulation of technological applications, such as the production of pharmaceuticals by using genetic engineering methods.

Research

With respect to research, the fundamental question is what limitations, if any, may be placed on the search for scientific knowledge. The primary applicable constitutional provision is the first amendment, which has been broadly interpreted by the Supreme Court to severely limit intrusion by the Government on all forms of expression.[1][2][3] Another constitutional safeguard, known as equal protection, is secondarily involved.

If the Supreme Court were to recognize a right of scientific inquiry, its boundaries would not exceed those for freedom of expression.[4] There is disagreement among commentators on this issue concerning the boundaries of the first amendment,[5] and certainly disagreement on the application of generally accepted principles to particular cases. Moreover, there have been no judicial decisions dealing with the precise issue at hand. However, it is possible to outline general principles derived from judicial decisions interpreting the first amendment, and indicate how they might be applied by the courts to attempts to regulate genetic research.

There are very few limitations on the written or spoken word. The prohibitions against obscenity or "fighting words"* clearly would be inapplicable here.

For many years, the Supreme Court has conceptualized the right of free expression in terms of a marketplace of ideas—through the open and full discussion of all ideas and related information, the valuable, valid, or useful ones will be accepted by society, while the ridiculous or even dangerous ones will be so demonstrated and discarded. This is a consensual process; no person, group, or institution has sufficient wisdom to prejudge ideas and deny them admittance to that intellectual marketplace, even if they threaten fundamental cultural values, for such values, if worthwhile, will survive. Under this concept, scientists would certainly have virtually unrestrained freedom to think, speak, and write.

Difficulties arise with actions, such as experimentation, which may be essential to the implementation of freedom of expression. Recent Supreme Court cases have recognized a limited protected interest of the media to gather information as an essential adjunct to freedom of publication. By analogy, it may be argued that scientists would also be protected in their research, as a necessary adjunct to freedom of expression. On the other hand, the information gathering cases usually involve access to Government facilities, such as courtrooms or prisons. They are based on the principle that actions by the Government should be open to public scrutiny—a concept not directly applicable to the present issue. More importantly, the Court has long recognized that actions related to expression can be regulated and that regulation may increase with the degree of the action's impact on people or the environment. The Court would probably apply what has been called a structured balancing test;[6] i.e., regulation would be deemed valid only when the Government sustains the burden of proving: 1) that there are "compelling reasons" for the regulation; and 2) that the objective cannot be achieved by "less drastic means," i.e., by more narrowly drafted regulations having less impact on first amendment rights.

The second part of the test is fairly straightforward. Governmental restrictions must be kept to a minimum. E.g., where possible, they should be regulatory rather than prohibitory, temporary rather than permanent, involve the least burden, and so on.

The difficult part of this test lies in determining

---

[1]Harold P. Green, "The Boundaries of Scientific Freedom" *Regulation of Scientific Inquiry: Societal Concerns With Research*, Keith M. Wulff (ed.) (Washington, D.C.: AAAS, 1979), pp. 139-143.

[2]Thomas I. Emerson, "The Constitution and Regulation of Research," *Regulation of Scientific Inquiry: Societal Concerns With Research*, Keith M. Wulff (ed.) (Washington, D.C.: AAAS, 1979), pp. 129-137.

[3]John A. Robertson, "The Scientists' Right to Research: A Constitutional Analysis," *Southern California Law Review* 51:1203, September 1978.

[4]Green, op. cit., p. 140.

[5]Emerson, op. cit., pp. 131-134.

*"Fighting words" are those provoking violent reaction or imminent disorder.

[6]Ibid., p. 134.

what is a compelling reason. The protection of health or the environment is the most clearly acceptable reason for regulation. In addition, the protection of individual rights and personal dignity is generally considered an acceptable reason. E.g., the National Research Act[7] requires that all biomedical and behavioral research involving human subjects supported under the Public Health Service Act be reviewed by an Institutional Review Board in order to protect the rights and welfare of the subjects.

The above discussion relates to protection from physical risks due to the process of research. Could the Government regulate or forbid experimentation solely because the product (knowledge) threatens cultural values or other intangibles such as the genetic inheritance of mankind? Religious or philosophical objections to research, based solely on the rationale that there are some things mankind should not know, conflict with the basic principles of freedom of expression and would not be sufficient reason on constitutional grounds to justify regulation. Even if the rationale underlying this objection were expanded to include situations where knowledge threatens fundamental cultural values about the nature of man, control of research for such a reason probably would not be constitutionally permissible. The rationale would again conflict with the marketplace of ideas concept that is central to freedom of expression. However, what if the knowledge were to provide the means to alter the human species in such a way that the physical, psychological, and emotional essence of what it is to be human could be changed? No precedent exists to provide guidance in determining an answer. Were the situation to arise, the Supreme Court might fashion another limitation on the concept of free expression in the same way it developed the obscenity or "fighting words" doctrines.

The discussion thus far has had as its premise a direct regulatory approach to research. There is a more indirect approach, which would be constitutionally permissible and could accomplish much of what direct regulation might attempt, including prevention of the acquisition of some forms of knowledge. This is the use of the funding power. The lifeblood of modern science in the United States is the Federal grant system. Yet it is generally agreed that Government has no constitutional duty to fund scientific research.[8] This is a benefit voluntarily provided to which many kinds of conditions may be attached. The only constitutional limitation on such an approach would be the concept of equal protection—any restrictions must apply to all or must not be ap-

plied in a discriminatory way without compelling reasons.

Congress could therefore, mandate by law that certain kinds of research not be funded or be conducted in certain ways. An example is the National Research Act, discussed previously. However, this approach may have some serious practical limitations because of the difficulty of determining which molecular biological research might lead to the proscribed knowledge. Much discretion would have to be left to the funding agency, which is likely to be unsympathetic or even hostile to such an approach, if it views its primary mission as fostering research.

Applications and products

Although fears have been expressed that current genetic technologies may lead to applications that would be detrimental, no one can reasonably conclude, at the present time, that this will actually occur. For this reason, the most constitutionally permissible approach in all probability will be to regulate the applications of the science. In such situations, whatever harms occur tend to be more tangible and the governmental interests, therefore, more clearly defined. Moreover, since fundamental constitutional rights are generally not involved, statutes and regulations are subjected to a lower level of scrutiny by the Federal courts.

The constitutional authority for Federal regulation of the applications of technologies such as genetic engineering lies in the commerce clause, article I, section 8 of the Constitution, which grants Congress the power "To Regulate Commerce with foreign Nations, and among the Several States." In contrast to situations involving fundamental rights, the Supreme Court has interpreted this clause as giving Congress extremely broad authority to regulate any activity in any way connected with commerce. It has been virtually impossible for Congress not to find some connection acceptable to the courts between commerce and the goals of a particular piece of legislation.* The standard of review of such legislation by the Federal courts is to determine if it bears a rational relationship to a valid legislative purpose. If so, the Court will uphold the legislation and will not second guess the legislators. This standard of review recognizes that a statute results from the balancing of competing interests and policies by the branch of Government created to function in that manner.

*See *Wickard* v. *Filburn*, 317 U.S. 111 (1942) in which the Supreme Court upheld civil penalties for violation of acreage allotments established by the Agricultural Adjustment Act of 1938, covering the amount of wheat that individual farmers could plant, even if the wheat was intended for self-consumption. The rationale was that even though the individual farmer's wheat had no measurable impact on interstate commerce, Congress could properly determine that all wheat of this category, if exempted from regulation, could undercut the purpose of the Act, which was to increase the price farmers received for their various crops.

[7]Public Law 93-348 (1974), 42 U.S.C. §289 l-3.
[8]Green, op. cit., p. 141.

Information on International Guidelines for Recombinant DNA

The following information is based largely on international surveys undertaken by The Committee on Genetic Experimentation of the International Council of Scientific Unions reported as of July 1979.[1]

I. Nations that had established guidelines for conduct of rDNA research or were using the guidelines of other nations:

Australia	Italy
Belgium	Japan
Brazil	Mexico
Bulgaria	Netherlands
Canada	New Zealand
Czechoslovakia	Norway
Denmark	Poland
German Democratic Republic	South Africa
Federal Republic of Germany	Sweden
	Switzerland
Finland	Taiwan
France	United Kingdom
Hungary	United States
Israel	U.S.S.R.
	Yugoslavia

II. Nations that had not established guidelines or had not responded with updated information:

Country	Yes	No
Austria	X	
Ghana	X	
India	X	
Iran	X	
Jamaica		X
Korea		X
Nigeria		X
Singapore		X
Sri Lanka		X
Sudan		X
Turkey	X	

III. Nations that had drafted their own guidelines:

Canada	Japan
Federal Republic of	United Kingdom

Germany
France
Italy

United States
U.S.S.R.

IV. Nations that had modified the guidelines of other, indicated, countries:

Australia (UK, U.S.)	Mexico (U.S.)
Belgium (UK, U.S.)	Netherlands (U.S.)
Brazil (U.S.)	New Zealand
Bulgaria (U.S.S.R., U.S.)	Norway (U.S.)
Czechoslovakia (U.S.S.R., U.S., Fed. Rep. Ger.)	Poland (U.S.)
	South Africa (U.S.)
Denmark (UK)	Sweden (U.S)
East German Democratic Republic (UK, U.S., Netherlands)	Switzerland (U.S.)
	Taiwan (U.S., UK)
	Yugoslavia
Finland (U.S. mainly)	(European Science
Hungary (U.S.)	Foundation)

V. Nations in which entirely voluntary guidelines have been adopted:

Finland

VI. Nations with guidelines that are enforceable through control of research funding:

Australia[a]	Japan
Canada	Netherlands[d]
Czechoslovakia[b]	Norway
Denmark	South Africa
Federal Republic of Germany[c]	Sweden
	Switzerland
France	Taiwan[e]
German Democractic Republic	United Kingdom[f]
	United States

[a]"Submissions may be made directly to the Academy of Science or through a granting agency. In the latter case, it is a requirement for the applicant to observe the recommendations of the Academy's Standing Committee if the agency makes a grant for the work. Otherwise, the guidelines are voluntary with the worker required to make an annual report on progress, or more frequently if conditions of the experiment (such as volumes) are changed appreciably."

[b]"Control through Academy of Sciences and Ministry of Health."

[c]Several research organizations require receivers of grants to apply the NIH guidelines until their own national guidelines are completed.

[d]The Netherlands Organization for the Advancement of Pure Research will only subsidize projects which have been given the committee's consent.

[e]"Waiting for response from National Advisory Committee."

[f]"Notification of proposals to GMAG became compulsory August 1, 1978. In addition, funding bodies require, as a condition of funding, GMAG's advice to be sought and followed."

[1]Report to COGENE from the working group on Recombinant DNA Guidelines, May 1980.

VII. Nations in which guidelines are legally enforceable:

Hungary

U.S.S.R.

Finland . . . . "At present, the guidelines are entirely voluntary, but in the near future, the intention is to include them in the law of infectious diseases when they will become legally enforceable."

South Africa . . . "At present the guidelines are not legally enforceable. They will only become so if regulations under the existing Health Act of 1977 and the Animal Diseases and Parasites Act of 1956 are promulgated; and none are intended at present."

United Kingdom "The regulation to notify GMAG does not strictly mean that the Williams Guidelines themselves are legally enforceable. But, under the Health and Safety at Work Act (within which the Regulations were introduced), it is expected that account will be taken of the relevant Codes of Practice and the advice given by GMAG."

VIII. Nations in which observance of the guidelines is monitored by a nationally-directed mechanism:

Australia	Norway
Czechoslovakia	South Africa
German Democratic	Sweden
Republic	United Kingdom
France	United States
Hungary	U.S.S.R.
Japan	Yugoslavia

IX. Nations in which a license or other authorization for recombinant DNA activity is granted:

—to an institution: U.S.S.R.

—to an indivdual laboratory: Hungary, Czechoslovakia

—to an individual scientist: Australia, Canada, German Democratic Republic, Federal Republic of Germany, Finland, France, Japan, Norway, South Africa, Sweden, United Kingdom[a], United States and U.S.S.R.

Netherlands: "There are gentlemen's agreements, signed by the individual scientist, the institution and the Committee." The reports of the Committee also recommend legislation that will require registration of research projects in this field and make binding the guidelines and supervision of their observance. (*Report of the Committee in Charge of the Control on Genetic Manipulation,* Amsterdam, March 1977, p. 54.)

Bulgaria, Switzerland: None of the above.

Taiwan: No response.

---

[a]The Group advises on proposals from individual workers, but considers them in the context of information about the 'centre' in which the work is to go on."

X. Nations in which special provisions for agriculture and/or industrial research and applications have been made:

Czechoslovakia . "10 liter maximum volume of the culture containing recombinant DNA"

German
Democratic
Republic . . . . . "The GDR Guidelines will be compulsory for industrial and agricultural applications. 10-liter maximum deviations may be allowed by the Minister of Health if suggested by the Committee."

Federal Republic
of Germany . . "Specification of containment of plants"

France . . . . "Industry, maximum volume of cell culture is set at 10 liters"

Norway . . . . . . "The Guidelines cover both agriculture and industry. Application of recombinant DNA research outside an approved laboratory is prohibited. Otherwise the Committee follows the NIH Guidelines."

United Kingdom "Agriculture, industry; see Williams Report, paragraphs 1.3, 2.7, 5.13 and appendix II, section 34."

United States . . "Agriculture. NIH Guidelines provide containment levels for cloning total plant DNA, plant virus DNA and plant organelle DNA in *E. coli* K-12, and provide general guidance for the use of plant host-vector systems. 10 liter maximum. A proposed Supplement to the Guidelines for voluntary compliance by the private sector is under consideration by RAC. Development of a monograph for large-scale applications has been proposed."

U.S.S.R. . . . . . . "Guidelines are compulsory for industrial and agricultural applications. 10 liter maximum. Deviation is allowed by the Recombinant DNA Commission."

Other
respondents . . No

XI. Number of laboratories currently engaged in recombinant DNA activities:

Country	Any labs?	How many?
Australia	yes	16
Austria	no[a]	
Belgium.	yes[a]	6
Brazil.	yes	5
Bulgaria.	yes[a]	no response
Canada	yes	10-15
Czechoslovakia	yes	3
Denmark.	yes[a]	several
German Democratic Republic	yes	5
Federal Republic of Germany	yes	10-20
Finland	yes	3 (3-4 planned)
France.	yes[a]	12
Ghana	no[a]	
Hungary	yes[a]	1-2
India	no[a]	
Iran.	no[a]	
Israel.	yes[a]	1
Jamaica	no[a]	
Japan	yes	35
Korea	no[a]	
Netherlands	yes	7
New Zealand	yes	2
Nigeria	no	
Norway.	yes	not stated
Philippines	no[a]	
Poland.	yes[a]	3
Singapore	no[a]	
South Africa	yes[a]	3
Sri Lanka.	no[a]	
Sudan	no[a]	
Sweden	yes[a]	2
Switzerland.	yes[a]	18
Taiwan	yes	2
Turkey	no[a]	
United Kingdom	yes	45
United States.	yes[a]	50
U.S.S.R.	yes[a]	6
Yugoslavia.	yes[b]	4

[a]Based on replies from previous Questionnaires.
[b]In preparation.

XII. Countries in which specific training for workers and safety officers in recombinant DNA activities is required by the guidelines:

Country	Yes	No	Other
Australia			[a]
Bulgaria	X		
Canada.		X	
Czechoslovakia.		X[b]	
German Democratic Republic.	X[c]		
Federal Republic of Germany . .	X[d]		
Finland.		X	
France	X		
Hungary.		X	
Japan		X	
Netherlands . . .	X[e]		
Norway		X[f]	
South Africa . . .	X		
Sweden		X	
Switzerland . . .			"recommended"
Taiwan.			"recommended"
United Kingdom	X[g]		
United States . .		X[h]	
U.S.S.R.	X		
Yugoslavia			no response

Other respondents: no or no response to question.

[a]Australia: "Require expertise through Biosafety Committee."
[b]Czechoslovakia: "Specific training is recommended."
[c]German Democratic Republic: "Training courses are organized by the Committees in cooperation with Akademie fur Arztliche Fortbildung der DDR."
[d]Federal Republic of Germany: "Experience as required by law on the control of communicable diseases."
[e]Netherlands: "The scientists should be trained in microbiology."
[f]Norway: The Committee certifies training and expertise of personnel are adequate."
[g]United Kingdom: "Details of training are required; the employer is legally obliged to provide suitable training."
[h]United States: "Specific training not required. However, local biohazards committees are required to certify to the NIH that the training and expertise of the personnel are adequate."

XIII. Countries in which the guidelines are applicable only to biological agents containing recombinant DNA, or also cover the recombinant DNA molecules themselves:

Country	Only to biological agents	Also recombinant DNA molecules
Australia........		X
Bulgaria	X	
Canada.........	(a)	(a)
Czechloslovakia ..		X
German Democratic Republic.......	X	
Federal Republic of Germany.......	X	
Finland.........		X
France	X	
Japan	X	
Netherlands		X
New Zealand.....		X
Norway		X
South Africa.....		X
Sweden.........		X
Switzerland		X
Taiwan.........	X	
United Kingdom ..		X
United States		X[b]
U.S.S.R.........		X

[a]Guidelines apply to all, but containment is not required for naked DNA.
[b]"The Guidelines apply to recombinant DNA experiments that are not exempt under Section I-E of the Guidelines. Recombinant DNA molecules that are not in organisms or viruses are exempt from the Guidelines (I-E-1)."

XIV. Groups/Committees responsible for carrying out monitoring of containment procedures:

Country	Group
Australia.......	Institutional Biosafety Committees.
Bulgaria	National Committee
Canada.........	"University and Medical Research Council Biohazards Committees"
Czechoslovakia...	"Under consideration of the National Institutes of Public Health."
German Democratic Republic.......	"Monitoring is carried out by local Biosafety Officers, who are representatives of the Committee in their institutions."
Federal Republic of Germany.......	Officers for Biological Safety monitor the health of employees

and compliance at laboratories; ZKBS (Zentrale Kommission fur die biologische Sicherheit) has overall responsibility.

France	"Local safety committees"
Hungary........	"National Institutes of Public Health"
Japan	"Principal Investigator and Safety Officer"
Netherlands	"Site Inspection Commission"
New Zealand.....	"Local controlling Committees are charged with monitoring observance of Guidelines. Biological Safety Officers are appointed to take immediate responsibility."
Norway	"Physical containment: Norwegian National Institute of Public Health. Biological containment: Committee."
South Africa.....	"Above P3, Biosafety Committee of Institute involved and SAGENE. Below P3, SAGENE only."
Sweden.........	Not applicable.
Switzerland	"At the responsibility of either the individual investigator or a local biohazards committee."
Taiwan.........	No response
United Kingdom ..	The Health and Safety Executive
United States	"Observance of containment is to be monitored by biohazards committees located in institutions in which the research is conducted. Effectiveness of containment procedures is to be monitored by the principal investigator who is to report problems to the NIH."
U.S.S.R.........	"Local biosafety commission, State Sanitary Inspection control group of Recombinant DNA Commission.

XV. *Countries in which the guidelines apply to all gene combinations constructed by cell-free methods, or only to molecules containing combinations of genes from different species:*

Country	All gene combinations constructed by cell-free methods	Molecules containing combinations of genes from different species
Australia........		X
Canada.........	X	
Czechoslovakia...		X
German Democratic Republic.......	X	
Federal Republic of Germany.......		X[a]
Finland.........	X	
France.........	X	
Japan..........		X
Netherlands.....		X[b]
New Zealand.....	X	
Norway........	X	
South Africa.....		X
Sweden........		X
Switzerland.....	X	
Taiwan.........		X
United Kingdom..		X[c]
United States....		X
U.S.S.R.........	X	

[a]Federal Republic of Germany Self-cloning experiments involving non-pathogenic donors and hosts shall be reported to ZKBS.

[b]Netherlands "The definition of recombinant DNA has recently been modified and includes the insertion of chemically synthesized DNA molecules into a vector."

[c]United Kingdom "The Group's provisional interpretation of their own remit is that they are concerned with work involving genetic manipulation, defined for these purposes as: the formation of new combinations of heritable materials by the insertion of nucleic acid molecules, produced by whatever means outside the cell, into any virus, bacterial plasmid, or other vector system so as to allow their incorporation into a host organism in which they do not naturally occur but in which they are capable of continued propagation."

XVI. *Countries in which the guidelines restrict the intentional dissemination into the environment of biological agents containing recombinant DNA:*

All respondents .. Yes[a]
Australia........ Not explicity so
German Democratic Republic....... "Exceptions have to be discussed by the Committee and require special permission by the Minister of Health."

New Zealand..... "Yes, with the approval of the National Committee."
United Kingdom .. "The question has not arisen."
Other respondents No

[a]Are there any circumstances under which such dissemination can be carried out?

XVII. *Countries in which the guidelines are restricted to recombinant DNA activities or also cover other areas of genetic experimentation:*

Country	Recombinant DNA activities	Other areas of genetic experimentation
Australia........	X[a]	
Bulgaria........	X	
Canada.........		X[b]
Czechoslovakia...	X	
German Democratic Republic.......	X	
Federal Republic of Germany.......	X	
Finland.........	X	
France.........	X	
Hungary........	X	
Japan..........	X	
Netherlands.....	X	
New Zealand.....		X[c]
Norway........	X	
South Africa.....		X[d]
Sweden........	X	
Switzerland.....	X	
United Kingdom..	X	
United States....	X	
U.S.S.R.........	X	

[a]"At present, the terms of reference of the Academy Committee refer only to in vitro experiments (i.e., the use of restriction enzymes and ligases). An *ad hoc* Academy Committee is about to investigate in vivo experimentation, with the following terms of reference:

1. Examine whether, other than by using the technique of *in vitro* recombinant DNA construction, new hybrid nucleic acid molecules can be produced that are potentially dangerous to humans, animals, or plants.

In so doing, the committee should give particular attention to the following possibilities:
—The use of mixed infections involving human or animal viruses, or the use of bacteria or fungi.
—The introduction of foreign DNA into plants and the production of new plant pathogens.

2. Consider whether there are certain classes of viral pathogens (e.g., polio) on which experimentation should not be carried out unless a special need is demonstrated."

[b]"work with animal viruses and cells"

[c]"i.e., cell fusion with approval of National Committee"

[d]"Other closely related areas are also covered."

XVIII. Countries in which the recombinant DNA advisory committee includes public representatives as well as scientists:

Country	Yes	No
Australia.......		X
Bulgaria		X
Canada	X	
Czechoslovakia...		X
Denmark		X
German Democratic Republic.......	X	
Federal Republic of Germany.......	X	
Finland.........	X	
France	X	
Hungary........		X
Italy		X
Japan	X	
Netherlands		X
New Zealand.....		X
Norway		X
South Africa	X	
Sweden.........	X	
Switzerland	X	
Taiwan.........		X
United Kingdom ..	X	
United States	X	
U.S.S.R.........		X

Composition of DNA advisory committees is as follows:

Australia........ 8 scientists

Canada .... 5 laymen (1 lawyer, 1 businessman, 3 generalists); 6 scientists (2 M.D.s, 3 virologists/cancer specialists, 1 recombinant DNA specialist)

Czechoslovakia... 6 members representing molecular biology, genetics, microbiology, medicine

Denmark .... 9 scientists and administrative representatives.

German Democratic Republic....... 3 geneticists, 1 biochemist, 2 bacteriologists, 2 virologists, 1 jurist, 1 representative of trade union of GDR.

Federal Republic of Germany....... 4 experts working in the field of recombinant DNA research; 4 experts who, though not working in the field of recombinant DNA

research, possess specific knowledge in the implementation of safety measures in biological research work, particularly however in microbiology, cytobiology, or hygiene and, in addition, 4 outstanding individuals, for example from the trade unions, industry, and the research-promoting organizations.

Finland......... 27 members: 6 molecular biology, 3 genetics, 3 microbiology, 1 virology, 1 plant physiology, 3 infectious diseases, 3 epidemiology, 2 enteric bacteria, 1 cell cultures, 3 public health, 1 occupational health.

France .... 13 members, 4 observers, 1 secretary

Hungary........ Scientists

Italy .... 8 molecular biologists, 4 microbiologists, 1 civil servant (Health Ministry).

Japan .... (Combines both Steering Committee and Advisory Group): 7 recombinant DNA scientists, 7 scientists in other fields, 6 specialists in medicine and biohazards, 2 lawyers, 2 specialists in physical containment, 3 public representatives.

Netherlands ..... 14 scientists representing genetics, molecular biology, bacteriology, virology, botany, medicine, ethics and social aspects of health and health-care. To be added: a committee composed of scientists and representatives of industry and trade unions.

New Zealand..... 1 molecular biologist, 1 microbial geneticist, 1 virologist, 1 botanist (molecular biologist), 1 human geneticist (medically qualified).

Norway .... 3 biochemists, 2 medicine, 1 veterinary medicine, 1 lawyer, 1 artist.

South Africa ..... One each from: Council for Scientific and Industrial Research, Medical Research Council, Department of Health, Department of Agricultural Technical Services. Three from universities, public and legal professions.

Sweden........ No response

Switzerland . . . . . 12 members representing medicine, microbiology, molecular biology, antibiotics, industry, university management, and 7 governmental departmental assessors.

United States . . . . Molecular biology: 6, Molecular Genetics: 5, Ethics: 3, Microbiology: 2, Plant Genetics: 2, Law: 2, Environmental Concerns, Laboratory Technician, Infectious Diseases, Occupational Health, Education: 1 each.

U.S.S.R. . . . . 8 scientists

Yugoslavia . . . . . . 3 geneticists

Planning Workshop Participants, Other Contractors and Contributors, and Acknowledgments

Planning workshop participants

Philip Bereano
  University of Washington
Ralph Hardy
  E. I. Du Pont de Nemours & Co., Inc.
Patricia King
  Georgetown University Law Center
Charles Lewis
  U.S. Department of Agriculture
Herman Lewis
  National Science Foundation
Pamela Lippe
  Friends of the Earth
Victor McKusick
  Johns Hopkins University Medical School
Elena O. Nightingale
  Institute of Medicine
Gilbert Omenn
  Office of Science and Technology Policy
Donna Parratt
  Cogressional Research Service
Walter Shropshire
  Smithsonian Radiation Biology Laboratory
Leroy Walters
  The Kennedy Institute

Other contractors and contributors

Betsy Amin-Arsala
Richard J. Auchus
  Massachusetts Institute of Technology
Fred Bergmann
  National Institutes of Health
Charles L. Cooney
  Massachusetts Institute of Technology
Richard Curtin
Robert A. Cuzick
  Massachusetts Institute of Technology
Roslyn Dauber
Arnold L. Demain
  Massachusetts Institute of Technology
Richard B. Emmitt
  F. Eberstadt & Co.
Emanuel Epstein
  University of California, Davis

Robert F. Fleischaker
  Massachusetts Institute of Technology
Odelia Funke
George E. Garrison
  F. Eberstadt & Co.
Reinaldo F. Gomez
  Massachusetts Institute of Technology
John Hamilton
Neal F. Jensen
Scott A. King
  F. Eberstadt & Co.
Harvey Lodish
  Massachusetts Institute of Technology
L. D. Nyhart
  Massachusetts Institute of Technology
ChoKyun Rha
  Massachusetts Institute of Technology
William Scanlon
Andrew Schmitz
  University of California, Berkeley
Michael J. Shodell
  The Sterling-Hobe Corp.
David Tse
  Massachusetts Institute of Technology
James Welsh
  Montana State University
George Whiteside
  Massachusetts Institute of Technology
Bernard Wolnak and Associates
Nancy Woods
  OTA intern

Acknowledgments

A large number of individuals provided valuable advice and assistance to OTA during this assessment. In particular, we would like to thank the following people:
John Adams
  Pharmaceutical Manufacturers Association
Rupert Amann
  Colorado State University
William Amon, Jr.
  Cetus Corp.
Daniel Azarnoff
  Searle Laboratories

A. L. Barr
  West Virginia University
K. J. Betteridge
  Animal Diseases Research Institute
Jerome Birnbaum
  Merck, Sharp & Dohme Research Laboratories
Gerald Bjorge
  U.S. Patent and Trademark Office
Hugh Bollinger
  Plant Resources Institute
G. Eric Bradford
  University of California
Robert Brackett
  Parke, Davis & Co.
Robert Byrnes
  Genentech, Inc.
Daniel Callahan
  The Hastings Center
Alexander M. Capron
  President's Commission for the Study of Ethical
    Problems in Medicine and Biomedical and
    Behavioral Research
Robert Church
  University of Calgary
H. Wallis Clark
  University of California
Gail Cooper
  Environmental Protection Agency
Joseph P. Dailey
  Revlon Health Care Group
Frank Dickinson
  Agricultural Research Center
Donald R. Dunner
  Finnegan, Henderson, Farabow, Garrett &
    Dunner
Roger B. Dworkin
  Indiana University School of Law
Richard P. Elander
  Bristol-Myers Co.
Peter Elsden
  Colorado State University
Haim Erder
  University of Pennsylvania
James F. Evans
  Pennsylvania Embryo Transfer Service
Kenneth Evans
  Plant Variety Protection Office
Richard Faust
  Hoffman-La Roche, Inc.
Herman Finke
  Sterling Systems
Robert H. Foote
  Cornell University
Orrie M. Friedman
  Collaborative Research, Inc.

William J. Gartland, Jr.
  National Institutes of Health
Kenneth Goertzen
  Seed Research, Inc.
Michael Goldberg
  Food and Drug Administration
Maxwell Gordon
  Bristol Laboratories
Lorance L. Greenlee
  Keil and Witherspoon
Ralph Hardy
  E. I. Du Pont de Nemours and Co., Inc.
W. C. D. Hare
  Animal Diseases Research Institute
Paul Harvey
  U.S. Department of Agriculture
Harold W. Hawk
  Agricultural Research Center
Richard L. Hinman
  Pfizer, Inc.
Peter Barton Hutt
  Covington & Burling
E. Keith Inskeep
  West Virginia University
Irving Johnson
  Lilly Research Laboratories
Charles Kiddy
  Agricultural Research Center
Thomas D. Kiley
  Genentech, Inc.
Carole Kitti
  National Science Foundation
Duane C. Kraemer
  Texas A&M University
Sheldon Krimsky
  Tufts University
W. W. Lampeter
  Lehn und Versuchtsgut
Earl Lasley
  Monsanto Farmers Hybrid
F. Douglas Lawrason
  Schering Corp.
Bernard Leese
  Plant Variety Protection Office
Stanley Leibo
  Oak Ridge National Laboratory
Morris Levin
  Environmental Protection Agency
Herman Lewis
  National Science Foundation
Peter Libassi
  Verner, Lipfert, Bernhard & MacPherson
Paul J. Luckern
  U.S. Department of Justice

Clement Markert
  Yale University
Ralph R. Maurer
  U.S. Meat Animal Research Center
Robert McKinnell
  University of Minnesota
Edward Mearns, Jr.
  Case Western Reserve University School
    of Medicine
Alan S. Michaels
  Stanford University
Elizabeth Milewski
  National Institutes of Health
Henry I. Miller
  Food and Drug Administration
Paul Miller
  W. R. Grace
A. V. Nalbandov
  University of Illinois
Claude H. Nash
  Smith Kline & French Laboratories
Dorothy Nelkin
  Resources for the Future
Gordon Niswender
  Colorado State University
Elena Ottolenghi-Nightengale
  National Academy of Medicine
David Padwa
  Agrigenetics, Inc.
Seth Pauker
  National Institute of Occupational Safety
    & Health
J. B. Peters
  West Virginia University
Nancy Pfund
  Stanford University School of Medicine
James Punch
  The Upjohn Co.
Neils Reimers
  Stanford University
Ira Ringler
  Abbott Laboratories
Roman Saliwanchik
  The Upjohn Co.
Robert B. Samuels
  Beckman Instruments
George E. Seidel, Jr.
  Colorado State University

Sarah M. Seidel
  Colorado State University
Thomas J. Sexton
  U.S. Department of Agriculture
Brian F. Shea
Ralph Silber
  Stanford University
Elizabeth L. Singh
Charles G. Smith
  Revlon Health Group
  Animal Diseases Research Institute
Davor Solter
  Wistar Institute of Anatomy and Biology
Mark Sorrells
  Cornell University
G. F. Sprague
  University of Illinois
Richard Staples
  Cornell University
Gerald G. Still
  U.S. Department of Agriculture
Charles W. Stuber
  North Carolina State University
Bernard Talbot
  National Institutes of Health
Rene Tegtmeyer
  U.S. Patent and Trademark Office
Clair E. Terrill
  U.S. Department of Agriculture
Robin Tervit
  Colorado State University
Stephen Turner
  Bethesda Research Laboratories, Inc.
L. D. Van Vleck
  Cornell University
Robert Walton
  W. R. Grace
Charles Weiner
  Massachusetts Institute of Technology
Ray W. Wright, Jr.
  Washington State University
Susan Wright
  University of Michigan
Oskar R. Zaborsky
  National Science Foundation
M. S. Zuber
  University of Missouri